Advancements in Cardiovascular and Antidiabetic Drug Therapy

Advancements in Cardiovascular and Antidiabetic Drug Therapy

Guest Editor

Alfredo Caturano

Basel • Beijing • Wuhan • Barcelona • Belgrade • Novi Sad • Cluj • Manchester

Guest Editor
Alfredo Caturano
Department of Human
Sciences and Promotion of the
Quality of Life
San Raffaele Roma Open
University
Rome
Italy

Editorial Office
MDPI AG
Grosspeteranlage 5
4052 Basel, Switzerland

This is a reprint of the Special Issue, published open access by the journal *Pharmaceuticals* (ISSN 1424-8247), freely accessible at: https://www.mdpi.com/journal/pharmaceuticals/special_issues/ND3N9D0P22.

For citation purposes, cite each article independently as indicated on the article page online and as indicated below:

Lastname, A.A.; Lastname, B.B. Article Title. *Journal Name* **Year**, *Volume Number*, Page Range.

ISBN 978-3-7258-3267-5 (Hbk)
ISBN 978-3-7258-3268-2 (PDF)
https://doi.org/10.3390/books978-3-7258-3268-2

© 2025 by the authors. Articles in this book are Open Access and distributed under the Creative Commons Attribution (CC BY) license. The book as a whole is distributed by MDPI under the terms and conditions of the Creative Commons Attribution-NonCommercial-NoDerivs (CC BY-NC-ND) license (https://creativecommons.org/licenses/by-nc-nd/4.0/).

Contents

About the Editor vii

Preface ix

Selin Genc, Bahri Evren, Onur Selcuk Yigit, Ibrahim Sahin, Ramazan Dayanan, Aleksandra Klisic, et al.
Evolving Clinical Features of Diabetic Ketoacidosis: The Impact of SGLT2 Inhibitors
Reprinted from: *Pharmaceuticals* 2024, 17, 1553, https://doi.org/10.3390/ph17111553 1

Gamze Camlik, Besa Bilakaya, Esra Küpeli Akkol, Adrian Joshua Velaro, Siddhanshu Wasnik, Adi Muradi Muhar, et al.
Oral Active Carbon Quantum Dots for Diabetes
Reprinted from: *Pharmaceuticals* 2024, 17, 1395, https://doi.org/10.3390/ph17101395 12

Teodor Salmen, Valeria-Anca Pietrosel, Delia Reurean-Pintilei, Mihaela Adela Iancu, Radu Cristian Cimpeanu, Ioana-Cristina Bica, et al.
Assessing Cardiovascular Target Attainment in Type 2 Diabetes Mellitus Patients in Tertiary Diabetes Center in Romania
Reprinted from: *Pharmaceuticals* 2024, 17, 1249, https://doi.org/10.3390/ph17091249 26

Božena Bradarić, Tomislav Bulum, Neva Brkljačić, Željko Mihaljević, Miroslav Benić and Božo Bradarić Lisić
The Influence of Dapagliflozin on Foot Microcirculation in Patients with Type 2 Diabetes with and without Peripheral Arterial Disease—A Pilot Study
Reprinted from: *Pharmaceuticals* 2024, 17, 1127, https://doi.org/10.3390/ph17091127 39

Emilia Błeszyńska-Marunowska, Kacper Jagiełło, Łukasz Wierucki, Marcin Renke, Tomasz Grodzicki, Zbigniew Kalarus and Tomasz Zdrojewski
Potentially Inappropriate Medications Involved in Drug–Drug Interactions in a Polish Population over 80 Years Old: An Observational, Cross-Sectional Study
Reprinted from: *Pharmaceuticals* 2024, 17, 1026, https://doi.org/10.3390/ph17081026 56

Alfredo Caturano, Roberto Nilo, Davide Nilo, Vincenzo Russo, Erica Santonastaso, Raffaele Galiero, et al.
Advances in Nanomedicine for Precision Insulin Delivery
Reprinted from: *Pharmaceuticals* 2024, 17, 945, https://doi.org/10.3390/ph17070945 65

Akshyaya Pradhan, Somya Mahalawat and Marco Alfonso Perrone
From the INVICTUS Trial to Current Considerations: It's Not Time to Retire Vitamin K Inhibitors Yet!
Reprinted from: *Pharmaceuticals* 2024, 17, 1459, https://doi.org/10.3390/ph17111459 92

Nicia I. Profili, Roberto Castelli, Antonio Gidaro, Roberto Manetti, Margherita Maioli and Alessandro P. Delitala
Sodium-Glucose Cotransporter-2 Inhibitors in Diabetic Patients with Heart Failure: An Update
Reprinted from: *Pharmaceuticals* 2024, 17, 1419, https://doi.org/10.3390/ph17111419 102

Teodor Salmen, Claudia-Gabriela Potcovaru, Ioana-Cristina Bica, Rosaria Vincenza Giglio, Angelo Maria Patti, Roxana-Adriana Stoica, et al.
Evaluating the Impact of Novel Incretin Therapies on Cardiovascular Outcomes in Type 2 Diabetes: An Early Systematic Review
Reprinted from: *Pharmaceuticals* 2024, 17, 1322, https://doi.org/10.3390/ph17101322 115

Silvius Alexandru Pescariu, Ahmed Elagez, Balaji Nallapati, Felix Bratosin, Adina Bucur, Alina Negru, et al.
Examining the Impact of Ertugliflozin on Cardiovascular Outcomes in Patients with Diabetes and Metabolic Syndrome: A Systematic Review of Clinical Trials
Reprinted from: *Pharmaceuticals* **2024**, *17*, 929, https://doi.org/10.3390/ph17070929 130

Mounica Vorla and Dinesh K. Kalra
Meta-Analysis of the Safety and Efficacy of Direct Oral Anticoagulants for the Treatment of Left Ventricular Thrombus
Reprinted from: *Pharmaceuticals* **2024**, *17*, 708, https://doi.org/10.3390/ph17060708 144

Marco Alexander Valverde Akamine, Beatriz Moreira Ayub Ferreira Soares, João Paulo Mota Telles, Arthur Cicupira Rodrigues de Assis, Gabriela Nicole Valverde Rodriguez, Paulo Rogério Soares, et al.
Role of Dapagliflozin in Ischemic Preconditioning in Patients with Symptomatic Coronary Artery Disease—DAPA-IP Study Protocol
Reprinted from: *Pharmaceuticals* **2024**, *17*, 920, https://doi.org/10.3390/ph17070920 157

Robert Krysiak, Marcin Basiak, Witold Szkróbka and Bogusław Okopień
Autoimmune Thyroiditis Mitigates the Effect of Metformin on Plasma Prolactin Concentration in Men with Drug-Induced Hyperprolactinemia
Reprinted from: *Pharmaceuticals* **2024**, *17*, 976, https://doi.org/10.3390/ph17080976 165

About the Editor

Alfredo Caturano

Alfredo Caturano, MD, PhD, graduated in Medicine and Surgery from the Second University of Naples (SUN) in 2016. He completed his Specialization in Internal Medicine in 2021 at the University of Campania "Luigi Vanvitelli". His specialization thesis, focusing on the impact of glycemic control on cardiovascular outcomes and mortality in a cohort of patients with type 2 diabetes mellitus, was awarded the 2021 Annapaola Braggio prize by the Italian Society of Diabetology as the best thesis on insulin resistance. He earned his PhD in January 2025 from the University of Campania "Luigi Vanvitelli" with a thesis titled *Impact of Empagliflozin and Metformin Combination Therapy vs. Metformin Monotherapy on MASLD Progression in Type 2 Diabetes: Findings from the IMAGIN Study*. Alfredo Caturano currently serves as an Assistant Professor at the Department of Human Sciences and Promotion of the Quality of Life at San Raffaele Roma Open University. He is also a Regional Counselor for the Campania-Basilicata region of the Italian Society of Diabetology and a member of the Early Career Academy Committee of the European Association for the Study of Diabetes (EASD). He has been awarded multiple grants from both Italian and European scientific societies and is actively involved in research focusing on diabetes and cardiovascular health.

Preface

This reprint, *Advancements in Cardiovascular and Antidiabetic Drug Therapy*, explores the transformative progress in pharmacological strategies addressing the intertwined challenges of cardiovascular diseases (CVDs) and diabetes. These conditions remain among the leading global health burdens, necessitating innovative therapeutic solutions to reduce morbidity and mortality. The scope of this Special Issue encompasses the latest advancements, including sodium–glucose cotransporter 2 inhibitors, novel incretin-based therapies, lipid-lowering agents, and groundbreaking insulin delivery systems.

The motivation behind this compilation stems from the urgent need to bridge gaps in knowledge and clinical practice, fostering a deeper understanding of these groundbreaking therapies. By assembling contributions from leading researchers and clinicians, this reprint serves as a comprehensive resource for healthcare professionals, researchers, and policymakers striving to enhance patient care and outcomes in these interrelated fields.

We extend our heartfelt gratitude to the authors for their invaluable contributions and to the reviewers for their critical insights and constructive feedback. Their dedication has ensured the highest quality of scientific content. We also acknowledge the support of the editorial team, whose expertise and guidance made this publication possible.

It is our hope that this reprint inspires further research and clinical advancements, paving the way for innovative strategies to combat the global burden of CVDs and diabetes.

Alfredo Caturano
Guest Editor

Article

Evolving Clinical Features of Diabetic Ketoacidosis: The Impact of SGLT2 Inhibitors

Selin Genc [1], Bahri Evren [2], Onur Selcuk Yigit [3], Ibrahim Sahin [4], Ramazan Dayanan [5], Aleksandra Klisic [6,7], Ayse Erturk [8,*] and Filiz Mercantepe [9,*]

1. Department of Endocrinology and Metabolism, Konya State Hospital, Konya 42250, Türkiye; selin.genc@saglik.gov.tr
2. Department of Endocrinology and Metabolism, Faculty of Medicine, Inonu University, Malatya 44280, Türkiye; bahri.evren@inonu.edu.tr
3. Department of Internal Medicine, Ordu State Hospital, Ordu 52200, Türkiye; onurselcuk.yigit@saglik.gov.tr
4. Department of Endocrinology and Metabolism, Memorial Sisli Hospital, Istanbul 34384, Türkiye; ibrahimsahin@memorial.com.tr
5. Department of Endocrinology and Metabolism, Batman Training and Research Hospital, Batman 72070, Türkiye; ramazan.dayanan@saglik.gov.tr
6. Faculty of Medicine, University of Montenegro, 81000 Podgorica, Montenegro; aleksandranklisic@gmail.com
7. Primary Health Care Center, Center for Laboratory Diagnostics, 81000 Podgorica, Montenegro
8. Department of Infection Disease, Faculty of Medicine, Recep Tayyip Erdogan University, Rize 53100, Türkiye
9. Department of Endocrinology and Metabolism, Faculty of Medicine, Recep Tayyip Erdogan University, Rize 53100, Türkiye
* Correspondence: ayse.erturk@erdogan.edu.tr (A.E.); filiz.mercantepe@saglik.gov.tr (F.M.)

Abstract: **Background/Objectives**: The antidiabetic effect of SGLT2 inhibitors (SGLT2-is) is based on their ability to increase glucose excretion through urine by inhibiting the kidney-resident SGLT2 protein. Euglycemic diabetic ketoacidosis (EuDKA) is an uncommon but potentially life-threatening adverse effect of these medications, which are notable for their antidiabetic, cardiovascular, and renal protective properties. This study aimed to clarify the impact of SGLT2-is on demographic, clinical, and biochemical characteristics in patients with DKA. **Methods**: A total of 51 individuals with a diagnosis of DKA were included in the trial; 19 of these patients were treated with SGLT2-is, while 32 were not. Patients diagnosed with DKA and treated with SGLT2-is were compared to those not treated with the medication in terms of clinical, biochemical, and laboratory characteristics. **Results**: The age of patients utilizing SGLT2-is was statistically considerably greater than that of non-users ($p < 0.001$). EuDKA was exclusively noted in the SGLT2-is cohort ($p = 0.005$). Urinary tract infections, vulvovaginitis, and genitourinary infections were substantially more prevalent among SGLT2-i users compared with non-users among both women and the overall patient group ($p = 0.036$, $p = 0.001$, $p = 0.005$, $p = 0.003$, respectively). Plasma glucose concentrations were significantly higher in SGLT2-i non-users ($p = 0.006$). Chloride (Cl$^-$) concentrations were elevated among SGLT2-i users ($p = 0.036$). **Conclusions**: The study findings indicate that SGLT2 inhibitors may substantially influence age, serum chloride, EuDKA, and the occurrence of genitourinary infections in individuals with DKA.

Keywords: diabetic ketoacidosis (DKA); euglycemic diabetic ketoacidosis (euDKA); genitourinary infection (GUI); latent autoimmune diabetes in adults (LADA); sodium–glucose co-transporter-2 inhibitor (SGLT2-i); type 1 diabetes mellitus (T1DM)

Citation: Genc, S.; Evren, B.; Yigit, O.S.; Sahin, I.; Dayanan, R.; Klisic, A.; Erturk, A.; Mercantepe, F. Evolving Clinical Features of Diabetic Ketoacidosis: The Impact of SGLT2 Inhibitors. *Pharmaceuticals* **2024**, *17*, 1553. https://doi.org/10.3390/ph17111553

Academic Editors: Agnieszka Sliwinska and Alfredo Caturano

Received: 28 September 2024
Revised: 7 November 2024
Accepted: 18 November 2024
Published: 20 November 2024

Copyright: © 2024 by the authors. Licensee MDPI, Basel, Switzerland. This article is an open access article distributed under the terms and conditions of the Creative Commons Attribution (CC BY) license (https://creativecommons.org/licenses/by/4.0/).

1. Introduction

Sodium–glucose co-transporter-2 inhibitors (SGLT2-is) are a class of recently developed oral antidiabetic medications that lower blood glucose levels by inhibiting the renal reabsorption of glucose in the proximal tubule [1]. The cardioprotective and renoprotective attributes of these medicines have broadened their application owing to their potential to diminish mortality in diabetic patients suffering from heart failure and chronic kidney disease [2–4]. Anti-inflammatory, antioxidant, and immunomodulatory effects may also be achieved by using SGLT2 inhibitors [5–7]. Despite the multiple beneficial therapeutic effects of SGLT2 inhibitors, fresh concerns have emerged regarding their potential to elevate the risk of euglycemic diabetic ketoacidosis (euDKA), particularly in individuals with type 1 diabetes mellitus (T1DM) and latent autoimmune diabetes (LADA) [8,9].

Diabetic ketoacidosis (DKA) is a severe acute metabolic complication of diabetes characterized by the excessive synthesis of ketone bodies due to insulin insufficiency, posing a life-threatening risk [10]. Hyperglycemia is a significant indicator in classic diabetic ketoacidosis and is frequently identified promptly. Nonetheless, hyperglycemia may be absent in ketoacidosis associated with the administration of SGLT2 inhibitors [9]. This condition is marked by elevated ketone synthesis despite generally normal blood glucose levels, complicating the diagnosis [11]. Prolonged diagnosis of EuDKA elevates the risk of consequences for patients [12].

The mechanism through which SGLT2 inhibitors enhance ketone synthesis remains inadequately elucidated. It is believed that by decreasing hyperglycemia, they also diminish insulin secretion, increase glucagon secretion, and ultimately promote lipolysis and enhance ketone generation [8,13]. Conditions like T1DM or LADA are believed to induce ketoacidosis in patients due to their reduction of insulin needs for the same rationale [11,14]. The literature underscores that the utilization of SGLT2 inhibitors elevates the likelihood of euDKA, necessitating caution from practitioners about this matter. SGLT2 inhibitors are linked to consequences, including genitourinary infections (GUIs), particularly urinary tract infections (UTIs) and vulvovaginitis, due to their promotion of glucosuria [15,16]. This circumstance markedly heightens the vulnerability to bacterial and fungal infections, particularly in women. The side effects diminish the drug's efficacy, which is an antidiabetic notable for its cardioprotective and renoprotective attributes, and adversely impact the patient's quality of life.

This study aimed to investigate the impact of SGLT2 inhibitor administration on clinical, biochemical, and electrolyte parameters in patients presenting to the emergency department with DKA. The potential effects of SGLT2 inhibitor usage on various parameters were assessed by comparing demographic characteristics, laboratory results, and serum electrolyte levels between the groups.

2. Results

The general characteristics of the study participants are shown in Table 1. Of the 51 DKA patients, 5 had EuDKA. Table 2 presents the comparative demographic and clinical data of patient groups utilizing and not utilizing SGLT2 inhibitors (SGLT2-is). The age of patients utilizing SGLT2 inhibitors was statistically considerably greater than that of non-users ($p < 0.001$). None of the T1DM patients utilized SGLT2 inhibitors; however, all four patients misdiagnosed as having T2DM at external facilities but subsequently identified as having LADA through comprehensive evaluations at our center were found to be utilizing SGLT2 inhibitors. EuDKA was exclusively noted in the SGLT2-is cohort, and this disparity was statistically significant ($p = 0.005$). Urinary tract infections, vulvovaginitis, and genitourinary system infections were substantially more prevalent among SGLT2 inhibitor users compared with non-users among both women and the overall patient group ($p = 0.036$, $p = 0.001$, $p = 0.005$, $p = 0.003$, respectively).

Table 3 presents the comparative biochemical and hematological laboratory data of patient groups utilizing and not utilizing SGLT2 inhibitors. Plasma glucose concentrations were significantly higher in SGLT2-i non-users ($p = 0.006$). Chloride (Cl^-) concentrations were elevated among SGLT2-i users ($p = 0.036$). Urine density was similar in both groups; urinary white blood cell count (WBC) was significantly lower in the SGLT2-is group ($p < 0.001$). No notable variation was detected in the other urine, hematological, and biochemical markers. These findings indicate that SGLT2 inhibitors may substantially influence age, serum chloride, and glucose concentrations, as well as the occurrence of urinary infections, in individuals with DKA.

Table 1. General characteristics of the study participants.

Variables	Categorization	n (n = 51)	Percentage (%)
Gender	Male	18	35.3
	Female	33	64.7
Smoking	Yes	14	27.5
	No	37	72.5
DM Type	T1DM	25	49
	T2DM	22	43.1
	LADA	4	7.8
SGLT2-is	Yes	19	37.3
	No	32	62.7
EuDKA	Yes	5	9.8
	No	46	90.2
UTI	Yes	13	25.5
	No	38	74.5
Vulvovaginitis	Yes	6	11.8
	No	45	88.2
HT	Yes	16	31.4
	No	35	68.6
Hyperlipidemia	Yes	3	5.9
	No	48	94.1
Mortality	Yes	4	7.8
	No	47	92.2
Intubation	Yes	5	9.8
	No	46	90.2
DM Treatment	No Treatment	1	2
	Insulin	28	54.9
	OAD	13	25.5
	Insulin + OAD	7	13.7
	Insulin Pump	2	3.9

Abbreviations: DM, diabetes mellitus; T1DM, type 1 DM; T2DM, type 2 DM; LADA, latent autoimmune diabetes in adults; SGLT2-is, sodium–glucose co-transporter-2 inhibitors; EuDKA, euglycemic diabetic ketoacidosis; UTI, urinary tract infection; HT, hypertension; OAD, oral antidiabetic drug.

Table 2. Comparative demographic and clinical parameters of patients according to SGLT2-i use.

Variables		SGLT2-is (+)	SGLT2-is (−)	Total	p–Value
Sex	Male	7	11	18	0.859
	Female	12	21	33	
Age (year)		59 ± 18	37 ± 17		<0.001
Height (cm)		164 ± 7	164 ± 9		0.795
Weight (kg)		70 ± 16	67 ± 14		0.525
BMI (kg/m^2)		26 ± 6	25 ± 5		0.429
Smoking (+)		5 (26.3)	9 (28.1)	14	0.889
DM Type	T1DM	0	25	25	<0.001
	T2DM	7	15	22	<0.001
	LADA	4	0	4	<0.001
EuDKA (+)		5	0	5	0.005 *
HT (+)		6 (31.6)	10 (31.3)	16	>0.05
UTI		8 (42.1)	5 (15.6)	13	0.036 *
Vulvovaginitis		6 (31.6)	0 (0)	6	0.001 *
GUS infection—female		10 (52.6)	5 (15.6)	15	0.005 *
GUS infection—total		12 (63.2)	7 (21.9)	19	0.003 *

* p < 0.05 is significant. Abbreviations: SGLT2-is, sodium–glucose co-transporter-2 inhibitors; BMI, body mass index; DM, diabetes mellitus; T1DM, type 1 DM; T2DM, type 2 DM; LADA, latent autoimmune diabetes in adults; EuDKA, euglycemic diabetic ketoacidosis; UTI, urinary tract infection; HT, hypertension; GUS, genitourinary system.

Table 3. Comparative laboratory parameters of patients according to SGLT2-i use.

Variables	SGLT2-is (+)	SGLT2-is (−)	p–Value
HbA1c (%)	11.2 ± 2.5	10.9 ± 2.2	0.623
Glucose (mg/dL)	398 ± 191	564 ± 206	0.006 *
BUN (mg/dL)	25 ± 13	27 ± 23	0.649
Creatinine (mg/dL)	1.3 ± 0.5	1.6 ± 1.2	0.247
Na^+ (mmol/L)	135 ± 5	130 ± 6	0.008 *
cNa^+ (mmol/L)	140 ± 4	137 ± 6	0.205
K^+ (mmol/L)	4.5 ± 0.8	4.8 ± 0.9	0.176
Cl^- (mmol/L)	103 (90–112)	98 (41–113)	0.036 *
Ca^{2+} (mmol/L)	8.9 ± 0.7	8.9 ± 0.8	0.996
Mg^{2+} (mEq/L)	2.3 ± 0.9	2 ± 0.5	0.201
P (mEq/L)	3.9 ± 3.4	4 ± 2.1	0.898
Albumin (mg/dL)	3.6 ± 0.5	3.6 ± 0.9	0.885
WBC (10^3 μL)	13.5 ± 6.5	14.5 ± 12	0.734
Neutrophil (10^3 μL)	11.7 ± 6	11.5 ± 10.4	0.935
Lymphocyte (10^3 μL)	1.4 ± 0.8	2 ± 1.8	0.121
Hemoglobin (gr/dL)	14.2 ± 3.2	13 ± 2.6	0.147
Platelet (10^3 μL)	261 ± 149	303 ± 115	0.262
CRP (mg/dL)	4.8 (0.3–36)	4.2 (0.3–30)	0.822
pH	7.17 ± 0.17	7.16 ± 0.16	0.853
HCO_3 (mmol/L)	12.5 ± 5.5	11.7 ± 4.5	0.796
Lactate mmol/L	3.3 ± 3.7	2.6 ± 1.9	0.344
PO_2 (mmHg)	66 ± 42	69 ± 37	0.733
O_2 Saturation (%)	97 ± 3	97 ± 2	0.232
Urine Density (g/cm^3)	1019 ± 7	1019 ± 7	0.831
Urine WBC (/HPF)	16 (1–70)	22 (0–617)	<0.001 *
Urine RBC (/HPF)	12 (0–51)	9 (0–75)	0.178

* $p < 0.05$ is significant. Abbreviations: SGLT2-is, sodium–glucose co-transporter-2 inhibitors; BUN, blood urea nitrogen; Na, sodium; cNa, corrected sodium; K, potassium; Cl, chloride; Ca, calcium; Mg, magnesium; P, phosphorus; WBC, white blood cell; CRP, C-reactive protein; HCO_3, bicarbonate; PO_2, partial pressure of oxygen; RBC, red blood cell.

3. Discussion

This study investigated the impact of SGLT2 inhibitors on patients with diabetic ketoacidosis, revealing a higher prevalence of euglycemic ketoacidosis in LADA and an increased incidence of genitourinary infections, particularly among female patients. Our findings corroborate analogous studies in the literature and appear to align with prevailing notions indicating that SGLT2-i use may facilitate the onset of EuDKA [17].

The correlation between SGLT2 inhibitors and DKA, as well as EuDKA, is found in the mechanism via which these medications diminish glucose reabsorption in the kidneys and augment glucose excretion in the urine. This may stimulate ketone body formation by enhancing fatty acid use for energy, hence elevating the risk of ketoacidosis [18,19]. It is important to note that glucose levels may be normal or only marginally raised in patients utilizing SGLT2 inhibitors, leading to euglycemic ketoacidosis, which presents a clinical profile distinct from classical diabetic ketoacidosis. The increased occurrence of euglycemic ketoacidosis among patients utilizing SGLT2 inhibitors in our study aligns with this mechanism, a phenomenon commonly highlighted in numerous publications in the literature [9,20–22].

The investigation revealed that the incidence of EuDKA escalated with the use of SGLT2 inhibitors in LADA patients. LADA is a form of diabetes characterized by autoimmune properties akin to type 1 diabetes, although it manifests in adults [8]. Endogenous insulin synthesis is diminished, and beta cell functionality is compromised in these patients. SGLT2 inhibitors enhance glucose excretion via urine by inhibiting glucose reabsorption in the kidneys, thereby lowering blood glucose levels through an insulin-independent mechanism. In LADA patients with restricted insulin production, the use of SGLT2 inhibitors may prompt cells to utilize fatty acids as an energy source, resulting in elevated ketone generation [20]. This is the primary mechanism that elevates the risk of developing EuDKA [9]. The elevated occurrence of euglycemic ketoacidosis can be attributed to SGLT2 inhibitors potentially inducing excessive ketone body formation while maintaining normal glucose levels. Symptoms of classical ketoacidosis may be disregarded due to acceptable glucose levels, complicating early identification in LADA patients. The restricted endogenous insulin reserves in LADA patients may render them more susceptible to SGLT2-i administration, hence elevating the likelihood of developing euDKA [8]. An internal medicine specialist rather than an endocrinologist diagnosed all LADA patients in the present study, and they were all under observation at locations other than our own. We believe that because LADA is a kind of diabetes typically observed in adult individuals, it was probably misclassified as T2DM, leading to the use of SGLT2 inhibitors. Considering the literature and this study's findings, we underscore the necessity of recognizing the potential for LADA and pursuing research in this area, particularly when assessing adult-onset diabetes at the initial diagnosis. Consequently, as demonstrated in the present study, diabetic patients may face lethal consequences from adverse effects arising from the improper use of antidiabetic medications. In conclusion, due to the elevated risk of euglycemic ketoacidosis in LADA patients utilizing SGLT2 inhibitors alongside insulin therapy, these patients require enhanced surveillance to mitigate the risk of ketoacidosis, necessitating more clinical oversight [13].

The higher incidence of genitourinary infections observed in this investigation, as well as in several studies in the literature, may be attributed to the propensity of SGLT2 inhibitors to heighten infection risk, particularly in women, by augmenting glucose excretion via urine [15,16]. Elevated glucose levels in urine foster an environment conducive to bacterial proliferation, hence heightening the risk of genitourinary infections [15]. The prevalence of these conditions, particularly among female patients, can be attributed to the anatomical characteristics of the female genitourinary system, which predispose them to such infections.

An interesting finding in present study is that serum chloride levels were higher in patients using SGLT2-is. SGLT2 inhibitors lower glucose levels by enhancing urinary glucose excretion while also influencing body fluid distribution and electrolyte equilibrium via osmotic diuresis and natriuresis actions [23–25]. Chloride is the second most prevalent serum electrolyte following sodium, crucial for regulating body fluids, maintaining electrolyte balance, and ensuring acid–base equilibrium. It is a fundamental component utilized in the assessment of numerous clinical conditions. Studies indicate that chloride is the principal electrolyte governing fluid distribution in the human body based on the likely biochemical nature of solutes. The hypothesis positing that chloride, rather than sodium, serves as the primary electrolyte governing plasma volume in the human body is referred to as the "Chloride theory" [26]. The impact of SGLT2 inhibitors on serum chloride levels remains incompletely elucidated; however, various mechanisms have been proposed, such as the stimulation of the renin–angiotensin–aldosterone system (RAAS), bicarbonate depletion, and proximal tubule effects [26].

SGLT2 inhibitors are believed to enhance urinary chloride reabsorption through the activation of the RAAS [26]. This may result in the maintenance or elevation of chloride levels. Nonetheless, the impact on the RAAS remains contentious, with several studies indicating that these drugs may suppress RAAS activity [26]. Secondly, SGLT2 inhibitors may diminish serum bicarbonate (HCO_3^-) concentrations as a result of buffering organic acids produced during ketone body metabolism. This may result in elevated chloride levels, as the rise in chloride through the bicarbonate–chloride equation could serve as a compensatory mechanism. An increase in blood chloride concentration is anticipated due to the decrease in serum HCO_3^- concentration caused by the buffering of strong organic acid metabolites, such as acetoacetic acid and 3-hydroxybutyric acid, after SGLT2-i therapy [23–25]. This also underscores a potential correlation between chloride and ketoacidosis. The study findings may indicate that elevated serum chloride levels in patients utilizing SGLT2 inhibitors corroborate this hypothesis. The limited sample size of the investigation inhibited a comparison of serum chloride levels between euglycemic diabetic ketoacidosis patients utilizing SGLT2 inhibitors and those not utilizing them. It is believed that SGLT2 inhibitors diminish NaHCO$_3$ reabsorption by inhibiting the Na$^+$/H$^+$ exchanger-3 (NHE3) in the renal proximal tubules, while perhaps enhancing chloride reabsorption [24,26,27]. In the investigation, no difference in Na$^+$ ion levels was seen between persons utilizing SGLT2-is and those not utilizing them, aligning with existing data [25,28]. The effect of SGLT2 inhibitors on electrolytes and glucose in the nephron is shown in Figure 1. Nevertheless, further research is required to conclusively ascertain which of these pathways is predominantly responsible for this phenomenon or if alternative mechanisms exist.

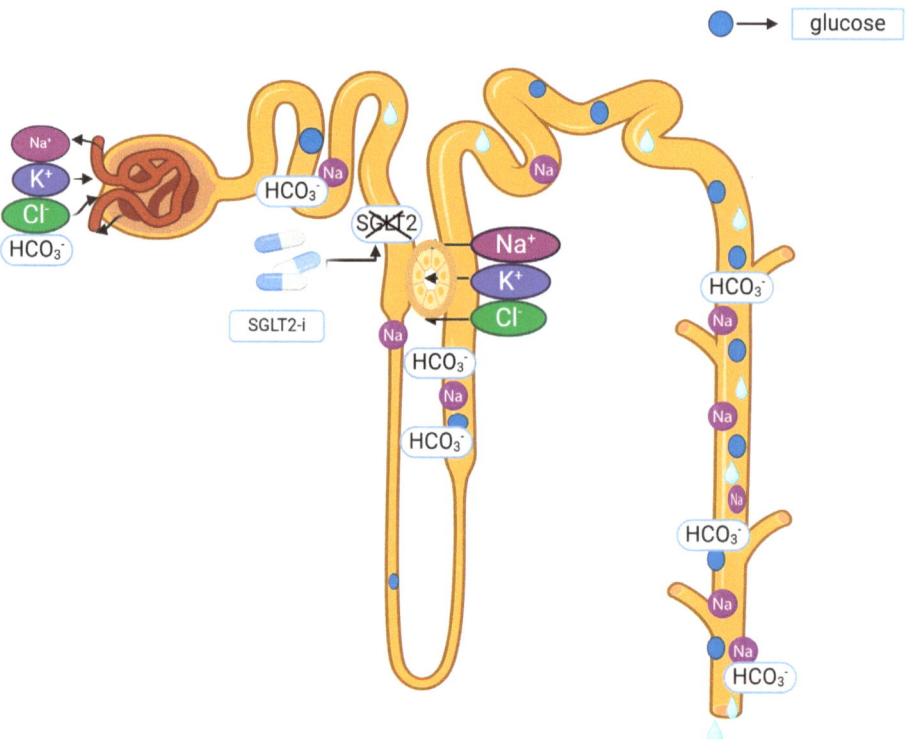

Figure 1. SGLT2-is inhibits SGLT2 in the S1 and S2 segments of the proximal tubule, preventing the reabsorption of sodium, glucose, bicarbonate, and water. SGLT2-is weakens the reabsorption of sodium, glucose, bicarbonate, and water, while enhancing the reabsorption of chloride and potassium. : sodium loss, : potassium gain, : chloride gain, : bicarbonate loss, : SGLT2 inhibitions by SGLT2-is.

While SGLT2 inhibitors are typically regarded as safe and advantageous for cardiorenal protection in the management of T2DM, they are linked to EuDKA, an uncommon yet severe consequence [18,29]. This condition is marked by an elevation of ketone bodies (notably β-hydroxybutyrate) and the onset of acidosis, despite the patient's generally normal blood glucose levels [22]. While elevated ketone generation from fatty acid use due to insulin insufficiency is attributed to the disease, alternative underlying processes may also be present [19,21,30]. In examining the impact of SGLT2-i administration on serum chloride levels in our investigation and the existing literature, we questioned whether this phenomenon could be associated with EuDKA. SGLT2 inhibitors enhance glucose excretion, stimulate fatty acid metabolism, and elevate ketone generation, resulting in decreased serum bicarbonate levels and acidosis [8], which prompted us to consider this condition as the cause of increased chloride levels. Chloride ions may serve as a compensatory response to address the prevailing acidosis through a bicarbonate–chloride balancing mechanism [26]. Alternatively, the converse may also be contemplated. It can be inferred that SGLT2 inhibitors diminish NaHCO3 reabsorption by inhibiting the Na^+/H^+ exchanger-3 (NHE3) in the renal proximal tubules while enhancing chloride reabsorption, leading to acidosis due to the reduction in NaHCO3 reabsorption, resulting in ketoacidosis in the absence of hyperglycemia. In EuDKA, the equilibrium of bodily fluids and electrolytes is disturbed [31]. A correlation may exist between the elevation of ketone bodies and chloride reabsorption. The impact of SGLT2 inhibitors on the RAAS system may also lead to electrolyte imbalance and elevate the risk of EuDKA. In conclusion, the capacity of SGLT2 inhibitors to sustain or elevate serum chloride levels may contribute to the acidosis that occurs during EuDKA. Nevertheless, comprehensive investigations examining all these pathways and additional factors are essential for a more definitive understanding.

The current study possesses several limitations, and its deficiencies must be acknowledged during evaluation. The limited sample size, retrospective methodology, insufficient assessment of additional risk factors potentially influencing the onset of EuDKA, variability in treatment protocols, and absence of long-term follow-up are significant factors impacting the results.

This study's retrospective design makes it difficult to infer cause–effect linkages with certainty. Furthermore, results may not be as robust as those from a prospective design, and the design is susceptible to bias. The results can only be applied to the specific population studied because the sample size was so small (51 patients). Results from studies that include more patients will show how effective SGLT2 inhibitors are. The metabolic differences among this study's diabetic participants, particularly between type 1 and type 2 diabetes, impact the onset of euDKA and the efficacy of SGLT2 inhibitors. The impact of SGLT2 inhibitors on patients with poor insulin reserves, including those with type 1 diabetes or living with LADA, may vary from that on those with type 2 diabetes. A crucial consideration in assessing the efficacy of SGLT2 inhibitors is, thus, the type of diabetes. Glucosuria, a side effect of SGLT2 inhibitors, raises the danger of genitourinary infections. In addition to altering the clinical course of DKA or its severity, these infections can impact patients' overall health. This is a significant consideration since it might change the results of this study. Patients on SGLT2 inhibitors had elevated chloride (Cl^-) levels, according to this research. Although the exact reason for these variations is still unclear, it is possible that other fluid–electrolyte imbalances or acid–base disorders influenced the findings. This condition needs to be clarified by additional study. The results can be skewed due to individual variances in the study's treatment regimens. How SGLT2 inhibitors work and how the condition progresses in each individual patient may depend on their unique therapy and care regimen. Unfortunately, we do not know much about the SGLT2 inhibitors' long-term consequences because this study did not follow up with patients for very long. How these medications' impact on EuDKA or other consequences will evolve over time requires more research.

4. Materials and Methods

4.1. Study Participants

This retrospective cohort study was performed in the emergency department of a tertiary hospital from April 2018 to November 2022. The Inonu University Scientific Research and Publication Ethics Committee approved this study (Approval Date: 2 May 2023, Approval Number: 2023/4560). During this research, we meticulously adhered to the guidelines established in the Declaration of Helsinki at every phase. The retrospective nature of this study obviates the need for a voluntary consent form. This study comprised 508 individuals aged 18 and older who presented to the emergency room with hyperglycemia. Among the 508 patients assessed by the endocrinologist in the emergency department for hyperglycemia, 51 were diagnosed with DKA. The patients were classified based on their utilization of SGLT2 inhibitors. This study did not include participants who were under the age of 18, breastfeeding mothers, or pregnant women. The data of all patients were retrospectively gathered from the hospital network system and the national health database. The sociodemographic data, diagnoses, diabetes types, age at admission, medical history, physical examination results, medications administered, and laboratory parameters at admission were processed through this system, anonymized, and documented. Individuals with unspecified diabetic management and missing laboratory results were excluded from this study. A flow diagram of the study design is shown in Figure 2.

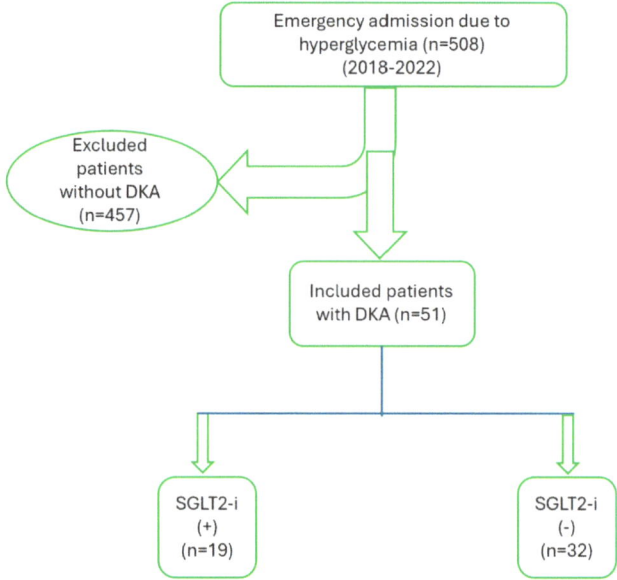

Figure 2. Flow diagram showing the study design. Abbreviations: DKA, diabetic ketoacidosis; SGLT2-is, sodium–glucose co-transporter-2 inhibitors.

4.2. Data Collection

When patients were admitted to the emergency department, blood samples were taken to measure plasma glucose, hemoglobin A1c (HbA1c), creatinine, uric acid, aspartate aminotransferase (AST), alanine aminotransferase (ALT), albumin, sodium, potassium, chloride, magnesium, calcium, phosphorus, blood gas, C-reactive protein (CRP), and complete blood count. A full blood count was conducted using a Sysmex XN-1000 automated hematological cell calculator device. A chemiluminescence spectrophotometric method (Beckman Coulter, Brea, CA, USA)) was used to measure the following biochemical parameters: albumin, sodium, potassium, chloride, magnesium, calcium, AST, and ALT, as well as plasma glucose, creatinine, uric acid, and blood urea nitrogen (BUN). HbA1c was measured by the high-performance liquid chromatography method (Adams A1c Ha-8160-BIODPC SN:10912002, Manufacturer: Arkray Factory Inc., 1480 Koji, Konan-Cho, Koka-Shi Shiga, Japan; European Representative Arkray Europe, Prof. J.H. Bavincklaan 51183 At Amstelveen, The Netherlands, made in Japan) [32]. Nephelometric analysis was used to detect CRP (Siemens Healthcare Diagnostics Products GmbH, 35041 Marburg, Germany, Type BN II System, SN: 202826). The emergency department's ABL 800 flex blood gas instrument was used to perform venous blood gas investigations. Digital imaging of the whole urinalysis flow cell and automated analysis of the urine's chemistry using dual-wavelength reflectance photometry were employed in the automated urine sediment analysis (BT URICELL 1280–1600 URINALYSIS, Gaziemir, İzmir, Turkey).

A health professional employed a Jadever-Türkter NLD-W 300 kg height and weight scale to collect additional measurements. BMI measurement was calculated as: BMI = weight (kg)/height (m)^2 [33]. Adjustment for sodium (cNa) was calculated as: cNa = measured Na + (glucose level − 100) × 0.016. Urinary tract infection and vulvovaginitis were confirmed by urine culture and genital examination.

4.3. Statistical Analysis

All analyses were conducted using SPSS 22.0 (SPSS Inc. of Chicago, IL, USA). A normality test was conducted before beginning the analyses. Frequency analysis was used to acquire descriptive information regarding this study's variables. The Student's t-test and the Mann–Whitney U Test were employed to compare normally and non-normally distributed data between two independent groups, respectively. In 2 × 2 comparisons of categorical variables, the Pearson Chi-square test was employed when the expected value exceeded 5, whereas the Chi-square Yates test was utilized when the expected value ranged from 3 to 5. Fisher's Exact test was employed when the anticipated value was less than 3. The Pearson Chi-square test was employed for comparisons of categorical variables greater than 2 × 2 when the expected value exceeded 5, whereas the Fisher–Freeman–Halton test was utilized when the expected value was less than 5. The threshold for statistical significance was established at $p < 0.05$.

5. Conclusions

The current study indicated that SGLT2 inhibitors may elevate the incidence of euglycemic ketoacidosis in diabetic patients and that genitourinary infections are more prevalent, particularly among women. Moreover, serum chloride concentrations were elevated in patients utilizing SGLT2 inhibitors. These data suggest that caution should be observed for the potential hazards and adverse effects of SGLT2 inhibitors. The prompt identification of euglycemic ketoacidosis and vigilant surveillance of patients predisposed to genitourinary infections is crucial for prevention. Furthermore, this research underscores the need to provide SGLT2 inhibitors to the appropriate patient demographic and accurately differentiate among the many kinds of diabetes. Additional study is required to comprehend the clinical implications of elevated serum chloride levels and their impact on ketoacidosis, as well as cardiovascular and renal consequences.

Author Contributions: All the authors contributed to the study conception and design. Material preparation, data collection, and laboratory analyses were performed by S.G., B.E., O.S.Y. and I.S. Statistical analysis was performed by F.M. and A.K. The first draft of the manuscript was written by F.M., A.E., R.D. and S.G. and all the authors commented on previous versions of the manuscript. All authors have read and agreed to the published version of the manuscript.

Funding: This research received no external funding.

Institutional Review Board Statement: This study received ethical approval from the local ethics commission (Approval Date: 2 May 2023, Approval Number: 2023/4560).

Informed Consent Statement: This is a retrospective study; an informed consent form was unnecessary.

Data Availability Statement: All data generated or analyzed during this study are included in this article. The data will be available upon reasonable request (contact person: filiz.mercantepe@saglik.gov.tr).

Conflicts of Interest: The authors declare that they have no known competing financial interests or personal relationships that could have appeared to influence the work reported in this paper.

References

1. Karamichalakis, N.; Kolovos, V.; Paraskevaidis, I.; Tsougos, E. A New Hope: Sodium-Glucose Cotransporter-2 Inhibition to Prevent Atrial Fibrillation. *J. Cardiovasc. Dev. Dis.* **2022**, *9*, 236. [CrossRef]
2. Santos-Gallego, C.G.; Ibanez, J.A.R.; Antonio, R.S.; Ishikawa, K.; Watanabe, S.; Picatoste Botija, M.B.; Salvo, A.J.S.; Hajjar, R.; Fuster, V.; Badimon, J. Empagliflozin Induces a Myocardial Metabolic Shift from Glucose Consumption to Ketone Metabolism that Mitigates Adverse Cardiac Remodeling and Improves Myocardial Contractility. *J. Am. Coll. Cardiol.* **2018**, *71*, A674. [CrossRef]
3. Anders, H.J.; Huber, T.B.; Isermann, B.; Schiffer, M. CKD in diabetes: Diabetic kidney disease versus nondiabetic kidney disease. *Nat. Rev. Nephrol.* **2018**, *14*, 361–377. [CrossRef]
4. Packer, M.; Anker, S.D.; Butler, J.; Filippatos, G.; Pocock, S.J.; Carson, P.; Januzzi, J.; Verma, S.; Tsutsui, H.; Brueckmann, M.; et al. Cardiovascular and Renal Outcomes with Empagliflozin in Heart Failure. *N. Engl. J. Med.* **2020**, *383*, 1413–1424. [CrossRef]
5. Schönberger, E.; Mihaljević, V.; Steiner, K.; Šarić, S.; Kurevija, T.; Majnarić, L.T.; Bilić Ćurčić, I.; Canecki-Varžić, S. Immunomodulatory Effects of SGLT2 Inhibitors—Targeting Inflammation and Oxidative Stress in Aging. *Int. J. Environ. Res. Public Health* **2023**, *20*, 6671. [CrossRef]
6. Yaribeygi, H.; Atkin, S.L.; Butler, A.E.; Sahebkar, A. Sodium–glucose cotransporter inhibitors and oxidative stress: An update. *J. Cell. Physiol.* **2019**, *234*, 3231–3237. [CrossRef]
7. Peng, W.K.; Chen, L.; Boehm, B.O.; Han, J.; Loh, T.P. Molecular phenotyping of oxidative stress in diabetes mellitus with point-of-care NMR system. *npj Aging Mech. Dis.* **2020**, *6*, 11. [CrossRef]
8. Vadasz, B.; Arazi, M.; Shukha, Y.; Koren, O.; Taher, R. Sodium-glucose cotransporter-2-induced euglycemic diabetic ketoacidosis unmasks latent autoimmune diabetes in a patient misdiagnosed with type 2 diabetes mellitus: A case report. *J. Med. Case Rep.* **2021**, *15*, 62. [CrossRef]
9. Morace, C.; Lorello, G.; Bellone, F.; Quartarone, C.; Ruggeri, D.; Giandalia, A.; Mandraffino, G.; Minutoli, L.; Squadrito, G.; Russo, G.T.; et al. Ketoacidosis and SGLT2 Inhibitors: A Narrative Review. *Metabolites* **2024**, *14*, 264. [CrossRef] [PubMed]
10. Ata, F.; Yousaf, Z.; Khan, A.A.; Razok, A.; Akram, J.; Ali, E.A.H.; Abdalhadi, A.; Ibrahim, D.A.; Al Mohanadi, D.H.S.H.; Danjuma, M.I. SGLT-2 inhibitors associated euglycemic and hyperglycemic DKA in a multicentric cohort. *Sci. Rep.* **2021**, *11*, 110293. [CrossRef] [PubMed]
11. Oriot, P.; Hermans, M.P. Euglycemic diabetic ketoacidosis in a patient with type 1 diabetes and SARS-CoV-2 pneumonia: Case-report and review of the literature. *Acta Clin. Belgica Int. J. Clin. Lab. Med.* **2022**, *77*, 113–117. [CrossRef] [PubMed]
12. Siebel, S.; Galderisi, A.; Patel, N.S.; Carria, L.R.; Tamborlane, W.V.; Sherr, J.L. Reversal of ketosis in type 1 diabetes is not adversely affected by SGLT2 inhibitor therapy. *Diabetes Technol. Ther.* **2019**, *21*, 101–104. [CrossRef] [PubMed]
13. Barski, L.; Eshkoli, T.; Brandstaetter, E.; Jotkowitz, A. Euglycemic diabetic ketoacidosis. *Eur. J. Intern. Med.* **2019**, *63*, 9–14. [CrossRef] [PubMed]
14. Kapila, V.; Topf, J. Sodium-Glucose Co-transporter 2 Inhibitor-Associated Euglycemic Diabetic Ketoacidosis After Bariatric Surgery: A Case and Literature Review. *Cureus* **2021**, *13*, e17093. [CrossRef] [PubMed]
15. Pishdad, R.; Auwaerter, P.G.; Kalyani, R.R. Diabetes, SGLT-2 Inhibitors, and Urinary Tract Infection: A Review. *Curr. Diab. Rep.* **2024**, *24*, 108–117. [CrossRef]
16. Arshad, M.; Hoda, F.; Siddiqui, N.; Najmi, A.; Ahmad, M. Genito Urinary Infection and Urinary Tract Infection in Patients with Type 2 Diabetes Mellitus Receiving SGLT2 Inhibitors: Evidence from a Systematic Literature Review of Landmark Randomized Clinical Trial. *Drug Res.* **2024**, *7*, 307–313. [CrossRef]
17. Somagutta, M.R.; Agadi, K.; Hange, N.; Jain, M.S.; Batti, E.; Emuze, B.O.; Amos-Arowoshegbe, E.O.; Popescu, S.; Hanan, S.; Kumar, V.R.; et al. Euglycemic Diabetic Ketoacidosis and Sodium-Glucose Cotransporter-2 Inhibitors: A Focused Review of Pathophysiology, Risk Factors, and Triggers. *Cureus* **2021**, *13*, e13665. [CrossRef]
18. Juneja, D.; Nasa, P.; Jain, R.; Singh, O. Sodium-glucose Cotransporter-2 Inhibitors induced euglycemic diabetic ketoacidosis: A meta summary of case reports. *World J. Diabetes* **2023**, *14*, 1314–1322. [CrossRef]
19. Bonora, B.M.; Avogaro, A.; Fadini, G.P. Euglycemic Ketoacidosis. *Curr. Diab. Rep.* **2020**, *20*, 25. [CrossRef]
20. He, Z.; Lam, K.; Zhao, W.; Yang, S.; Li, Y.; Mo, J.; Gao, S.; Liang, D.; Qiu, K.; Huang, M.; et al. SGLT-2 inhibitors and euglycemic diabetic ketoacidosis/diabetic ketoacidosis in FAERS: A pharmacovigilance assessment. *Acta Diabetol.* **2023**, *60*, 401–411. [CrossRef]
21. Puls, H.A.; Haas, N.L.; Franklin, B.J.; Theyyunni, N.; Harvey, C.E. Euglycemic diabetic ketoacidosis associated with SGLT2i use: Case series. *Am. J. Emerg. Med.* **2021**, *44*, 11–13. [CrossRef] [PubMed]
22. El Ess, M.S.; ElRishi, M.A. Severe euglycemic diabetic ketoacidosis secondary to sodium-glucose co-transporter 2 inhibitor: Case report and literature review. *Ann. Med. Surg.* **2023**, *85*, 2097–2101. [CrossRef] [PubMed]
23. Cianciolo, G.; De Pascalis, A.; Capelli, I.; Gasperoni, L.; Di Lullo, L.; Bellasi, A.; La Manna, G. Mineral and Electrolyte Disorders with SGLT2i Therapy. *JBMR Plus* **2019**, *3*, e10242. [CrossRef] [PubMed]
24. Cianciolo, G.; De Pascalis, A.; Gasperoni, L.; Tondolo, F.; Zappulo, F.; Capelli, I.; Cappuccilli, M.; Manna, G. La The Off Target Effects, Electrolyte and Mineral Disorders of SGLT2i. *Molecules* **2020**, *25*, 2757. [CrossRef] [PubMed]
25. Zhang, J.; Huan, Y.; Leibensperger, M.; Seo, B.; Song, Y. Comparative Effects of Sodium-Glucose Cotransporter 2 Inhibitors on Serum Electrolyte Levels in Patients with Type 2 Diabetes: A Pairwise and Network Meta-Analysis of Randomized Controlled Trials. *Kidney360* **2022**, *3*, 477–487. [CrossRef]

26. Kataoka, H.; Yoshida, Y. Enhancement of the serum chloride concentration by administration of sodium-glucose cotransporter-2 inhibitor and its mechanisms and clinical significance in type 2 diabetic patients: A pilot study. *Diabetol. Metab. Syndr.* **2020**, *12*, 5. [CrossRef]
27. León Jiménez, D.; Gómez Huelgas, R.; Miramontes González, J.P. The mechanism of action of sodium–glucose co-transporter 2 inhibitors is similar to carbonic anhydrase inhibitors. *Eur. J. Heart Fail.* **2018**, *20*, 409. [CrossRef]
28. Stamatiades, G.A.; D'Silva, P.; Elahee, M.; Viana, G.M.; Sideri-Gugger, A.; Majumdar, S.K. Diabetic Ketoacidosis Associated with Sodium-Glucose Cotransporter 2 Inhibitors: Clinical and Biochemical Characteristics of 29 Cases. *Int. J. Endocrinol.* **2023**, *2023*, 6615624. [CrossRef]
29. Koufakis, T.; Pavlidis, A.N.; Metallidis, S.; Kotsa, K. Sodium-glucose co-transporter 2 inhibitors in COVID-19: Meeting at the crossroads between heart, diabetes and infectious diseases. *Int. J. Clin. Pharm.* **2021**, *43*, 764–767. [CrossRef]
30. Fernandez Felix, D.A.; Madrigal Loria, G.; Sharma, S.; Sharma, S.; Arias Morales, C.E. A Rare Case of Empagliflozin-Induced Euglycemic Diabetic Ketoacidosis Obscured by Alkalosis. *Cureus* **2022**, *14*, e25818. [CrossRef]
31. Banakh, I.; Kung, R.; Gupta, S.; Matthiesson, K.; Tiruvoipati, R. Euglycemic diabetic ketoacidosis in association with dapagliflozin use after gastric sleeve surgery in a patient with type II diabetes mellitus. *Clin. Case Rep.* **2019**, *7*, 1087–1090. [CrossRef] [PubMed]
32. Penttilä, I.; Penttilä, K.; Holm, P.; Laitinen, H.; Ranta, P.; Törrönen, J.; Rauramaa, R. Methods, units and quality requirements for the analysis of haemoglobin A 1c in diabetes mellitus. *World J. Methodol.* **2016**, *6*, 133–142. [CrossRef] [PubMed]
33. Mercantepe, F.; Baydur Sahin, S.; Cumhur Cure, M.; Karadag, Z. Relationship Between Serum Endocan Levels and Other Predictors of Endothelial Dysfunction in Obese Women. *Angiology* **2022**, *74*, 948–957. [CrossRef] [PubMed]

Disclaimer/Publisher's Note: The statements, opinions and data contained in all publications are solely those of the individual author(s) and contributor(s) and not of MDPI and/or the editor(s). MDPI and/or the editor(s) disclaim responsibility for any injury to people or property resulting from any ideas, methods, instructions or products referred to in the content.

 *pharmaceuticals*

Article

Oral Active Carbon Quantum Dots for Diabetes

Gamze Camlik [1,2], Besa Bilakaya [1,2], Esra Küpeli Akkol [3], Adrian Joshua Velaro [4,5,6], Siddhanshu Wasnik [7], Adi Muradi Muhar [4], Ismail Tuncer Degim [1,2,*] and Eduardo Sobarzo-Sánchez [8,9,*]

1. Department of Pharmaceutical Technology, Faculty of Pharmacy, Biruni University, Istanbul 34015, Türkiye; gcamlik@biruni.edu.tr (G.C.); bbilakaya@biruni.edu.tr (B.B.)
2. Biruni University Research Center (B@MER), Biruni University, Istanbul 34015, Türkiye
3. Department of Pharmacognosy, Faculty of Pharmacy, Gazi University, Ankara 06330, Türkiye; esrak@gazi.edu.tr
4. Department of Surgery, Faculty of Medicine, Universitas Sumatera Utara, Medan 20155, Indonesia; ajoshuav@gmail.com (A.J.V.); adi.muradi@usu.ac.id (A.M.M.)
5. Artisan Karya Abadi Research, Medan 20155, Indonesia
6. Department of Surgery, Dr. Djasamen Saragih Regional Public Hospital, Pematang Siantar 21121, Indonesia
7. Faculty of Medicine, Government Medical College and Hospital, Miraj 416410, Maharashtra, India; anshuwasnik@gmail.com
8. Instituto de Investigación y Postgrado, Facultad de Ciencias de la Salud, Universidad Central de Chile, Lord Cochrane 417, Santiago 8330507, Chile
9. Department of Organic Chemistry, Faculty of Pharmacy, University of Santiago de Compostela, 15782 Santiago de Compostela, Spain
* Correspondence: tdegim@biruni.edu.tr (I.T.D.); eduardo.sobarzo@ucentral.cl (E.S.-S.)

Abstract: Background/Objectives: Metformin (Met), an oral drug used to treat type II diabetes, is known to control blood glucose levels. Metformin carbon quantum dots (MetCQDs) were prepared to enhance the bioavailability and effectiveness of metformin. Several studies have shown that carbon quantum dots (CQDs) have attractive properties like small particle size, high penetrability, low cytotoxicity, and ease of synthesis. CQDs are made from a carbon source, namely, citric acid, and a heteroatom, such as nitrogen. The active molecule can be a carbon source or a heteroatom, as reported here. Methods: This study aims to produce MetCQDs from an active molecule. MetCQDs were successfully produced by microwave-based production methods and characterized. The effect of the MetCQDs was tested in Wistar albino rats following a Streptozocin-induced diabetic model. Results: The results show that the products have a particle size of 9.02 ± 0.04 nm, a zeta potential of -10.4 ± 0.214 mV, and a quantum yield of $15.1 \pm 0.045\%$. Stability studies and spectrophotometric analyses were carried out and the effectiveness of MetCQDs evaluated in diabetic rats. The results show a significant reduction in blood sugar levels (34.1–51.1%) compared to the group receiving only metformin (37.1–55.3%) over a period of 30 to 360 min. Histopathological examinations of the liver tissue indicate improvement in the liver health indicators of the group treated with MetCQDs. Conclusions: Based on these results, the products have potential therapeutic advantages in diabetes management through their increased efficacy and may have reduced side effects compared to the control group.

Keywords: carbon quantum dots; metformin; oral administration; type II diabetes; nanotechnology

1. Introduction

Metformin HCl (Met), an antidiabetic agent, is a biguanide class of drug used as the first step in pharmacological therapy for the treatment of type II diabetes. In addition, it was approved by the US Food and Drug Administration (FDA) in 1994 [1]. Met is a basic hydrophilic drug with a pKa value of 11.5. Animal and human studies have shown that it acts in the liver and inhibits gluconeogenesis by blocking mitochondrial redox shuttles [2–5]. It also increases insulin-mediated glucose use in peripheral tissue such as the liver and muscles, reduces glucose absorption in the small intestines, and lessens gluconeogenesis by reducing free fatty acid in plasma. The drug essentially activates the AMPk-activated protein kinase, which plays an important role in the anti-hyperglycemic effect [6]. Met is not a hypoglycemic drug, but an euglycemic drug, thereby reducing the potential for hypoglycemia. Several studies have shown that Met treatment is preferred for patients with type II diabetes due to its benefits, such as long-term safety and efficacy data, low risk of hypoglycemia, cardiovascular benefits, mortality benefits, additive or synergistic effects in combination therapy, low cost, and wide availability [7].

In recent years, carbon-based nanomaterials, including fullerenes, graphene, nanodiamonds, carbon nanotubes (CNTs), and carbon quantum dots (CQDs), have garnered significant attention due to their unique structural characteristics and exceptional chemical and physical properties [8]. However, the synthesis and purification of nanodiamonds present challenges, while other nanomaterials, such as graphene, fullerenes, and CNTs, suffer from poor water solubility and lack strong visible-region fluorescence [9]. As mentioned in the literature, CQDs can be produced with simple and easy microwave-assisted technology at a low temperature and power (100–200 W). The production is simple: if the color changes to dark brown, it can be accepted as an indication of the presence of graphene quantum dots. A clear solution obtained after synthesis can indicate the presence of CQDs [10]. Although semiconductor quantum dots (QDs) show good fluorescent properties, these materials are toxic and are obtained from heavy metals. Consequently, their biological application in biosensors, bioimaging, and drug delivery are limited [11,12]. To overcome these limitations, CQDs have been reported to have attractive properties, such as high photostability, low cytotoxicity, ease of synthesis, good water solubility, a high photo-response, easy surface functionalization, good catalysis, and tunable excitation–emission [13,14].

According to previous studies, CQDs can be used in imaging as well as in drug delivery systems due to their nature. The drug delivery system protects against degradation (enzymatic) as well as reduces the number of doses required and their side effects, leading to increased effectiveness and bioavailability [15].

In this study, MetCQD formulations were produced and tested for better results. The formulations were prepared using citric acid as a carbon source and Met was used for dual-action purposes as a heteroatom and an active molecule. The mitochondrial accumulation properties of Met were also considered. The initial hypothesis considers the action of MetCQDs on mitochondria, which is essential for the treatment of diabetes and many other diseases related to mitochondrial dysfunction.

2. Results
2.1. The One-Step Production of MetCQDs

MetCQDs were successfully synthesized and the process proved to be straightforward and easily applicable. A total of 0.2 g of citric acid monohydrate was used as a carbon source, with 0.5 g of Met used as both a heteroatom and an active pharmaceutical ingredient. The method was found to be cost-effective and rapid. A blue-green fluorescent emission was observed under UV light (365 nm) after the reaction (Figure 1). This served as a rapid initial confirmation and the presence of the blue emission showed the completion of a successful reaction, as failure would result in the absence of this emission.

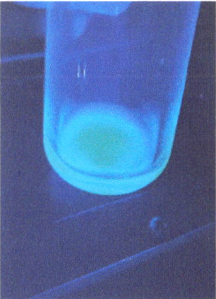

Figure 1. The physical appearance of MetCQDs under UV light (365 nm).

2.2. Characterization of MetCQDs

2.2.1. Physical Appearance, Particle Size and Distribution, and Zeta Potential

The analysis of MetCQDs involved an evaluation of their particle size (PS), distribution, and zeta potential (ZP) using the Lite Sizer 500 from Anton Paar (Anton Paar, St. Albans Hertfordshire, AL4 0LA, UK) (Table 1).

Table 1. Particle size, zeta potential, and polydispersity index of prepared MetCQDs.

Formulation	Particle Size (nm)	Zeta Potential (mV)	Polydispersity Index (%) (PDI %)
MetCQDs	9.02 ± 0.04	−10.4 ± 0.214	15.1 ± 0.045

The prepared MetCQDs were diluted with water then imaged using a TEM (Figure 2). It was observed that the MetCQDs were of a very small size, measuring less than 10 nm.

Figure 2. Physical appearance of MetCQDs under TEM.

2.2.2. Fluorescence

The MetCQDs were used to examine fluorescence, showing successful preparation through the observation of vivid blue emissions. Exciting this fluorescence with 365 nm of light verified the optimal emissions, as confirmed by a spectrofluorometer (RF-6000 fluorescence spectrophotometer, Shimadzu, Kyoto, Japan). However, using a longer wavelength of UV light (420 nm) led to a shift in the emissions toward the red spectrum. Factors such as the selection of the carbon source and pH prove to have an influence on both surface property and doped atom selection. The prepared MetCQDs appeared as a blue-green fluorescence under UV light.

2.2.3. Quantum Yield

Quinine sulfate solution served as the established standard for comparison. The emissions obtained from the MetCQDs exhibited a satisfactory intensity leading to a calculated quantum yield deemed sufficiently high. Graphs plotting the intensity against the wavelength for both the quinine sulfate and the MetCQDs were generated. In the experiment, the slope of the area under the curve (AUC)–intensity relationship for the quinine sulfate measured 895.72, while the CQDs registered at 720.34 with an AUC value of 0.804. Consequently, the quantum yield was determined to be 80.3%.

2.2.4. Stability

The characterization parameters (particle size and distribution and zeta potential) of the prepared formulations were kept at different temperatures (−20 °C, 5 °C, 25 °C, and 40 °C) during stability testing and the shelf life was calculated at more than 2 years. The relative humidity values of the formulations were 60% and 75% at 25 °C and 40 °C, respectively. There was no significant change in the results obtained.

2.3. Assay

The amount of Met was successfully determined by the spectrophotometric method and was found to be linear, repeatable, and stable. The dose given to the animals was kept constant by controlling the quantification.

2.3.1. In Vivo Experiments

Animals were grouped, and the experiment proceeded according to the protocol in the Materials and Methods Section. No animal died and no animal was excluded from the experiment.

2.3.2. Determination of Blood Glucose Levels

The animals' blood glucose levels were measured successfully by following the protocol specified in the Materials and Methods Section. There were no problems related to the measurement and method used.

2.3.3. Effects on Normal and Glucose-Loaded Rats (NG-OGTT)

A significant decrease in blood glucose level was achieved after metformin administration. The test material was found to show stronger hypoglycemic activity compared to metformin (Table 2 and Figure 3).

Table 2. Acute effect of test samples on blood glucose levels in normal and glucose-loaded hyperglycemic (NG-OGTT) rats.

Group	Blood Glucose Concentration ± S.E.M. (Inhibition %)						
	0 min	30 min	60 min (+Glucose)	90 min	120 min	240 min	360 min
Control	102.5 ± 4.8	109.2 ± 6.3	124.7 ± 6.1	183.0 ± 5.2	151.3 ± 4.8	130.2 ± 4.4	114.6 ± 4.9
Metformin	106.2 ± 5.7	71.2 ± 3.9 *	80.6 ± 3.1 **	89.4 ± 3.6 **	73.6 ± 3.2 **	71.6 ± 3.5 **	70.8 ± 4.2 *
MetCQD	99.3 ± 4.1	45.1 ± 3.7 **	49.4 ± 3.2 **	53.8 ± 3.3 **	46.3 ± 3.1 **	45.5 ± 3.2 **	43.7 ± 3.0 **

*: $p < 0.01$; **: $p < 0.001$; S.E.M.: standard error of the mean.

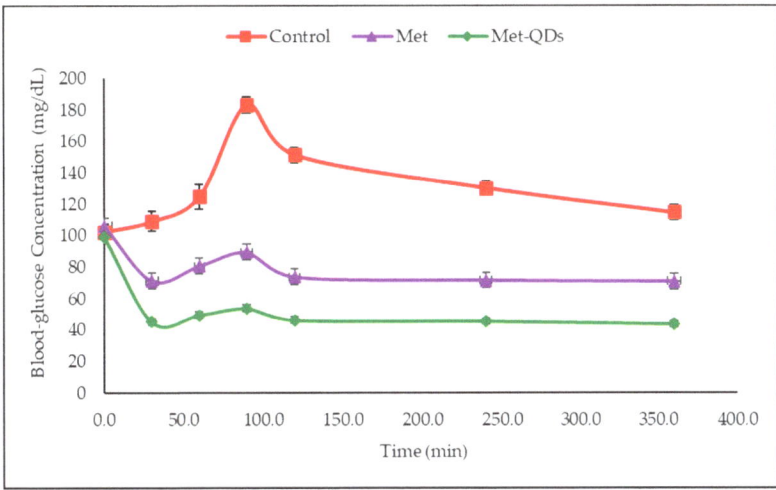

Figure 3. Blood glucose concentration graph in healthy animals.

2.3.4. Acute Antidiabetic Effect

A total of 60 min after oral administration, a 23.9–31.1% decrease in blood glucose was observed in the metformin-administered group and a 28–35.2% decrease in the test-material-applied group (Table 3 and Figure 4).

Table 3. Effect of test samples on blood glucose levels in diabetic rats with STZ.

Groups	Blood Glucose Concentration ± S.E.M. (Inhibition %)						
	0 min	30 min	60 min	90 min	120 min	240 min	360 min
Diabetic Control	314.3 ± 9.9	347.1 ± 10.8	356.3 ± 9.1	331.5 ± 8.2	322.7 ± 8.4	310.1 ± 6.3	285.5 ± 4.6
STZ + Metformin	315.9 ± 8.5	338.6 ± 8.2	271.2 ± 6.3 *	234.2 ± 5.1 *	244.5 ± 5.0 *	213.7 ± 5.1 **	218.2 ± 4.9
STZ + MetCQDs	309.7 ± 10.4	229.9 ± 4.9	168.6 ± 4.1 **	150.9 ± 3.3 **	152.5 ± 3.5 **	132.1 ± 3.4 **	131.3 ± 2.2 **

*: $p < 0.05$; **: $p < 0.01$; S.E.M.: standard error of the mean.

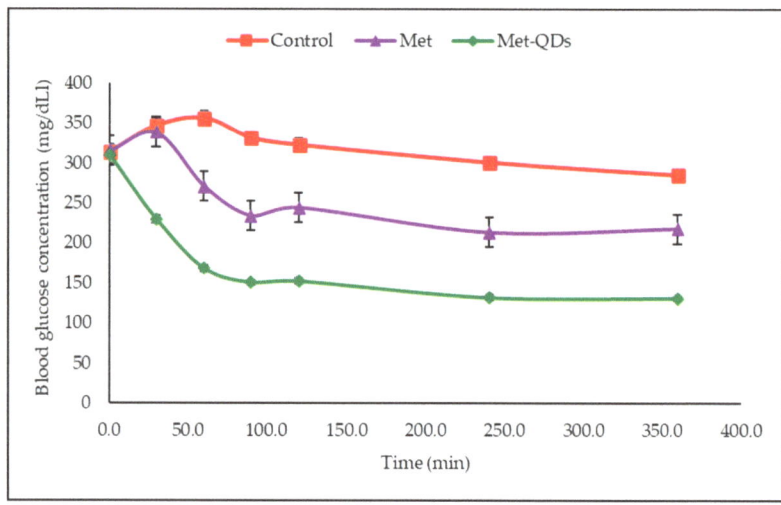

Figure 4. Blood glucose concentration graph in diabetic animals.

2.4. Histopathological Examinations

In this experimental study, the pathological results from the pancreata and livers of rats with type II diabetes and the changes observed as a result of the application of the test samples were examined, and the macroscopic and microscopic pathological results detected given below.

2.4.1. Macroscopic Observations

The pancreata and livers of the rats, necropsied after euthanasia, were examined to determine whether there were any macroscopic lesions. A remarkable finding was observed.

2.4.2. Histopathological Observations

Histopathological Results of the Pancreatic Tissue

The pancreatic tissue of rats in the control group exhibited a normal histological structure (Figure 5A). Rats in the STZ-induced diabetic group showed degenerative and necrotic changes in the endocrine portion of the pancreas, identified by the atrophic appearance of the islets of Langerhans in the endocrine pancreata due to degenerative and necrotic changes (Figure 5B). In the STZ + Metformin group, it was observed that the necrotic and degenerative changes decreased compared to the other experimental groups; a new pancreatic exocrine structure formed and regeneration occurred in the islets of Langerhans containing the endocrine portion of the pancreas (Figure 5C). In the STZ + Test material group, regeneration was observed in the exocrine gland epithelium, in the islets of Langerhans containing the endocrine portion of the pancreas, and in the connective tissue separating the pancreatic lobules (Figure 5D).

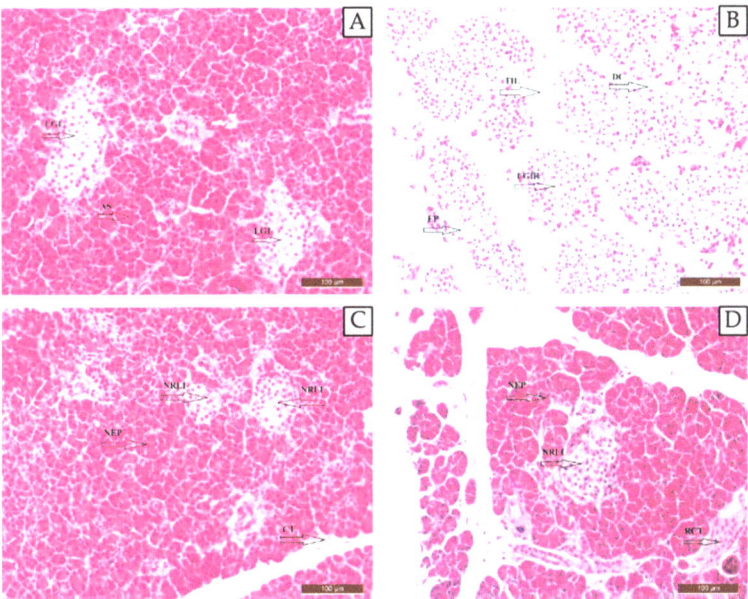

Figure 5. A histopathologic view of pancreatic tissue in the control group (healthy animals) (**A**), diabetic control group (**B**), metformin-administered group (**C**), and MetCQD-administered group (**D**). **AS**: acinus (exocrine pancreas), **CT**: normal connective tissue structure, **DC**: destruction of the connective tissue that separates the pancreatic lobules from each other, **EP**: exocrine pancreatic structure degeneration, **LGI**: Langerhans islet (pancreatic islet cells, endocrine pancreas), **LGIR**: Langerhans islets ruptured, **NEP**: new exocrine pancreatic structure, **NRLI**: new regeneration of the Langerhans islet, and **RCT**: regeneration of the connective tissue structure that separates the pancreatic lobules from each other. 40X.

Histopathological Results of the Liver Tissue

The liver tissue from rats in the control group was observed to have a normal histological structure (Figure 6A). In the diabetic control group, fatty degeneration, widespread hepatocyte enlargement, irregular trabecular structure, periacinar necrosis, and sinusoidal constriction were observed (Figure 6B). In the STZ + Metformin group, improvement in liver fat content, regeneration in hepatocytes, regeneration in the bile duct, improvement in the sinusoidal structure, a regular trabecular network, and a normal structure of the hepatic artery, central vein, and portal vein were observed (Figure 6C). In the STZ + MetCQDs group, regeneration, a regular trabecular network, and decreased liver fat content were identified due to increased mitotic activity in the hepatocytes (Figure 6D).

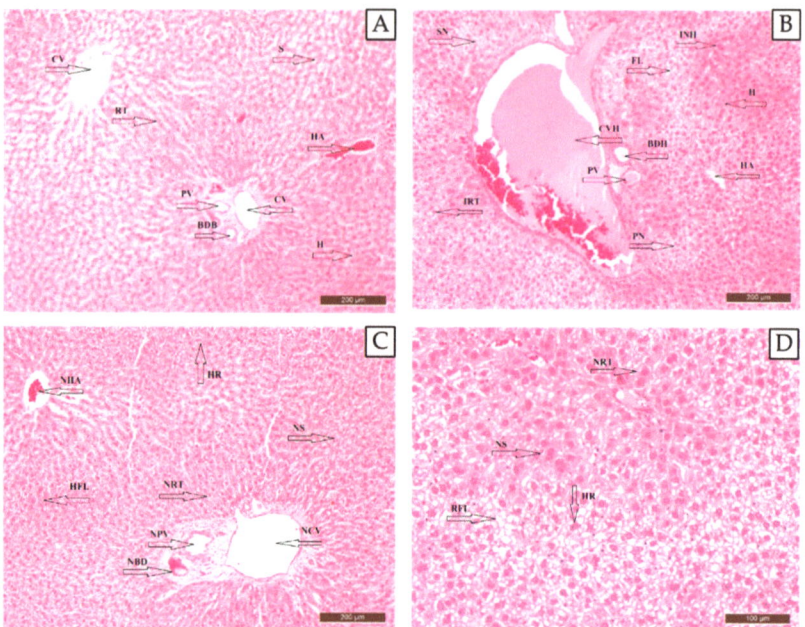

Figure 6. A histopathologic view of liver in the control group (healthy animals) (**A**), diabetic control group (**B**), STZ + Metformin-administered group (**C**), and STZ + MetCQD-administered group (**D**). 200X. **BDB:** bile duct branch, **BDH:** bile duct hyperplasia, **CV:** central vein, **CVH:** central vein hyperplasia, **FL:** fatty liver, **H:** hepatocytes, **HA:** hepatic artery, **HFL:** healing fatty liver, **HR:** hepatocytes regeneration (binucleated hepatocytes), **INH:** increase in number of large hepatocytes, **IRT:** irregular trabeculae, **NBD:** normal bile duct branch, **NCV:** normal central vein, **NHA:** normal hepatic artery, **NPV:** normal portal vein, **NRT:** new regular trabeculae formation, **NS:** new sinusoid structure formation, **PN:** periacinar necrosis, **PV:** portal vein, **RFL:** reduction in fatty liver, **RT:** regular trabeculae, **S:** sinusoid, and **SN:** sinusoid narrowing.

3. Discussion

CQDs, known as excellent fluorescent nanomaterials, have a particle size of less than 10 nm, representing no optical vibration, a high fluorescence stability, a continuous and tunable emission wavelength, a wide excitation spectrum, good biocompatibility, and low toxicity. Due to these properties, they are widely used in biological systems for detection, in vivo and in vitro imaging, drug delivery, and diagnostic medicine [16,17]. Additionally, while they can be prepared from a carbon source, it is known that the active ingredient itself can be used as a carbon source.

The CQDs in this study were obtained from the parent active substance metformin HCl. The particle size and polydispersity index, zeta potential, and quantum yield were found to be 9.02 ± 0.04 nm, $19.2\% \pm 0.102$, 15.1 ± 0.045 mV, and 80.3%, respectively. In addition, TEM images showed homogeneous dispersion and no aggregation, proven by the PDI. This was also reported in the literature; homogeneous dispersion was obtained when the PDI was less than 20% [18]. This obtained value showed that the prepared formulations were homogeneous.

Met is an oral antidiabetic agent used to treat type II diabetes and is used as a first-line treatment. Metformin is shown to improve glycemic control, reduce hyperglycemia, and prevent the occurrence of diabetes-related complications [19]. Additionally, its mechanism of action can include the inhibition of glucose production in the liver and increased insulin sensitivity. The study proposed in this article suggests that it can increase the effectiveness of metformin in the treatment of type II diabetes, with potential benefits for its management, particularly through the use of nanotechnology [20].

Characterization parameters such as particle size and distribution and zeta potential of the prepared MetCQD formulations were continuously maintained at temperatures of −20 °C, 5 °C, 25 °C, and 40 °C during stability testing and the shelf life was calculated at more than 2 years. The relative humidity values of the formulations were identified as 60% and 75% for 25 °C and 40 °C, respectively. This indicates there was no significant change for all formulations.

The active substance Met was successfully measured at 232 nm and the method was determined to be linear, specific, selective, repeatable, and stable.

Met was slowly and incompletely absorbed after oral administration and resulted in some significant side effects when administered orally [2]. The most serious complication was metformin-associated lactic acidosis (MALA), with a mortality rate ranging from 25% to 50%. This term, classically, included any metformin user with increased lactate. In 2017, Lalau et al. described a series of clinical conditions such as metformin-induced lactic acidosis (MILA) and metformin-unrelated lactic acidosis (MULA). This occurred when high metformin serum levels were the underlying cause of patient death [21]. Therefore, the dose must be lowered to avoid unwanted side effects. The MetCQDs were prepared to increase their bioavailability at lower doses and to reduce their side effects. It was observed that although the same dose was administered, the effect with the MetCQDs was greater; to have the same effect, the dose should be minimized. In this way, the side effects could be lessened.

In this study, Wistar albino rats were used to test the effect of MetCQDs. In an environment where the temperature, relative humidity, and light/dark cycles were controlled, the rats were fed with the standard feed and clean drinking water. In the experiment, the blood glucose levels of the rats were measured and then divided into different groups, namely: Group I (control), Group II (STZ, diabetic control), Group III (STZ + Met), and Group IV (STZ + test sample (MetCQD)). An oral glucose tolerance test was performed and fasting blood sugar levels were determined.

A significant reduction in blood glucose levels after Met administration was observed (Table 1 and Figure 3) when healthy animals were used. However, a stronger hypoglycemic effect was observed in rats administered with MetCQDs than with Met alone. The same, stronger effect was observed with the MetCQDs when diabetic animals were subjected to the test (Table 2 and Figure 4).

At the end of the experiment, the rats were euthanized under anesthesia and the pancreas and liver tissue were prepared and evaluated for histopathological examination. As a result of these evaluations, it was determined that the liver tissue of the rats in the control group (Group I) had a normal histological structure (Figure 5A). In the diabetic control group (Group II), fatty liver, hepatocyte expansion, irregular trabecular structure, periacinar necrosis, and sinusoid narrowing were observed (Figure 5B). In the STZ + Metformin group (Group III), improvement in liver steatosis, hepatocyte regeneration, improvement in the bile duct and sinusoid structure, a regular trabecular network structure, and normal hepatic artery and portal vein structures were observed (Figure 5C). In the STZ + Test Material (MetCQDs) group (Group IV), increased mitotic activity in hepatocytes, along with regeneration, a regular trabecular network, and decreased fatty liver were observed (Figure 5D).

Liver tissue and cells were found to be normal in the control group (Figure 6A), whereas some fatty degeneration, enlarged hepatocytes with irregular trabecular structures, some necrosis, and sinusoidal constrictions were observed in the diabetic animals (Figure 6B). This was found to be proof of the STZ effect in the diabetic animal model. Met was found to be effective in improving liver fat content and some regeneration was also observed. Some regeneration in the bile duct, improvement in the sinusoidal structure, a regular trabecular network, and normal structures of the hepatic artery, central vein, and portal vein were observed (Figure 6C). However, even more regeneration, a regular trabecular network, accepted to be an indication of good regeneration, and decreased liver fat content were identified, and increased mitotic activity in hepatocytes was observed in the MetCQD-administered group of diabetic animals (Figure 6D). All these were attributed to the effects of CQDs previously described [12].

Literature reports demonstrate the ability of CQDs to traverse multiple membrane barriers [22]. Many reports also show that positively charged quantum dots can penetrate cell membranes much faster. MetCQDs represent a negative surface charge and penetrate biological membranes, as well. This shows that the penetration mechanism is more complex than just a charge; some other mechanism may be involved. The cellular uptake and subsequent cellular toxicity may also differ from the type of quantum dot. Furthermore, positively charged quantum dots have been shown to exhibit enhanced cellular membrane permeability. A 2020 study reported the unintentional formation of CQDs while cooking salmon in an oven at 200 °C. The same research indicated no histopathological effects in rats at a dose of 2 g/kg [23]. A 2015 study investigating pancreatic targeting with quantum dots observed reduced blood glucose levels three to four weeks post-administration [24]. Observations of enhanced metabolic activity in the mitochondria were reported. This may explain why some regeneration was observed in the tissue of the diabetic animals.

Cases of diabetes and related diseases have risen and the number of patients has grown exponentially around the world. Impaired glucose tolerance could result in high blood glucose levels, which severely endanger people. Therefore, controlling blood glucose levels is essential to therapy, and there are several drugs available to reduce blood glucose levels on the market. Biguanides or sulfonylureas can be used to successfully treat high blood glucose levels, but have some side effects [25,26]. The CQDs presented here appear to be a good alternative and are a subclass of carbon material used in various biological applications because of their low toxicity and good biocompatibility [15,27,28]. A carbon atom can have four bonds with other atoms; hence, it can be a chiral atom. Chirality is a common phenomenon and plays a critical role in physiological activities [29,30]. The CQDs in this study were made of carbon and nitrogen atoms and used as heteroatoms. Although it is not generally known and there are limited sources available in the literature, MetCQDs are more effective at reducing blood glucose levels, as mentioned here.

In studies conducted with CQDs, it was found that CQDs increased metabolic activity by affecting the energy pathways of the cells and by going to the mitochondria of the cells; therefore, they had a hypoglycemic effect [12].

4. Materials and Methods

4.1. Materials

Citric acid monohydrate and polyethylene glycol (PEG 3350) were purchased from Sigma Aldrich (St. Louis, MO, USA). The metformin HCl was donated by SANOVEL, and a microwave reactor (Anton Paar Microwave 300, Anton Paar, St. Albans Hertfordshire, AL4 0LA, UK), was used for the synthesis of the MetCQDs. Additionally, particle size, size distribution, and zeta potential were determined using an Anton Paar LiteSizer 500. The optical properties of the MetCQDs were characterized using a spectrofluorometer (Model 229129, Agilent Technologies, Santa Clara, CA, USA). For use in this experimental in vivo study, Wistar albino rats weighing 200–250 g were obtained from the Kobay Animal Study Center (Ankara, Turkey). The experiments were conducted with the approval of the ethics committee of the Ankara University Faculty of Pharmacy in the Department of Pharmacology (Local Ethics Committee for Animal Experiments, decision no. 044). This study adhered to the Animal Study Reporting In Vivo Experiments (ARRIVE) guidelines. Fasting blood glucose levels were measured using a Bayer Glucometer Elite. Streptozocin (St. Louis, MO, USA) (dissolved in 0.1 M of citrate buffer with a pH of 4.50) was used to induce diabetes in the animals, and sodium carboxymethylcellulose (Sigma-Aldrich, St. Louis, MO, USA) was administered to the rats in the STZ groups. Furthermore, a formalin solution was used for the macroscopic and histopathological examinations. An automatic tracking device (SHANDON Citadel 1000, Brentwood, TN 37027, United States) and microtome (Reichert rotary, Buffalo City, NY 14043, United States) devices were used for the histopathology examinations and the hematoxylin–eosin method was used for section staining. Microscopic examinations were performed with an Olympus BX51 (Olympus, Tokyo, Japan) (DP72 camera attachment) light microscope.

The SPSS 20.0 program was used for the statistical evaluation, and the Kruskal–Wallis and Mann–Whitney U tests were used for the histopathological examination.

4.2. Methods

4.2.1. The One-Step Production of MetCQDs

The effective conditions and parameters impacted the dimensions and surface properties of the MetCQDs and lead to the optimization of the production method. Additionally, to create the MetCQDs, 0.2 g of citric acid monohydrate served as a carbon source, while 0.5 g of Met acted as the active drug source. All chemicals were dissolved in 1 mL of distilled water and transferred to the microwave reactor. The heating program adhered to the following steps:

- Heat up to 160 °C for 5 min;
- Maintain a temperature of 160 °C for 20 min;
- Cool gradually to reach room temperature.
- Finally, the MetCQDs were successfully created and checked under UV light (365 nm) for rapid confirmation. After the reaction, surface modification was achieved by adding 0.02 g of PEG 3350.

4.2.2. Characterization of the MetCQDs

Physical Appearance, Particle Size and Distribution, and Zeta Potential

Characterization studies were conducted on the MetCQDs and a transmission electron microscope (TEM, Carl Zeiss EM 900, Jena, Germany) was used to ascertain their particle shape and morphological traits. Particle size and distribution were determined using an Anton Paar LiteSizer 500. All measurements were performed with 6 replicates for size, distribution, and zeta potential determination.

Fluorescence

A spectrofluorometer was used to determine the optical characteristics of the MetCQDs. Excitation of the MetCQDs allowed for the recording of their emissions while their response under UV light was observed and documented.

Quantum Yield

The calculation method used was adapted from a previously conducted study in the literature [11,14]. First, the absorbance of quinine sulfate in a sulfuric acid solution was gauged at 350 nm. Subsequently, the complete fluorescence spectrum, depicting fluorescent intensity across various wavelengths, was generated for the solution. Dilution steps were undertaken in increments of 0%, 20%, 40%, 60%, 80%, and 100% of the original solution, with fluorescence spectra plotted for each. The areas beneath these curves (AUC) were calculated using the trapezoidal method and the resultant AUC values were correlated with the intensity to determine the slope value. This entire procedure was replicated for the MetCQD solution, yielding a comparable fluorescence spectrum. Quantum yield was calculated as the following:

$$\text{Quantum yield} = \{m(\text{test})/m\ (\text{quinine sulphate})\} * A,$$

where A = (refractive index-water/refractive index-0.1 M sulfuric acid) = 0.99939985

[Refractive index-water = 1.3325, refractive index-0.1 M sulfuric acid = 1.3333] and m is the slope of the AUC-intensity graphic.

4.2.3. Assay

A Met standard solution was prepared and a UV spectrophotometer was used. The maximum wavelength (λ max) was 232 nm and the calibration curve was obtained using standard solutions. Additionally, the curve was found to be linear and the method reproducible. The other validation parameters were suitable, such as repeatability, specificity, stability, etc. [31].

Stability

Stability tests were performed to determine chemical and physical stability and the shelf life of the MetCQDs.

The In Vivo Experiments

Wistar albino rats weighing 200–250 g were obtained from the Guinea pig laboratory and housed in a room at a temperature of 22 ± 2 °C with approximately 60% relative humidity and a 12 h light/dark cycle. The experimental animals were fed with standard rat chow and their drinking water bottles were changed daily. Animals were randomly selected and groupings of 7 animals in each experimental group were created.

The Determination of Blood Glucose Levels

The blood glucose concentration of each rat was measured from a fasting state. Blood samples were collected from their tails and the Glucometer Elite commercial test (Bayer), based on the glucose oxidase method, was used to determine their blood glucose levels.

The Effects on Normal and Glucose-Loaded Rats (NG-OGTT)

Animals that received the Met solution or the MetCQCs were administered orally. Blood glucose levels were measured at the beginning and after 30 and 60 min, followed by an oral glucose tolerance test. The rats were orally loaded with glucose at a dose of 2 g/kg and their blood glucose levels determined at the 90th, 120th, 240th, and 360th min [32,33].

The Diabetes Model

The rats were made diabetic by administering a streptozotocin solution (STZ, 60 mg/kg, i.p.) (St. Louis, MO, USA) dissolved in 0.1 M of a citrate buffer solution with a pH of 4.50. The control group was given a citrate buffer solution that did not contain STZ. After 3 weeks of streptozotocin injections, a drop of blood was collected from the tail vein and the fasting blood glucose level was measured using a blood glucose monitoring system (One Touch® UltraMini®, Life Scan, Inc., Milpitas, CA, USA). Rats with a blood glucose concentration level of ≥300 mg/dL were considered diabetic [34–36].

Groups:

Group I: control group; the group without any experimental procedure ($n = 7$).

Group II (STZ, Diabetic control): animals in which diabetes was induced with STZ (60 mg/kg, single dose i.p., 1 mL) ($n = 7$).

Group III (STZ + Met): diabetic animals given a pure metformin solution (60 mg/kg, single dose i.p., 1 mL, and Met 250 mg/kg; per os) ($n = 7$).

Group IV (STZ + MetCQDs): diabetic animals given new formulations of Metformin HCl (60 mg/kg, single dose i.p., 1 mL, and MetCQDs 250 mg/kg; per os) ($n = 7$).

The administered dose of Met was calculated relative to the clinically relevant human dose based on body surface area.

Metformin dissolved in 0.5% sodium carboxymethylcellulose (250 mg/kg/day, Sigma-Aldrich, St. Louis, MO, USA) was administered via gastric gavage. Rats in the STZ groups were administered 0.5% sodium carboxymethylcellulose.

The acute antidiabetic effect of the test samples was determined from a fasting state and the blood sugar levels were measured at 30, 60, 90, 120, 240, and 360 min after the application of Met and the MetCQDs. At the end of the experiment, the rats were euthanized with a high dose of anesthesia. Pancreas and liver tissue were taken and placed in a 5% formalin solution until use.

Histopathological Examination

Tissue pieces prepared in a block size 0.5 cm thick were passed through the alcohol and xylol series for dehydration, clearing, and paraffinization processes with an automatic tracking device (SHANDON Citadel 1000). A total of 5 micron-thick sections were taken from the prepared paraffin blocks using a microtome (Reichart rotary), and all sections were stained using the hematoxylin–eosin staining method. Microscopic examination of the preparations were performed with an Olympus BX51 (DP72 camera attachment) light microscope. Histopathological results were observed in sections from 10 random areas examined under the light microscope and were evaluated according to the atrophic appearance of the islets of Langerhans in the pancreas, normal (−), mild (+), moderate (++), and severe (+++); in the liver, (−), mild (+), moderate (++), and severe (++); and in terms of necrotic and degenerative change, evaluated as (+).

Statistical Analysis

The SPSS 20.0 program was used for statistical evaluation. In histopathological examinations, the difference between the groups in the data regarding the atrophic appearance of the islets of Langerhans in the pancreas and the necrotic-degenerative changes in the liver was determined by the Kruskal–Wallis test. The detection of the groups creating the difference was determined by the Mann–Whitney U test (a $p < 0.05$ value is considered statistically significant).

5. Conclusions

In conclusion, MetCQDs were prepared using citric acid as a carbon source and Met was used for dual-action purposes as a heteroatom and an active molecule for the first time. The MetCQDs formulations were successfully produced and exhibited a hypoglycemic effect; the liver fat was also found to have decreased. This study concludes that the preparation of quantum dot formulations of active substances can open the door to new treatment alternatives that use drug molecules in both roles. Finally, further studies are necessary to determine the possible long-term effects.

Author Contributions: Conceptualization, I.T.D.; methodology, G.C., B.B., E.K.A., A.J.V., S.W. and A.M.M.; software, G.C. and B.B.; validation, G.C., B.B. and I.T.D.; formal analysis, G.C., B.B., E.K.A. and A.J.V.; investigation, G.C., B.B., E.K.A. and I.T.D.; resources, I.T.D.; data curation, G.C., B.B., E.K.A., A.J.V., S.W., A.M.M. and I.T.D.; writing—original draft preparation, G.C., B.B., E.K.A. and I.T.D.; writing—review and editing, E.K.A., I.T.D. and E.S.-S.; visualization, I.T.D.; supervision, I.T.D. All authors have read and agreed to the published version of the manuscript.

Funding: This research received no external funding.

Institutional Review Board Statement: This study was conducted in accordance with international guidelines for animal experiments and biodiversity rights, as per the ethical standards of the NESA Animal Breeding Laboratory, Ethical Council Project Number: 44.

Informed Consent Statement: Not applicable.

Data Availability Statement: Dataset available on request from the authors.

Conflicts of Interest: The authors declare that the research was conducted in the absence of any commercial or financial relationships that could be construed as a potential conflict of interest.

References

1. Bailey, C.J. Metformin: Historical Overview. *Diabetologia* **2017**, *60*, 1566–1576. [CrossRef] [PubMed]
2. Cetin, M.; Sahin, S. Microparticulate and Nanoparticulate Drug Delivery Systems for Metformin Hydrochloride. *Drug Deliv.* **2016**, *23*, 2796–2805. [CrossRef] [PubMed]

3. Flory, J.; Lipska, K. Metformin in 2019. *JAMA* **2019**, *321*, 1926. [CrossRef] [PubMed]
4. Hundal, R.S.; Inzucchi, S.E. Metformin. *Drugs* **2003**, *63*, 1879–1894. [CrossRef]
5. Nasri, H.; Rafieian-Kopaei, M. Metformin: Current Knowledge. *J. Res. Med. Sci.* **2014**, *19*, 658–664.
6. Hernández-Velázquez, E.D.; Alba-Betancourt, C.; Alonso-Castro, Á.J.; Ortiz-Alvarado, R.; López, J.A.; Meza-Carmen, V.; Solorio-Alvarado, C.R. Metformin, a Biological and Synthetic Overview. *Bioorg. Med. Chem. Lett.* **2023**, *86*, 129241. [CrossRef]
7. Foretz, M.; Guigas, B.; Viollet, B. Metformin: Update on Mechanisms of Action and Repurposing Potential. *Nat. Rev. Endocrinol.* **2023**, *19*, 460–476. [CrossRef]
8. Ilbasmiş-Tamer, S.; Yilmaz, Ş.; Banoğlu, E.; Değim, I.T. Carbon Nanotubes to Deliver Drug Molecules. *J. Biomed. Nanotechnol.* **2010**, *6*, 20–27. [CrossRef]
9. Wang, R.; Lu, K.-Q.; Tang, Z.-R.; Xu, Y.-J. Recent Progress in Carbon Quantum Dots: Synthesis, Properties and Applications in Photocatalysis. *J. Mater. Chem. A Mater.* **2017**, *5*, 3717–3734. [CrossRef]
10. Saud, A.; Oves, M.; Shahadat, M.; Arshad, M.; Adnan, R.; Qureshi, M.A. Graphene-Based Organic-Inorganic Hybrid Quantum Dots for Organic Pollutants Treatment. In *Graphene Quantum Dots*; Elsevier: Amsterdam, The Netherlands, 2023; pp. 133–155.
11. Camlik, G.; Bilakaya, B.; Uyar, P.; Degim, Z.; Degim, I.T. New Generation of Composite Carbon Quantum Dots for Imaging, Diagnosing, and Treatment of Cancer. In *Functionalized Nanomaterials for Cancer Research*; Elsevier: Amsterdam, The Netherlands, 2024; pp. 543–557.
12. Camlik, G.; Ozakca, I.; Bilakaya, B.; Ozcelikay, A.T.; Velaro, A.J.; Wasnik, S.; Degim, I.T. Development of Composite Carbon Quantum Dots-Insulin Formulation for Oral Administration. *J. Drug Deliv. Sci. Technol.* **2022**, *76*, 103833. [CrossRef]
13. Camlik, G.; Bilakaya, B.; Ozsoy, Y.; Degim, I.T. A New Approach for the Treatment of Alzheimer's Disease: Insulin-Quantum Dots. *J. Microencapsul.* **2024**, *41*, 18–26. [CrossRef] [PubMed]
14. Yadav, P.K.; Chandra, S.; Kumar, V.; Kumar, D.; Hasan, S.H. Carbon Quantum Dots: Synthesis, Structure, Properties, and Catalytic Applications for Organic Synthesis. *Catalysts* **2023**, *13*, 422. [CrossRef]
15. Lim, S.Y.; Shen, W.; Gao, Z. Carbon Quantum Dots and Their Applications. *Chem. Soc. Rev.* **2015**, *44*, 362–381. [CrossRef] [PubMed]
16. Yang, H.-L.; Bai, L.-F.; Geng, Z.-R.; Chen, H.; Xu, L.-T.; Xie, Y.-C.; Wang, D.-J.; Gu, H.-W.; Wang, X.-M. Carbon Quantum Dots: Preparation, Optical Properties, and Biomedical Applications. *Mater. Today Adv.* **2023**, *18*, 100376. [CrossRef]
17. Camlik, G.; Kupeli Akkol, E.; Degim, Z.; Degim, I.T. Can Carbon Quantum Dots (CQDs) or Boron Compounds Be an Ultimate Solution for COVID-19 Therapy? *Iran. J. Pharm. Res.* **2021**, *20*, 9–20.
18. Hoseini, B.; Jaafari, M.R.; Golabpour, A.; Momtazi-Borojeni, A.A.; Karimi, M.; Eslami, S. Application of Ensemble Machine Learning Approach to Assess the Factors Affecting Size and Polydispersity Index of Liposomal Nanoparticles. *Sci. Rep.* **2023**, *13*, 18012. [CrossRef]
19. Chen, Y.; Shan, X.; Luo, C.; He, Z. Emerging Nanoparticulate Drug Delivery Systems of Metformin. *J. Pharm. Investig.* **2020**, *50*, 219–230. [CrossRef]
20. Kenechukwu, F.C.; Nnamani, D.O.; Duhu, J.C.; Nmesirionye, B.U.; Momoh, M.A.; Akpa, P.A.; Attama, A.A. Potential Enhancement of Metformin Hydrochloride in Solidified Reverse Micellar Solution-Based PEGylated Lipid Nanoparticles Targeting Therapeutic Efficacy in Diabetes Treatment. *Heliyon* **2022**, *8*, e09099. [CrossRef]
21. Rivera, D.; Onisko, N.; Cao, J.D.; Koyfman, A.; Long, B. High Risk and Low Prevalence Diseases: Metformin Toxicities. *Am. J. Emerg. Med.* **2023**, *72*, 107–112. [CrossRef]
22. Zhang, M.; Wang, H.; Wang, B.; Ma, Y.; Huang, H.; Liu, Y.; Shao, M.; Yao, B.; Kang, Z. Maltase Decorated by Chiral Carbon Dots with Inhibited Enzyme Activity for Glucose Level Control. *Small* **2019**, *15*, 1901512. [CrossRef]
23. Song, Y.; Wu, Y.; Wang, H.; Liu, S.; Song, L.; Li, S.; Tan, M. Carbon Quantum Dots from Roasted Atlantic Salmon (*Salmo salar* L.): Formation, Biodistribution and Cytotoxicity. *Food Chem.* **2019**, *293*, 387–395. [CrossRef] [PubMed]
24. Liu, Z.; Lin, Q.; Huang, Q.; Liu, H.; Bao, C.; Zhang, W.; Zhong, X.; Zhu, L. Semiconductor Quantum Dots Photosensitizing Release of Anticancer Drug. *Chem. Commun.* **2011**, *47*, 1482–1484. [CrossRef]
25. Inzucchi, S.E. Oral Antihyperglycemic Therapy for Type 2 Diabetes. *JAMA* **2002**, *287*, 360. [CrossRef] [PubMed]
26. Nathan, D.M.; Buse, J.B.; Davidson, M.B.; Ferrannini, E.; Holman, R.R.; Sherwin, R.; Zinman, B. Medical Management of Hyperglycemia in Type 2 Diabetes: A Consensus Algorithm for the Initiation and Adjustment of Therapy. *Diabetes Care* **2009**, *32*, 193–203. [CrossRef] [PubMed]
27. Du, J.; Xu, N.; Fan, J.; Sun, W.; Peng, X. Carbon Dots for In Vivo Bioimaging and Theranostics. *Small* **2019**, *15*, 1805087. [CrossRef]
28. Zhu, S.; Meng, Q.; Wang, L.; Zhang, J.; Song, Y.; Jin, H.; Zhang, K.; Sun, H.; Wang, H.; Yang, B. Highly Photoluminescent Carbon Dots for Multicolor Patterning, Sensors, and Bioimaging. *Angew. Chem. Int. Ed.* **2013**, *52*, 3953–3957. [CrossRef]
29. Viedma, C. Chiral Symmetry Breaking During Crystallization: Complete Chiral Purity Induced by Nonlinear Autocatalysis and Recycling. *Phys. Rev. Lett.* **2005**, *94*, 065504. [CrossRef]
30. Barron, L.D. Chirality and Life. *Space Sci. Rev.* **2008**, *135*, 187–201. [CrossRef]
31. Rote, A.R.; Ravindranath, B. Saudagar Estimation of Metformin Hydrochloride by UV Spectrophotometric Method in Pharmaceutical Formulation. *World J. Pharm. Sci.* **2014**, *2*, 1841–1845.
32. Vijayaraghavan, K.; Iyyam Pillai, S.; Subramanian, S.P. Design, Synthesis and Characterization of Zinc-3 Hydroxy Flavone, a Novel Zinc Metallo Complex for the Treatment of Experimental Diabetes in Rats. *Eur. J. Pharmacol.* **2012**, *680*, 122–129. [CrossRef]

33. Kesari, A.N.; Gupta, R.K.; Singh, S.K.; Diwakar, S.; Watal, G. Hypoglycemic and Antihyperglycemic Activity of Aegle Marmelos Seed Extract in Normal and Diabetic Rats. *J. Ethnopharmacol.* **2006**, *107*, 374–379. [CrossRef] [PubMed]
34. Yu, J.-W.; Deng, Y.-P.; Han, X.; Ren, G.-F.; Cai, J.; Jiang, G.-J. Metformin Improves the Angiogenic Functions of Endothelial Progenitor Cells via Activating AMPK/ENOS Pathway in Diabetic Mice. *Cardiovasc. Diabetol.* **2016**, *15*, 88. [CrossRef] [PubMed]
35. Li, Z.-P.; Xin, R.-J.; Yang, H.; Jiang, G.-J.; Deng, Y.-P.; Li, D.-J.; Shen, F.-M. Diazoxide Accelerates Wound Healing by Improving EPC Function. *Front. Biosci.* **2016**, *21*, 1039–1051.
36. Szkudelski, T. The Mechanism of Alloxan and Streptozotocin Action in B Cells of the Rat Pancreas. *Physiol. Res.* **2001**, *50*, 537–546. [CrossRef] [PubMed]

Disclaimer/Publisher's Note: The statements, opinions and data contained in all publications are solely those of the individual author(s) and contributor(s) and not of MDPI and/or the editor(s). MDPI and/or the editor(s) disclaim responsibility for any injury to people or property resulting from any ideas, methods, instructions or products referred to in the content.

 *pharmaceuticals*

Article

Assessing Cardiovascular Target Attainment in Type 2 Diabetes Mellitus Patients in Tertiary Diabetes Center in Romania

Teodor Salmen [1], Valeria-Anca Pietrosel [2], Delia Reurean-Pintilei [3,4], Mihaela Adela Iancu [5,*], Radu Cristian Cimpeanu [6], Ioana-Cristina Bica [1], Roxana-Ioana Dumitriu-Stan [1], Claudia-Gabriela Potcovaru [1], Bianca-Margareta Salmen [1], Camelia-Cristina Diaconu [7], Sanda Maria Cretoiu [8] and Anca Pantea Stoian [9]

1. Doctoral School, "Carol Davila" University of Medicine and Pharmacy, 050474 Bucharest, Romania; teodor.salmen@drd.umfcd.ro (T.S.)
2. DiabetMed Clinic, 052034 Bucharest, Romania
3. Department of Medical-Surgical and Complementary Sciences, Faculty of Medicine and Biological Sciences, "Ștefan cel Mare" University, 720229 Suceava, Romania; delia.pintilei@usm.ro
4. Department of Diabetes, Nutrition and Metabolic Diseases, Consulted Medical Centre, 700544 Iasi, Romania
5. Department of Internal, Family and Occupational Medicine, "Carol Davila" University of Medicine and Pharmacy, 020021 Bucharest, Romania
6. Doctoral School, University of Medicine and Pharmacy, 200349 Craiova, Romania
7. 5th Department, "Carol Davila" University of Medicine and Pharmacy, 050474 Bucharest, Romania; camelia.diaconu@umfcd.ro
8. Department of Morphological Sciences, Cell and Molecular Biology and Histology, "Carol Davila" University of Medicine and Pharmacy, 050474 Bucharest, Romania; sanda.cretoiu@umfcd.ro
9. Diabetes, Nutrition and Metabolic Diseases Department, "Carol Davila" University of Medicine and Pharmacy, 050474 Bucharest, Romania; anca.stoian@umfcd.ro
* Correspondence: adela.iancu@umfcd.ro

Citation: Salmen, T.; Pietrosel, V.-A.; Reurean-Pintilei, D.; Iancu, M.A.; Cimpeanu, R.C.; Bica, I.-C.; Dumitriu-Stan, R.-I.; Potcovaru, C.-G.; Salmen, B.-M.; Diaconu, C.-C.; et al. Assessing Cardiovascular Target Attainment in Type 2 Diabetes Mellitus Patients in Tertiary Diabetes Center in Romania. *Pharmaceuticals* **2024**, *17*, 1249. https://doi.org/10.3390/ph17091249

Academic Editor: Alfredo Caturano

Received: 6 August 2024
Revised: 20 September 2024
Accepted: 21 September 2024
Published: 23 September 2024

Copyright: © 2024 by the authors. Licensee MDPI, Basel, Switzerland. This article is an open access article distributed under the terms and conditions of the Creative Commons Attribution (CC BY) license (https://creativecommons.org/licenses/by/4.0/).

Abstract: Introduction: Type 2 diabetes mellitus (T2DM) and cardiovascular disease (CVD) share a bidirectional link, and the innovative antidiabetic molecules GLP-1 Ras and SGLT-2is have proven cardiac and renal benefits, respectively. This study aimed to evaluate CV risk categories, along with lipid-lowering and antidiabetic treatments, in patients with T2DM from a real-life setting in Romania. Material and Methods: A cross-sectional evaluation was conducted on 405 consecutively admitted patients with T2DM in an ambulatory setting, assessing them according to the 2019 ESC/EAS guidelines for moderate, high, and very high CV risk categories. Results: The average age of the group was 58 ± 9.96 years, with 38.5% being female. The mean HbA1C level was 7.2 ± 1.7%. Comorbidities included HBP in 88.1% of patients, with a mean SBP and DBP of 133.2 ± 13.7 mm Hg and 79.9 ± 9 mm Hg, respectively, and obesity in 66.41%, with a mean BMI of 33 ± 6.33 kg/m^2. The mean LDL-C levels varied by CV risk category: 90.1 ± 34.22 mg/dL in very high risk, 98.63 ± 33.26 mg/dL in high risk, and 105 ± 37.1 mg/dL in moderate risk. Prescribed treatments included metformin (100%), statins (77.5%), GLP-1 Ras (29.4%), and SGLT-2is (29.4%). Conclusions: In Romania, patients with T2DM often achieve glycemic control targets but fail to meet composite targets that include glycemic, BP, and lipid control. Additionally, few patients benefit from innovative glucose-lowering therapies with proven cardio-renal benefits or from statins.

Keywords: cardiovascular disease; type 2 diabetes mellitus; LDL-cholesterol; risk; cardio-renal protection; statin; glucagon-like peptide-1 receptor agonists; sodium-glucose co-transporter-2 inhibitors

1. Introduction

The incidence of Type 2 Diabetes Mellitus (T2DM) is rising steadily, surpassing the predicted prevalence rates and becoming one of the most significant pandemics of this century. Additionally, this condition is a leading cause of death, disability, and lost years of life, posing a substantial burden to healthcare systems [1–3]. The PREDATORR study conducted in Romania in 2016 reported an undiagnosed DM prevalence of 2.4% among the population and a diagnosed prevalence of 11.6% among individuals aged 20 to 79 years [4].

The lack of proper glycemic and metabolic control can result in the development and progression of complications associated with T2DM, with high mortality rates. To mitigate these risks, the American and European Diabetes Associations have issued guidelines recommending the avoidance of therapeutic inertia and the use of innovative antidiabetic agents with cardioprotective and renal protective effects, such as GLP-1 Ras and SGLT-2is [2,5–7]. The efficacy of these agents was demonstrated through large-scale CVOTs, despite the heterogeneity of the various agents within the two pharmacological classes. These trials demonstrated improved metabolic control, reflected by a better management of weight, glycemic levels, BP, and lipid profiles. Additionally, they provided CV protection, as indicated by a reduction in major CVEs, including stroke and myocardial infarction, as well as a decreased risk of HF hospitalization or CV mortality [8,9].

The most widely prescribed medication for the treatment of T2DM globally is metformin, a biguanide derivative known as dimethylbiguanide. It is commonly prescribed as a first-line therapy, either alone or in combination with other medications. However, current guidelines recommend that innovative medication classes, such as GLP-1 Ras and SGLT-2is, should be used in T2DM patients regardless of their glycemic control, with this approach aiming to reduce CV and renal risks. The CV effects of metformin are limited, but it is known to reduce oxidative stress and inflammation and improve endothelial function [10]. There are also some nutritional supplements that have been shown to reduce oxidative stress [11]. Additionally, physical activity can help reduce oxidative stress and contribute to weight loss [12]. Given these factors, it is nevertheless still important to focus on patient education and lifestyle advice for those with T2DM and other CV risk factors. Education should include guidance on physical activity, understanding the impact of carbohydrates on blood glucose levels, and overall dietary management. DM can impact patients' functionality and lead to disability, often due to complications like diabetic neuropathy or dietary restrictions. These issues can be addressed through effective assessments, education, and comprehensive rehabilitation programs [13].

CVD includes ischemic coronary disease, stroke, and peripheral arterial disease, and constitutes the leading cause of mortality globally, greatly impairing individuals' quality of life [1,2,14]. The most frequent manifestation of CVD, CHD, was reported in 2019 with 197 million cases of incidence and 9.14 million fatalities by the Global Burden of Disease Study [15]. The 2022 update of the AHA reports a 7.2% CHD prevalence in individuals above 20 years old in the United States [16]. In the EU, in 2020, 1.70 million deaths resulted from diseases of the circulatory system, equivalent to 32.7% of all deaths as compared to 22.5% for the second most prevalent cause of death, cancer. These are alarming percentages, with more than half of all deaths registered in Central-East countries (60.6% in Bulgaria, which leads the ranking, shortly followed by Romania with 55.1% deaths) and Baltic States Lithuania and Latvia [17]. The 2019 report from the ESC Atlas stated that 15% of CVD deaths in Europe are due to poorly controlled DM [18].

CVD and T2DM are part of cardio-reno-metabolic syndrome; therefore, these entities have a bidirectional link [19]. The level of risk is further enhanced in the case of associations with additional standard risk factors, such as excess weight, high BP (HBP), dyslipidemia, smoking status, and physical inactivity, especially when they overlap with a hyperglycemic state [20]. The emergence of CVEs is more likely to occur in individuals with elevated levels of total cholesterol, particularly high levels of LDL-C. A reduction of one mmol/L in LDL-C levels can substantially decrease the risk of CVEs, regardless of the baseline level [21]. The long-term exposure to blood glucose levels, specifically the HbA1C level, is closely associated with CVEs. An increase of 1% in HbA1C is linked to a 20% increase in the risk of atherosclerotic CVEs. Conversely, maintaining the HbA1C level below 7% for over 10 years significantly decreases the risk of CVD [22,23].

To enhance CV risk stratification, various predictive models have been developed to enable a multifactorial approach to CVD. These models comprehensively identify and quantify risk factors. The early identification of high risk individuals is critical, particularly since DM frequently coexists with CVD and is a significant risk factor itself. Therefore, in trials, risk stratification has been continuously refined to include the most relevant variables for optimal estimation, supporting personalized care management, meaning targeted therapy and specific treatment goals [23].

Between 2019 and 2023, continuous efforts were made to improve CV risk stratification in patients with DM. The 2019 ESC/EAS guidelines categorized T2DM patients as very high, high, or moderate risk, eliminating the need for the SCORE assessment. In 2021, the ESC Guidelines recommended using the ADVANCE or DIAL models for CV risk assessment in DM patients. The 2023 update introduced the SCORE2-Diabetes algorithm, which utilizes factors like age, gender, smoking status, BP, cholesterol levels, and renal health to classify CV risk into low, moderate, high, or very high categories. [11,16,24,25]. These risk factors are also addressed by the American Diabetes Association (ADA) guidelines, which outline the four pillars of DM management: glucose, BP, lipid, and weight control. The guidelines recommend using glucose-lowering medications that offer cardiac or renal benefits [5,14].

Despite these advances in treatment guidelines and risk stratification models that provide clear guidance on selecting therapeutic agents according to the CV risk groups, a significant gap remains in understanding how many patients with T2DM achieve the recommended therapeutic targets and how closely their medications are prescribed according to these guidelines. This research aims to bridge this gap by offering updated insights into the management of T2DM patients as it evaluates the achievement of specific therapeutic targets—HbA1C, LDL-C, and BP—both individually and collectively. The study also examines the utilization rates of statins and novel antidiabetic therapies with established cardio-renal benefits, including GLP-1 Ras and SGLT-2is. Specifically, it assesses the proportion of patients classified according to the 2019 ESC/EAS CV risk categories and determines the percentage of patients prescribed statins and novel antidiabetic agents with proven cardio-renal protective effects. By analyzing these factors, the study aims to enhance awareness of a critical public health issue and provide guidance for future clinical decisions.

2. Results

The demographic characteristics of the 405 patients included a mean age of 58 ± 9.96 years, with 38.5% females, and a median duration of T2DM of 6 (0, 59) years.

Our cohort's unmodifiable risk factors consisted of age, gender, and duration of DM, while the modifiable risk factors included poorly controlled DM, obesity, dyslipidemia, and BP. Additional patient characteristics are detailed in Table 1.

Table 1. Patients' characteristics.

Characteristic	n = 405
Demographics	
Age (years), mean (SD)	58 ± 9.96
Women, %, (n)	38.5% (156)
DM mean duration, median (25–75% IQR)	6 (2, 12)
Risk factors	
BMI (kg/m^2), mean (SD)	33 ± 6.33
Obesity, %, (n)	66.41% (269)
SBP (mm Hg), mean (SD)	132.3 ± 13.7
DBP (mm Hg), mean (SD)	79.9 ± 9
HBP, %, (n)	88.1% (357)
Smoking status, %, (n)	20% (81)

Table 1. Cont.

Characteristic	n = 405
Demographics	
HbA1C (%), mean (SD)	7.2 ± 1.7
Total-C (mg/dL), mean (SD)	168.8 ± 42.6
HDL-C (mg/dL), mean (SD)	44.7 ± 14.2
TGs (mg/dL), median (25–75% IQR)	156 (106, 206)
LDL-C (mg/dL), mean (SD)	91.5 ± 34.1
Atherosclerotic CVD, %, (n)	34.32% (139)
eGFR (mL/min/1.73 m^2)	97.6 ± 16.8
LDL-C in very high CV risk category, (mg/dL), mean (SD)	90.1 ± 34.22
LDL-C in high CV risk category, (mg/dL), mean (SD)	98.63 ± 33.26
LDL-C in moderate CV risk category, (mg/dL), mean (SD)	105 ± 37.1
Glucose-lowering medication usage	
Insulin, %, (n)	23.9% (97)
Metformin, %, (n)	100% (405)
GLP-1 Ras, %, (n)	29.4% (119)
SGLT-2is, %, (n)	29.4% (119)
Other therapies	
ACEi/ARBs, %, (n)	69.1% (280)
Statins, %, (n)	77.5% (314)
Beta-blockers, %, (n)	60.7% (246)
Calcium channel blockers, %, (n)	24.4% (99)
Diuretics, %, (n)	37.8% (153)

Among the patients, 66.41% were obese, with a mean body mass index of 33 ± 6.33 kg/m^2; 88.1% had high BP, with a mean SBP of 132.3 ± 13.7 mm Hg and a mean DBP of 79.9 ± 9 mm Hg. Smoking was reported by 20% of the patients, and 34.32% had atherosclerotic CVD. The mean LDL-C values were distributed as follows: 90.1 ± 34.22 mg/dL in the very high CV risk category, 98.63 ± 33.26 mg/dL in the high CV risk category, and 105 ± 37.1 mg/dL in the moderate CV risk category. The mean HbA1C level was 7.2 ± 1.7%, and the lipid profile showed a mean total-C of 168.8 ± 42.6 mg/dL, a mean HDL-C of 44.7 ± 14.2 mg/dL, a mean LDL-C of 91.5 ± 34.1 mg/dL, and a median TG level of 156 mg/dL (ranging from 38 to 1080).

Table 1 provides a summary of the prescribed glucose-lowering therapies by category, including metformin (100%), insulin (23.9%), GLP-1 Ras (29.4%), and SGLT-2is (29.4%). Additionally, the table details the use of other therapies targeting the CV risk factors, such as ACEi/ARBs (69.1%), statins (77.5%), beta-blockers (60.7%), calcium channel blockers (24.4%), and diuretics (37.8%).

Tables 2–4 illustrate the subsequent distribution of the 405 patients into the CV risk categories, with 340 patients (83.9%) in the very high CV risk category, 62 patients (15.3%) in the high CV risk category, and 3 patients (0.8%) in the moderate CV risk category. A concise description of our results is shown in the tables, with the number of patients who achieved their targets for LDL-C, HbA1C, and BP, both separately and in combination. They also summarize the use of innovative antidiabetic and lipid-lowering medications. According to the 2019 ESC/EAS Guidelines for LDL-C and the 2019 ADA Guidelines for HbA1C and BP, data about the patients with very high, high, and moderate CV risk are presented in Table 2, Table 3, and Table 4, respectively.

Table 2. Very high CV risk category.

Treatment Target for Patients with Very High CV Risk Category (n = 340)	Patients Achieving Target	Patients with SGLT-2is Prescription	Patients with GLP-1 Ras Prescription	Patients with Statin Prescription
LDL-C < 55 mg/dL, %, (n)	13.23% (45)	5% (17)	3.53% (12)	9.7% (33)
HbA1C < 7%, %, (n)	40.59% (138)	5% (17)	19.7% (67)	27.65% (94)
BP < 130/80 mmHg, %, (n)	14.42% (49)	4.71% (16)	5.3% (18)	10.9% (37)
LDL-C < 55 mg/dL + HbA1C < 7%, %, (n)	5.88% (20)	0.59% (2)	2.06% (7)	3.53% (12)
HbA1C < 7% + BP < 130/80 mmHg, %, (n)	6.76% (23)	0.88% (3)	3.53% (12)	4.18% (14)
LDL-C < 55 mg/dL + BP < 130/80 mmHg, %, (n)	2.06% (7)	0.3% (1)	0.88% (3)	1.47% (5)
LDL-C < 55 mg/dL + HbA1C < 7% + BP < 130/80 mmHg, %, (n)	1.18% (4)	0	0.3% (1)	0.59% (2)

Table 3. High CV risk category.

Treatment Target for Patients with High CV Risk Category (n = 62)	Patients Achieving Target	Patients with SGLT-2is Prescription	Patients with GLP-1 Ras Prescription	Patients with Statin Prescription
LDL-C < 70 mg/dL, %, (n)	20.96% (13)	6.45% (4)	6.45% (4)	14.51% (9)
HbA1C < 7%, %, (n)	38.7% (24)	6.45% (4)	14.51% (9)	27.42% (17)
BP < 130/80 mmHg, %, (n)	16.13% (10)	8.06% (5)	3.23% (2)	11.29% (7)
LDL-C < 70 mg/dL + HbA1C < 7%, %, (n)	9.68% (6)	1.61% (1)	4.84% (3)	4.84% (3)
HbA1C < 7% + BP <130/80 mmHg, %, (n)	4.84% (3)	0	3.23% (2)	4.84% (3)
LDL-C < 70 mg/dL + BP < 130/80 mmHg, %, (n)	1.61% (1)	1.61% (1)	0	1.61% (1)
LDL-C < 70 mg/dL + HbA1C < 7% + BP < 130/80 mmHg, %, (n)	0	0	0	0

Table 4. Moderate CV risk category.

Treatment Target for Patients with Moderate CV Risk Category (n = 3)	Patients Achieving Target	Patients with SGLT-2is Prescription	Patients with GLP-1 Ras Prescription	Patients with Statin Prescription
LDL-C < 100 mg/dL, %, (n)	66.6% (2)	0	0	0
HbA1C < 7%, %, (n)	0	0	0	0
BP < 130/80 mmHg, %, (n)	33.3% (1)	0	0	0
LDL-C < 100 mg/dL + HbA1C < 7%, %, (n)	0	0	0	0
HbA1C < 7% + BP < 130/80 mmHg, %, (n)	0	0	0	0
LDL-C < 100 mg/dL + BP < 130/80 mmHg, %, (n)	33.3% (1)	0	0	0
LDL-C < 100 mg/dL + HbA1C < 7% + BP < 130/80 mmHg, %, (n)	0	0	0	0

3. Discussion

This study provided an illustrative image of the CV risk categories for T2DM patients from a 2019 ambulatory setting in a Romanian tertiary care center. These categories were defined using the 2019 ESC/EAS guidelines for LDL-C and the 2019 ADA guidelines for HbA1C and BP, in effect at the time of the cohort's assessment. Additionally, we evaluated the prescription rates of statins and innovative cardio-renal protective antidiabetic drugs, specifically GLP-1 Ras and SGLT-2is, across these CV risk categories.

The cohort's mean age was 58 ± 9.9 years, highlighting that these molecules were prescribed early. With a median DM duration of 6 years, the patients' metabolic control was borderline, evidenced by a mean HbA1C of $7.2 \pm 1.7\%$. This data aligns with other cohort studies of T2DM patients, such as Vintila et al. [26], who reported a mean age of 71 years and mean HbA1C of 7.2%; Reurean-Pintilei et al. [23], with a mean age of 62.9 ± 7.7 years and mean HbA1C of $7.1 \pm 1.3\%$; and Cokolic et al. [27], with a mean age of 63.5 ± 10.7 years, a mean DM duration of 8.9 ± 7.1 years, and a mean HbA1C of $7.3 \pm 1.5\%$.

Obesity, a key risk factor for developing DM, affects a substantial portion of the population, accounting for 66.41% with an average BMI of 33 ± 6.33 kg/m^2, similar to other Romanian reports [23] and higher than other Eastern European countries and the USA, which report a 20–30% prevalence [28,29].

HBP was recorded in an overwhelming percentage in our cohort (88.1% of patients), similar to a recent report by Reurean-Pintilei et al. [23]. The percentage is almost double than the SEPHAR III study's reports of 45.1%, but this was expected, because in patients with T2DM, usually higher rates of HBP as compared to the general population are encountered [30,31].

When compared with reports from other standard of care approaches from similar time frames in Romania [23], Scotland [32], and Denmark [33], the included patients were younger (58 years old as compared to 62 years old, 67 years old, and 72 years old, respectively), suggesting that the risk factors appear early, especially in Eastern European populations. Moreover, the patients had a shorter DM duration (6 years versus 9 years and 7.8 years). In terms of CVEs, the 34.32% of CVEs is similar to the one reported for Scotland (32%), but bigger than other regions in Romania (13.9%) or Denmark (21.4%) report. On the other hand, the mean levels of HbA1C, SBP, and BMI are similar.

When comparing the CV risk category percentages according to the 2019 ESC/EASD guidelines with a similar Romanian study by Reurean-Pintilei et al. [23], we observed fewer patients in the very high CV risk category, 83.9% versus 92.7%, and more in the high CV risk category, 15.3% versus 1.12%, with a lower moderate CV risk category percentage of 0.8% versus 6.7%. When comparing with the Santorini study [34], which included over 9000 patients from 14 European but non-eastern countries, we observed a lower median LDL-C level of 82 mg/dL and larger percentages for the very high CV risk of 91% and the high CV risk of 6.5% as compared to our results. Because only four patients (1.18%) achieved all of the HbA1C, LDL-C, and BP targets in the very high CV risk category, it is important to emphasize that ASCVD was present in almost one of three enrolled patients, similar to the data reported by McGurnaghan et al. [33], but almost triple compared to Reurean-Pintilei et al. [23]. On the other hand, if taken individually, 49 patients (14.42%) achieved the BP target alone, 138 patients (40.59%) achieved the HbA1C target, and 45 patients (13.23%) achieved the LDL-C target. Another important aspect found in the patients that reached all of the three targets simultaneously is the low rate of prescription of innovative cardio-renal protective medication; respectively, one patient received GLP-1 Ras and two patients received statins. It can be inferred that the patients included in the study exhibited a commendable level of metabolic control, which suggests that these individuals were proactive in managing their health by consistently attending medical check-ups and maintaining a healthy lifestyle. Therefore, the low prescription rate of the innovative pharmaceuticals may be deceptively minimal, particularly given that these medications were prescribed exclusively in accordance with the stringent and inflexible criteria established by the national prescription guidelines in 2019, when the HbA1C levels exceeded 7%.

An important aspect in our study is that 13.23% patients from the very high CV risk category met the 2019 recommended guidelines for the LDL-C targets, more than double the previously reported data from a Romanian study, as well [23] as the DA VINCI study [35], where the percentages were 5% and 4%, respectively. The previous rates of achieved LDL-C targets, according to the 2016 ESC/EAS guidelines, were similar to our reported data but higher than the ones reported in the DA VINCI study, which included over 2000 Romanian patients; therefore, we can conclude that the physician's underestimation of the patient's risk and fear of escalating lipid-lowering treatment in groups of patients other than the very high CV risk category are aspects to be taken into account [31]. Still, in our cohort, 77.5% of patients were prescribed statins, more than the 67.8% reported in other regions of the same country [23], with both percentages being higher than the 48.4% reported by the Santorini study [34]. Delving deeper, only 9.7% of the patients in the very high CV risk category treated with statins met the LDL-C target, a figure that lies between the 15% reported by Morieri et al. [36] and the 2.7% reported by Reurean-Pintilei et al. [23]. Despite the recommendations from several medical societies, including the American Association of Clinical Endocrinologists, the ESC/EAS guidelines for dyslipidaemia management, and the American College of Cardiology/American Heart Association Task Force [15,37,38], which advocate for an LDL-C target below 55 mg/dL for those at very high or extreme CV risk, the consistently low achievement rates highlight an urgent need to increase awareness and adopt more aggressive lipid management strategies in these populations [23,26,34–36]. Expanding our analysis to the high and moderate CV risk categories, we find that no patients simultaneously met the LDL-C, BP, and HbA1C targets.

Statin use was lower than in other Romanian data, but with similarities with Scotland and Denmark, while ACEi/ARBs were lower in both Romanian studies as compared to the other two countries [23,32,33]. Moreover, even if in our cohort the rates of SGLT-2is and GLP-1 Ras were low, they were recommended more than in previously reported data on similar cohorts, 29.4% versus 3.9% and 29.4% versus 8.1%, respectively, and also in higher proportion for the very high and high CV risk categories. The trend of under-prescribing these innovative medications is similar to the reported data of Vencio et al. [39], where 15% of patients received SGLT-2is and 9% received GLP-1 Ras, or the Discover study by Arnold et al. [40], where almost 16.1% of patients received these types of medications at follow-up, but with variable values between the included countries. A rate over 63% was, on the contrary, reported by countries with smaller economic power from Southern and Eastern Europe, as reported by Banach et al. [41]. In Denmark and Scotland, SGLT-2is and GLP-1 Ras were also under-prescribed, with 4% and 2% and 8% and 5.4%, respectively [32,33].

An interesting fact is the high percentage of metformin prescription that was present in all patients from our cohort, in comparison with the data from Scotland, Denmark, or Romania, with rates of 57%, 54%, and 87%, respectively; for insulin, the 23.9% rate was similar to the data reported in Romania of 25%, but higher than the ones from Scotland or Denmark of, respectively, 11% and 19.5% [23,32,33].

In a broader context, simultaneously with the development of these novel molecules and the updates of the medical guidelines on their use, medical research showed more clearly that DM is evolving silently and leads to the appearance of CVEs and high mortality rates; hence, a more accurate stratification for CV risk is essential. Because of DM and CVD's silent progression, complementary investigations such as magnetic resonance imaging and coronary computed tomography angiography are needed both in primary and secondary prevention, in order to be able to include specific therapies or interventions in the management of very high risk patients [42]. Another approach to refining CV risk involves the early assessment of atherosclerosis to detect progression towards plaque instability or rupture using biological markers. In this context, Lp(a) is recognized as both an independent and causal risk factor for ASCVD due to its inflammatory, prothrombotic, and atherogenic effects. It is a primary target of the latest lipid-lowering therapies, including proprotein convertase subtilisin/kexin type 9 inhibitors, small interfering RNA, and inhibitory antisense oligonucleotides. However, the implementation of Lp(a) screening and treatment faces significant challenges, including limited physician awareness, the absence of consensus guidelines amid ongoing research, and high costs, with cost-effectiveness data being both scarce and inconsistently assessed [15,43–45].

This study's strengths include providing important insights into T2DM among European patients, particularly focusing on the underrepresented Romanian demographic, utilizing a cross-sectional design. This approach allows for a comprehensive snapshot of current disease management practices and CV risks at a specific point in time, facilitating an immediate comparison with other countries. The utilization of the 2019 ESC/EAS and ADA guidelines in use at the time of the evaluation enhances the depiction of the CV risk categories, while the detailed data collection underscores significant areas of underutilization due to factors such as physician unawareness and medical inertia. Moreover, the comparative analysis not only underscores the regional variations that may influence disease management outcomes, but also identifies potential areas for improvement in local healthcare strategies. This is essential for acknowledging disparities compared to other regions or populations, particularly given the underrepresentation of Romanian patients in studies that often focus on Western European populations. The fact that our cohort consisted of younger patients can be also considered a strength, as it adds important data on the real-life clinical status of the working-age population with DM.

The cross-sectional design of our study serves as both a strength and a limitation. So, the study's main limitations are the cross-sectional design, the exclusively Romanian cohort from a single center, the relatively small sample, and the lack of data about physician inertia levels and cost perceived barriers in prescribing the innovative molecules. This design limits our ability to infer causality or track changes over time, preventing an evaluation of T2DM progression, the evolution of prescription patterns, and the long-term effectiveness of treatment strategies. Additionally, with a focus predominantly on a Romanian cohort, our study may have a limitation on the generalizability of the findings to other regions or populations. The specific healthcare setting, different cultural factors, and economic conditions in different geographical regions of Romania may not be representative of other European or global contexts. This study included participants from a single tertiary care center, leading to a potential selection bias, as we are unable to accurately reflect the broader population of T2DM patients in Romania or elsewhere. This aspect could influence the observed prevalence of CV risk factors and the reported prescription rates of medication. Lastly, while providing data on prescription patterns (although not including data on dipeptidyl-peptidase 4 inhibitors or sulphonylurea) and risk categorization, our study does not offer the explanations behind physician inertia or the barriers to the adoption of innovative treatments beyond cost and awareness. It should be noted that in 2019, prescription practices were subject to restrictions under the Romanian health insurance system.

In order to mitigate the issues identified in our study and enhance T2DM management, several measures could be recommended as possible perspectives. This should start with future longitudinal studies to aid in policy adjustments and to improve educational and policy interventions. A more detailed exploration of these factors could offer clearer guidance for interventions designed to enhance T2DM management. Simplifying the national prescribing criteria would facilitate easier access to innovative antidiabetic drugs for high risk patients. Targeted educational programs could increase healthcare providers' awareness of the latest guidelines and emphasize comprehensive treatment strategies. Establishing regular audits of prescription patterns, coupled with feedback mechanisms, would potentially encourage guideline adherence and improve prescribing practices. Additionally, patient engagement and education would empower individuals to actively participate in their treatment, thereby enhancing adherence and outcomes.

4. Materials and Methods

4.1. Study Design and Patients

This single-center, consecutive-case investigation was cross-sectional and population-based, directed as part of a sub-analysis derived from a retrospective investigation. The principles outlined in the Declaration of Helsinki were followed and protocol number 5591, dated 17 November 2022 from the Institutional Ethics Committee of the N Paulescu National Institute for Diabetes Mellitus, Nutrition, and Metabolic Disorders in Bucharest, Romania, was obtained. Initially, between January and July in 2019, 477 patients with T2DM who met the inclusion criteria were sequentially invited to participate during their routine visits, during which their medical records were gathered. Subsequently, data analysis was performed on the records of the 405 patients who provided informed consent. The following criteria were used to determine eligibility for the study: adults over the age of 18, the provision of informed consent, a confirmed diagnosis of T2DM for at least six months, and treatment with the standard of care maximum tolerated doses for the associated conditions for at least six months prior to the study. The study also excluded patients who were under 18 years of age, those who were diagnosed with other types of DM than T2DM (such as Type 1 DM or secondary DM), those who did not provide signed informed consent, and those who had severe or acute heart failure, renal insufficiency, or hepatic insufficiency.

The patient data were collected from hospital reports and included demographics (age, gender, and residential background), anthropometrics (height, weight, and BMI), concurrent conditions (e.g., smoking status, HBP, dyslipidemia, and atherosclerotic CV disease), laboratory test results (HbA1C, LDL-C, total-C, high-density lipoprotein cholesterol (HDL-C), triglycerides, uric acid, UACR, and eGFR), and treatment specifics (such as antidiabetics, BP, and lipid-lowering medications).

4.2. Statistical Analysis

The categorical data were expressed as numbers and percentages. Continuous variables were evaluated using the Kolmogorov–Smirnov test, and depending on their distribution, were expressed as either the median (IQR) or the mean ± SD. We divided the population data into three different groups according to the antidiabetic treatments (metformin, metformin + SGLT-2is, and metformin + GLP-1 Ras, respectively). This information was organized into Excel spreadsheets and analyzed using Excel 2019 software.

4.3. Patients' Stratification

The patients were stratified into CV risk categories following the 2019 ESC/EAS guidelines, namely the moderate, high, and very high CV risk groups.

The low risk category was not included due to the presence of T2DM in all patients, which represented that they were in the most favorable scenario in the moderate risk category. The moderate CV risk category included the patients with T2DM younger than 50 years of age, with a DM duration of less than 10 years, and without additional risk factors. The high risk category encompassed the patients with T2DM without TOD, with a DM duration of at least ten years, or presenting an additional risk factor.

The very high risk category included the patients with T2DM and TOD, T2DM and at least three major risk factors, or documented ASCVD. Documented ASCVD included a history of ACS, stable angina, coronary revascularization procedures (PCI, CABG, and other arterial revascularization), stroke, TIA, or peripheral arterial disease. Unequivocal documentation on imaging involved findings predictive of clinical events, such as significant plaque on a coronary angiography or CT scan or notable carotid ultrasound results.

The goals for each CV risk category were established based on the 2019 ESC/EAS standards for LDL-C and the 2019 ADA guidelines for HbA1C and BP:

i. Moderate CV risk category: LDL-C < 100 mg/dL, HbA1C < 7%, and BP < 130/80 mmHg.
ii. High CV risk category: LDL-C < 70 mg/dL, HbA1C < 7%, and BP < 130/80 mmHg.
iii. Very high CV risk category: LDL-C < 55 mg/dL, HbA1C < 7%, and BP < 130/80 mmHg.

Upon assessing the biological and paraclinical parameters in accordance with the guidelines, we evaluated the use of statins and novel T2DM cardio-renal protective therapies, specifically the GLP-1 Ras and SGLT-2is classes, which were accessible in Romania in 2019.

5. Conclusions

Romanian patients typically fall into very high or high CV risk categories. Despite recent scientific advances, the strict criteria of the national prescribing protocols and prevalent medical inertia mean that in real-world settings, these patients rarely achieve the optimal prescriptions of lipid-lowering and innovative antidiabetic drugs, such as SGLT-2is and GLP-1 Ras. Conversely, they often meet BP or glycemic targets individually. Given that patients with T2DM frequently exhibit additional CV risk factors, improving their prognosis requires a comprehensive treatment evaluation that simultaneously targets BP, HbA1C, and LDL-C. The failure to meet the guideline recommendations for cardio-renal benefits is a critical issue, particularly for those with T2DM and high CV risk. A longitudinal follow-up of this cohort will determine whether there are improvements in guideline adherence, better target achievement, and enhanced overall care and potential guidance.

Abbreviation List

ACEi/ARBs	Angiotensin-Converting Enzyme Inhibitors/Angiotensin Receptor Blockers
ACS	Acute Coronary Syndrome
AHA	American Heart Association
ASCVD	Atherosclerotic Cardiovascular Disease
BMI	Body Mass Index
BP	Blood Pressure
CHD	Coronary Heart Disease
CV	Cardiovascular
CVD	Cardiovascular Disease
CVEs	Cardiovascular Events
CVOTs	Cardiovascular Outcome Trials
DBP	Diastolic Blood Pressure
DM	Diabetes Mellitus
EAS	European Atherosclerosis Society
ESC	European Society of Cardiology
EU	European Union
GLP-1 Ras	Glucagon-Like Peptide-1 Receptor Agonists
HbA1C	A1c Haemoglobin
HBP	High Blood Pressure
HDL-C	High-Density Lipoprotein Cholesterol
HF	Heart Failure
IQR	Interquartile Range
LDL-C	Low-Density Lipoprotein Cholesterol
Lp(a)	Lipoprotein(a)
SBP	Systolic Blood Pressure
SD	Standard Deviation
SGLT-2is	Sodium-Glucose Co-Transporter-2 Inhibitors
T2DM	Type 2 Diabetes Mellitus
TGs	Triglycerides
TIA	Transient Ischemic Attack
TOD	Target Organ Damage
Total-C	Total Cholesterol

Author Contributions: Conceptualization, T.S., V.-A.P. and D.R.-P.; methodology, M.A.I. and C.-C.D.; software, T.S.; validation, A.P.S., S.M.C. and I.-C.B.; formal analysis, R.C.C.; investigation, S.M.C. and R.-I.D.-S.; resources, T.S., B.-M.S. and D.R.-P.; data curation, V.-A.P.; writing—original draft preparation, T.S. and R.-I.D.-S.; writing—review and editing, C.-G.P. and V.-A.P.; visualization, A.P.S. and M.A.I.; supervision, A.P.S. and C.-C.D.; project administration, A.P.S. and M.A.I.; funding acquisition, R.C.C. and R.-I.D.-S. All authors have read and agreed to the published version of the manuscript.

Funding: This research received no external funding.

Institutional Review Board Statement: The study was conducted in accordance with the Declaration of Helsinki and approved by the Institutional Ethics Committee of the N Paulescu National Institute for Diabetes Mellitus, Nutrition, and Metabolic Disorders, Bucharest, Romania (protocol number 5591, from 17 November 2022).

Informed Consent Statement: Not applicable.

Data Availability Statement: The data supporting this study's findings are available upon request from the authors. Please note that the data are also being utilized in an ongoing doctoral project evaluating the novel antidiabetic drugs.

Conflicts of Interest: The authors declare no conflicts of interest.

References

1. Sun, H.; Saeedi, P.; Karuranga, S.; Pinkepank, M.; Ogurtsova, K.; Duncan, B.B.; Stein, C.; Basit, A.; Chan, J.C.N.; Mbanya, J.C.; et al. IDF Diabetes Atlas: Global, regional and country-level diabetes prevalence estimates for 2021 and projections for 2045. *Diabetes Res. Clin. Pract.* **2022**, *183*, 109119. [CrossRef] [PubMed]
2. Salmen, T.; Pietroșel, V.-A.; Mihai, B.-M.; Bica, I.C.; Teodorescu, C.; Păunescu, H.; Coman, O.A.; Mihai, D.-A.; Pantea Stoian, A. Non-Insulin Novel Antidiabetic Drugs Mechanisms in the Pathogenesis of COVID-19. *Biomedicines* **2022**, *10*, 2624. [CrossRef] [PubMed]
3. Khan, M.A.B.; Hashim, M.J.; King, J.K.; Govender, R.D.; Mustafa, H.; Al Kaabi, J. Epidemiology of Type 2 Diabetes—Global Burden of Disease and Forecasted Trends. *J. Epidemiol. Glob. Health* **2020**, *10*, 107–111. [CrossRef] [PubMed]
4. Mota, M.; Popa, S.G.; Mota, E.; Mitrea, A.; Catrinoiu, D.; Cheta, D.M.; Guja, C.; Hancu, N.; Ionescu-Tirgoviste, C.; Lichiardopol, R.; et al. Prevalence of diabetes mellitus and prediabetes in the adult Romanian population: PREDATORR study. *J. Diabetes* **2016**, *8*, 336–344. [CrossRef] [PubMed]
5. ElSayed, N.A.; Aleppo, G.; Aroda, V.R.; Bannuru, R.R.; Brown, F.M.; Bruemmer, D.; Collins, B.S.; Hilliard, M.E.; Isaacs, D.; Johnson, E.L.; et al. 9. Pharmacologic approaches to glycemic treatment: Standards of care in diabetes—2023. *Diabetes Care* **2023**, *46*, S140–S157. [CrossRef]
6. Castro Conde, A.; Marzal Martín, D.; Campuzano Ruiz, R.; Fernández Olmo, M.R.; Morillas Ariño, C.; Gómez Doblas, J.J.; Gorriz Teruel, J.L.; Mazón Ramos, P.; García-Moll Marimon, X.; Soler Romeo, M.J.; et al. Comprehensive Cardiovascular and Renal Protection in Patients with Type 2 Diabetes. *J. Clin. Med.* **2023**, *12*, 3925. [CrossRef]
7. Salmen, T.; Bobirca, F.-T.; Bica, I.-C.; Mihai, D.-A.; Pop, C.; Stoian, A.P. The Safety Profile of Sodium-Glucose Cotransporter-2 Inhibitors and Glucagon-like Peptide 1 Receptor Agonists in the Standard of Care Treatment of Type 2 Diabetes Mellitus. *Life* **2023**, *13*, 839. [CrossRef]
8. Scheen, A.J. Cardiovascular outcome studies in type 2 diabetes: Comparison between SGLT2 inhibitors and GLP-1 receptor agonists. *Diabetes Res. Clin. Pract.* **2018**, *143*, 88–100. [CrossRef]
9. Acharya, T.; Deedwania, P. Cardiovascular outcome trials of the newer anti-diabetic medications. *Prog. Cardiovasc. Dis.* **2019**, *62*, 342–348. [CrossRef]
10. Nesti, L.; Natali, A. Metformin Effects on the Heart and the Cardiovascular System: A Review of Experimental and Clinical Data. *Nutr. Metab. Cardiovasc. Dis.* **2017**, *27*, 657–669. [CrossRef]
11. Manolescu, B.N.; Berteanu, M.; Cinteză, D. Effect of the Nutritional Supplement ALAnerv® on the Serum PON1 Activity in Post-Acute Stroke Patients. *Pharmacol. Rep.* **2013**, *65*, 743–750. [CrossRef]
12. Huang, C.-J.; McAllister, M.J.; Slusher, A.L.; Webb, H.E.; Mock, J.T.; Acevedo, E.O. Obesity-Related Oxidative Stress: The Impact of Physical Activity and Diet Manipulation. *Sports Med.-Open* **2015**, *1*, 32. [CrossRef] [PubMed]
13. Potcovaru, C.-G.; Salmen, T.; Bîgu, D.; Săndulescu, M.I.; Filip, P.V.; Diaconu, L.S.; Pop, C.; Ciobanu, I.; Cinteză, D.; Berteanu, M. Assessing the Effectiveness of Rehabilitation Interventions through the World Health Organization Disability Assessment Schedule 2.0 on Disability—A Systematic Review. *J. Clin. Med.* **2024**, *13*, 1252. [CrossRef] [PubMed]
14. ElSayed, N.A.; Aleppo, G.; Bannuru, R.R.; Bruemmer, D.; Collins, B.S.; Das, S.R.; Ekhlaspour, L.; Hilliard, M.E.; Johnson, E.L.; Khunti, K.; et al. 10. Cardiovascular Disease and Risk Management: Standards of Care in Diabetes—2024. *Diabetes Care* **2024**, *47*, 179–218.

15. Mach, F.; Baigent, C.; Catapano, A.L.; Koskinas, K.C.; Casula, M.; Badimon, L.; Chapman, M.J.; De Backer, G.G.; Delgado, V.; Ference, B.A.; et al. 2019 ESC/EAS Guidelines for the management of dyslipidaemias: Lipid modification to reduce cardiovascular risk. *Eur. Heart J.* **2020**, *41*, 111–188. [CrossRef] [PubMed]
16. Tsao, C.W.; Aday, A.W.; Almarzooq, Z.I.; Alonso, A.; Beaton, A.Z.; Bittencourt, M.S.; Boehme, A.K.; Buxton, A.E.; Carson, A.P.; Commodore-Mensah, Y.; et al. Heart disease and stroke statistics—2022 update: A report from the American Heart Association. *Circulation* **2022**, *145*, e153–e639.
17. Eurostat. Deaths from Cardiovascular Diseases. Cardiovascular Diseases Statistics. 2023. Available online: https://ec.europa.eu/eurostat/statistics-explained/index.php?title=Cardiovascular_diseases_statistics#Deaths_from_cardiovascular_diseases (accessed on 13 May 2024).
18. European Society of Cardiology. Fact Sheets for Press–CVD in Europe and ESC Congress Figures. 2022. Available online: https://www.escardio.org/The-ESC/Press-Office/Fact-sheets (accessed on 13 May 2024).
19. Handelsman, Y.; Butler, J.; Bakris, G.L.; DeFronzo, R.A.; Fonarow, G.C.; Green, J.B.; Grunberger, G.; Januzzi, J.L.; Klein, S.; Kushner, P.R.; et al. Early intervention and intensive management of patients with diabetes, cardiorenal, and metabolic diseases. *J. Diabetes Complicat.* **2023**, *37*, 108389. [CrossRef] [PubMed]
20. Marx, N.; Federici, M.; Schütt, K.; Müller-Wieland, D.; Ajjan, R.A.; Antunes, M.J.; Christodorescu, R.M.; Crawford, C.; Di Angelantonio, E.; Eliasson, B.; et al. 2023 ESC Guidelines for the management of cardiovascular disease in patients with diabetes. *Eur. Heart J.* **2023**, *44*, 4043–4140. [PubMed]
21. Wang, N.; Fulcher, J.; Abeysuriya, N.; Park, L.; Kumar, S.; Di Tanna, G.L.; Wilcox, I.; Keech, A.; Rodgers, A.; Lal, S. Intensive LDL cholesterol-lowering treatment beyond current recommendations for the prevention of major vascular events: A systematic review and meta-analysis of randomised trials including 327,037 participants. *Lancet Diabetes Endocrinol.* **2020**, *8*, 36–49. [CrossRef]
22. Chen, J.; Dong, Y.; Kefei, D. Intensified glycemic control by HbA1c for patients with coronary heart disease and type 2 diabetes: A review of findings and conclusions. *Cardiovasc. Diabetol.* **2023**, *22*, 146. [CrossRef]
23. Reurean-Pintilei, D.; Potcovaru, C.-G.; Salmen, T.; Mititelu-Tartau, L.; Cinteză, D.; Lazăr, S.; Pantea Stoian, A.; Timar, R.; Timar, B. Assessment of Cardiovascular Risk Categories and Achievement of Therapeutic Targets in European Patients with Type 2 Diabetes. *J. Clin. Med.* **2024**, *13*, 2196. [CrossRef] [PubMed]
24. Visseren, F.L.J.; Mach, F.; Smulders, Y.M.; Carballo, D.; Koskinas, K.C.; Bäck, M.; Benetos, A.; Biffi, A.; Boavida, J.M.; Capodanno, D.; et al. 2021 ESC Guidelines on cardiovascular disease prevention in clinical practice. *Eur. Heart J.* **2021**, *42*, 3227–3337. [CrossRef] [PubMed]
25. Kannel, W.B. Risk stratification in hypertension: New insights from the Framingham Study. *Am. J. Hypertens.* **2000**, *13*, 3S–10S. [CrossRef] [PubMed]
26. Vintila, A.M.; Horumba, M.; Cimpu, C.; Dumitrescu, D.; Miron, P.; Alucai, A.; Cristea, G.; Tudorica, C.C.; Vintila, V.D. Target Achievement in Very High Risk Patients in Light of the New Dyslipidemia Guidelines. *J. Hypertens.* **2021**, *39*, e375. [CrossRef]
27. Cokolic, M.; Lalic, N.M.; Micic, D.; Mirosevic, G.; Klobucar Majanovic, S.; Lefterov, I.N.; Graur, M. Patterns of diabetes care in Slovenia, Croatia, Serbia, Bulgaria and Romania: An observational, non-interventional, cross-sectional study. *Wien. Klin. Wochenschr.* **2017**, *129*, 192–200. [CrossRef]
28. Janssen, F.; Bardoutsos, A.; Vidra, N. Obesity prevalence in the long-term future in 18 European countries and in the USA. *Obes. Facts* **2020**, *13*, 514–527. [CrossRef]
29. Stival, C.; Lugo, A.; Odone, A.; van den Brandt, P.A.; Fernandez, E.; Tigova, O.; Soriano, J.B.; Lopez, M.J.; Scaglioni, S.; Gallus, S. Prevalence and Correlates of Overweight and Obesity in 12 European Countries in 2017–2018. *Obes. Facts* **2022**, *15*, 655–665. [CrossRef]
30. Pop, C.; Fronea, O.F.G.; Pop, L.; Iosip, A.; Manea, V.; Dorobantu, L.; Cotoraci, C.; Bala, C.; Pop, D.; Dorobantu, M. High-normal blood pressure and related cardiovascular risk factors prevalence in the Romanian adult population: Insights from the SEPHAR III study. *J. Hum. Hypertens.* **2021**, *35*, 884–895. [CrossRef]
31. Shariq, O.A.; McKenzie, T.J. Obesity-related hypertension: A review of pathophysiology, management, and the role of metabolic surgery. *Gland. Surg.* **2020**, *9*, 80–93. [CrossRef]
32. McGurnaghan, S.; Blackbourn, L.A.K.; Mocevic, E.; Haagen Panton, U.; McCrimmon, R.J.; Sattar, N.; Wild, S.; Colhoun, H.M. Cardiovascular disease prevalence and risk factor prevalence in Type 2 diabetes: A contemporary analysis. *Diabet. Med.* **2019**, *36*, 718–725. [CrossRef]
33. Rungby, J.; Schou, M.; Warrer, P.; Ytte, L.; Andersen, G.S. Prevalence of cardiovascular disease and evaluation of standard of care in type 2 diabetes: A nationwide study in primary care. *Cardiovasc. Endocrinol.* **2017**, *6*, 145–151. [CrossRef] [PubMed]
34. Ray, K.K.; Haq, I.; Bilitou, A.; Manu, M.C.; Burden, A.; Aguiar, C.; Arca, M.; Connolly, D.L.; Eriksson, M.; Ferrieres, J.; et al. Treatment gaps in the implementation of LDL cholesterol control among high- and very high-risk patients in Europe between 2020 and 2021: The multinational observational SANTORINI study. *Lancet Reg. Health Eur.* **2023**, *29*, 100624. [CrossRef] [PubMed]
35. Vrablik, M.; Seifert, B.; Parkhomenko, A.; Banach, M.; Jóźwiak, J.J.; Kiss, R.G.; Gaita, D.; Rašlová, K.; Zachlederova, M.; Bray, S.; et al. Lipid-lowering therapy use in Central and Eastern Europe primary and secondary care: DA VINCI observational study. *Atherosclerosis* **2021**, *334*, 66–75. [CrossRef] [PubMed]
36. Morieri, M.L.; Avogaro, A.; Fadini, G.P. Cholesterol-lowering therapies and achievement of targets for primary and secondary cardiovascular prevention in type 2 diabetes: Unmet needs in a large population of outpatients at specialist clinics. *Cardiovasc. Diabetol.* **2020**, *19*, 190. [CrossRef] [PubMed]

37. Jellinger, P.S.; Handelsman, Y.; Rosenblit, P.D.; Bloomgarden, Z.T.; Fonseca, V.A.; Garber, A.J.; Grunberger, G.; Guerin, C.K.; Bell, D.S.H.; Mechanick, J.I.; et al. American Association of Clinical Endocrinologists and American College of Endocrinology Guidelines for Management of Dyslipidemia and Prevention of Cardiovascular Disease. *Endocr. Pract.* **2017**, *23*, 1–87. [CrossRef]
38. Grundy, S.M.; Stone, N.J.; Bailey, A.L.; Beam, C.; Birtcher, K.K.; Blumenthal, R.S.; Braun, L.T.; De Ferranti, S.; Faiella-Tommasino, J.; Forman, D.E.; et al. 2018 guideline on the management of blood cholesterol: A report of the American College of Cardiology/American Heart Association Task Force on clinical practice guidelines. *J. Am. Coll. Cardiol.* **2019**, *73*, e285–e350. [CrossRef]
39. Vencio, S.; Alguwaihes, A.; Leon, J.L.A.; Bayram, F.; Darmon, P.; Dieuzeide, G.; Hettiarachchige, N.; Hong, T.; Kaltoft, M.S.; Lengyel, C.; et al. Contemporary use of diabetes medications with a cardiovascular indication in adults with type 2 diabetes: A secondary analysis of the multinational CAPTURE study. *Diabetologia* **2020**, *63*, A945.
40. Arnold, S.V.; Tang, F.; Cooper, A.; Chen, H.; Gomes, M.B.; Rathmann, W.; Shimomura, I.; Vora, J.; Watada, H.; Khunti, K.; et al. Global use of SGLT2 inhibitors and GLP-1 receptor agonists in type 2 diabetes. Results from DISCOVER. *BMC Endocr. Disord.* **2022**, *22*, 111. [CrossRef]
41. Banach, M.; Gaita, D.; Haluzik, M.; Janez, A.; Kamenov, Z.; Kempler, P.; Nebojsa, L.; Ales, L.; Dimitri, P.M.; Aleksandra, N.; et al. Cardio-Metabolic Academy Europe East. Adoption of the ADA/EASD guidelines in 10 Eastern and Southern European countries: Physician survey and good clinical practice recommendations from an international expert panel. *Diabetes Res. Clin. Pract.* **2021**, *172*, 108535. [CrossRef]
42. Perone, F.; Bernardi, M.; Redheuil, A.; Mafrica, D.; Conte, E.; Spadafora, L.; Ecarnot, F.; Tokgozoglu, L.; Santos-Gallego, C.G.; Kaiser, S.E.; et al. Role of Cardiovascular Imaging in Risk Assessment: Recent Advances, Gaps in Evidence, and Future Directions. *J. Clin. Med.* **2023**, *12*, 5563. [CrossRef]
43. Reyes-Soffer, G.; Ginsberg, H.N.; Berglund, L.; Duell, P.B.; Heffron, S.P.; Kamstrup, P.R.; Lloyd-Jones, D.M.; Marcovina, S.M.; Yeang, C.; Koschinsky, M.L.; et al. Lipoprotein (a): A genetically determined, causal, and prevalent risk factor for atherosclerotic cardiovascular disease: A scientific statement from the American Heart Association. *Arterioscler. Thromb. Vasc. Biol.* **2022**, *42*, e48–e60. [CrossRef] [PubMed]
44. Di Fusco, S.A.; Arca, M.; Scicchitano, P.; Alonzo, A.; Perone, F.; Gulizia, M.M.; Gabrielli, D.; Oliva, F.; Imperoli, G.; Colivicchi, F. Lipoprotein(a): A risk factor for atherosclerosis and an emerging therapeutic target. *Heart* **2022**, *109*, 18–25. [CrossRef] [PubMed]
45. Alonso, R.; Mata, P. What are the controversies and appropriate guidance for cascade screening for lipoprotein (a)? *Expert. Rev. Cardiovasc. Ther.* **2023**, *21*, 241–243. [CrossRef] [PubMed]

Disclaimer/Publisher's Note: The statements, opinions and data contained in all publications are solely those of the individual author(s) and contributor(s) and not of MDPI and/or the editor(s). MDPI and/or the editor(s) disclaim responsibility for any injury to people or property resulting from any ideas, methods, instructions or products referred to in the content.

Article

The Influence of Dapagliflozin on Foot Microcirculation in Patients with Type 2 Diabetes with and without Peripheral Arterial Disease—A Pilot Study

Božena Bradarić [1], Tomislav Bulum [1,2,*], Neva Brkljačić [1], Željko Mihaljević [3], Miroslav Benić [3] and Božo Bradarić Lisić [4]

[1] Vuk Vrhovac University Clinic for Diabetes, Endocrinology and Metabolic Diseases, Merkur University Hospital, 10000 Zagreb, Croatia
[2] School of Medicine, University of Zagreb, 10000 Zagreb, Croatia
[3] Croatian Veterinary Institute, 10000 Zagreb, Croatia
[4] Professional Study Program in Physiotherapy, University of Applied Health Sciences, 10000 Zagreb, Croatia
* Correspondence: tomobulum@gmail.com

Abstract: The results of large cardiovascular studies indicate that SGLT-2 inhibitors may increase the risk of leg amputations. This study aims to investigate whether dapagliflozin therapy affects peripheral vascular oxygenation, i.e., microcirculation in the foot, as measured by transcutaneous oxygen pressure (TcPO2) in patients with type 2 diabetes (T2DM) and peripheral arterial disease (PAD) compared to patients without PAD. The patients with PAD were randomized into two groups. In the first 35 patients with PAD, dapagliflozin was added to the therapy; in the other 26 patients with PAD, other antidiabetic drugs were added to the therapy. Dapagliflozin was added to the therapy in all patients without PAD. TcPO2 measurement, Ankle Brachial Index (ABI), anthropometric measurements, and laboratory tests were performed. After a follow-up period of 119.35 days, there was no statistically significant difference in the reduction of mean TcPO2 values between the group with T2DM with PAD treated with dapagliflozin and the group with T2DM with PAD treated with other antidiabetic drugs (3.88 mm Hg, SD = 15.13 vs. 1.48 mm Hg, SD = 11.55, $p = 0.106$). Patients with control TcPO2 findings suggestive of hypoxia (TcPO2 < 40 mm Hg) who were treated with dapagliflozin had a clinically significant decrease in mean TcPO2 of 10 mm Hg or more (15.8 mm Hg and 12.90 mm Hg). However, the aforementioned decrease in TcPO2 was not statistically significantly different from the decrease in TcPO2 in the group with PAD treated with other diabetic medications ($p = 0.226$, $p = 0.094$). Based on the available data, dapagliflozin appears to affect tissue oxygenation in T2DM with PAD. However, studies with a larger number of patients and a longer follow-up period are needed to determine the extent and significance of this effect.

Keywords: microcirculation; transcutaneous oxygen pressure (TcPO2); peripheral arterial disease (PAD) diabetes mellitus; SGLT2–inhibitors

1. Introduction

Patients with diabetes have a two to four times higher risk of developing peripheral arterial disease (PAD) than people without diabetes [1–3]. About 15–20% of patients with PAD and diabetes end up with limb amputation [4–7]. Non-traumatic lower limb amputations are 15 times more common in patients with diabetes than in patients without diabetes [4–7]. The factors that increase the risk of amputation are complex and include not only macrovascular disease (PAD) but also diabetic neuropathy and microvascular foot disease, which are also the main etiopathogenetic factors in the development of the diabetic foot. Diabetic sensorimotor neuropathy leads to atrophy and dysfunction of the small muscles of the foot, resulting in deformities and changes in pressure distribution in the foot. Together with reduced touch sensation and proprioception, this leads to reduced capillary blood flow to the foot and increased ischemia in areas of increased pressure, as well as an increased risk of trauma. Due to diabetic neuropathy, patients with PAD often do not have typical claudication symptoms, which further increases the risk of developing wounds. Microvascular foot disease results from hyperglycemia-induced damage to the capillary endothelium, leading to disruption of endothelial permeability and induction of a proinflammatory state, along with disruption of nutrient and oxygen supply to the tissue. An imbalance between pro- and anti-neovascularization responses occurs, resulting in reduced formation of new blood vessels in the skin, decreased repair and regenerative capacity, and induction of a procoagulant state leading to microthrombosis and subsequent tissue ischemia and damage [2,8,9]. Transcutaneous tissue oximetry (TcPO2) is a non-invasive method for measuring oxygen pressure on the skin surface to assess peripheral vascular oxygenation and microcirculation. Values below 40 mm Hg are considered pathologically low or hypoxic, which can slow or reduce wound healing, while values above 40 mm Hg are considered normal [10–13]. PAD increases the risk of cardiovascular disease and death, and these risks are even higher when diabetes is associated with PAD [14].

SGLT-2 inhibitors are now considered one of the essential antidiabetic drugs in patients with type 2 diabetes (T2DM). In clinical trials, SGLT2 inhibitors reduce the number of hospitalizations due to heart failure and slow the progression of kidney complications in patients with T2DM. According to the joint ADA/EASD guidelines, these drugs are recommended for patients with T2DM and established atherosclerotic cardiovascular disease and/or heart failure and/or chronic kidney disease and for patients at increased risk for these conditions [15–22]. However, in the CANVAS study (Canagliflozin Cardiovascular Assessment Study), patients taking SGLT-2 inhibitor canagliflozin had a twofold higher risk of amputations compared to placebo [17]. In the DECLARE-TIMI 58 study (Dapagliflozin and Cardiac, Kidney, and Limb Outcomes in Patients With and Without Peripheral Artery Disease in DECLARE-TIMI 58), a non-significantly higher number of amputations was observed in the group of patients with PAD treated with dapagliflozin compared to placebo [18].

Nowadays, diabetes is a leading cause of non-traumatic lower-extremity amputation. Since results from large cardiovascular trials indicated that SGLT-2 inhibitors might increase the risk of leg amputations, this study aims to determine whether dapagliflozin therapy affects peripheral vascular oxygenation, as measured by transcutaneous tissue oximetry, in patients with type 2 diabetes and PAD compared to patients without PAD.

2. Results

At baseline, we found significant differences between the groups in BMI values ($p = 0.012$), prevalence of hyperlipidemia ($p = 0.015$), prevalence of diabetic polyneuropathy ($p = 0.001$), frequency of smoking ($p = 0.009$), history of ischemic heart disease ($p = 0.006$), myocardial infarction ($p = 0.004$) and cerebrovascular disease ($p = 0.001$), and in the proportion of patients taking antiaggregation and/or anticoagulation therapy ($p = 0.001$) and lipid-lowering therapy ($p = 0.048$).

At baseline, the group of patients with PAD treated with dapagliflozin had a significantly higher BMI (28.999 vs. 26.582 kg/m^2, $p = 0.0075$), a significantly higher prevalence of hyperlipidemia (33 vs. 17 patients, $p = 0.006$), and a larger proportion taking lipid-lowering therapy (27 vs. 13 patients, $p = 0.027$), as well as a significantly higher frequency of previous amputations (9 vs. 1 patient, $p = 0.034$), compared to the group with PAD treated with other antidiabetic drugs. The average follow-up period for all groups of patients was 119.35 days. The intergroup comparisons of all patients with and without PAD at the beginning of the follow-up are presented in Tables 1 and 2.

Table 1. The intergroup comparisons of all patients at the beginning of the follow-up.

	Baseline Characteristics of All Patients (n = 97)				
	Without Peripheral Artery Disease	With Peripheral Artery Disease			
	n = 36	n = 61			
	Dapagliflozin	Dapagliflozin n = 35	Other Antidiabetics n = 26	p Value Group 2 vs. 3	p Value All Groups
Age, median	67.5	70	71.5	p = 0.780	p = 0.817
Female sex (n)	10	8	7	p = 0.770	p = 0.877
Body mass index, mean (SD)	30.387 (5.717)	28.999 (7.453)	26.571 (3.742)	p = 0.0075	p = 0.012
History hypertension, (n) (%)	31 (31.96)	34 (35.05)	23 (23.71)	p = 0.303	p = 0.290
History hyperlipidemia, n (%)	30 (30.93)	33 (34.02)	17 (17.53)	p = 0.006	p = 0.015
History diabetic polyneuropathy, n (%)	16 (16.46)	29 (29.90)	22 (22.68)	p = 1	p = 0.001
History diabetic retinopathy, n (%)	7 (7.22)	14 (14.43)	8 (8.29)	p = 0.459	p = 0.166
Smoker, n (%)	16 (16.49)	26 (26.80)	20 (20.62)	p = 0.813	p = 0.009
Hemoglobin A1c mean (SD),	7.350 (1.289)	7.314 (1.506)	7.062 (1.452)	p = 0.320	p = 0.200
Duration of diabetes mellitus, (y) mean (SD)	8.694 (7.920)	16.829 (9.724)	12.462 (7.393)	p = 0.248	p = 0.216
Estimated GFR (CKD-EPI) mL/min/1.73 m^2, n (%)	79.028 (19.191)	73.400 (18.753)	73.692 (22.922)	p = 0.455	p = 0.476
History of ischemic heart disease, n (%)	3 (3.09)	14 (14.43)	6 (6.19)	p = 0.164	p = 0.006
History of myocardial infarction, n (%)	1 (1.03)	11 (11.34)	4 (4.12)	p = 0.230	p = 0.004
History of cerebrovascular disease, n (%)	4 (4.12)	17 (17.53)	13 (13.4)	p = 0.912	p = 0.001
History of cerebral infarction, n (%)	3 (3.09)	7 (7.22)	5 (5.15)	p = 0.940	p = 0.314
Antiaggregation and/or anticoagulation therapy, n (%)	9 (9.28)	32 (32.99)	21 (21.65)	p = 0.268	p = 0.001
Lipid-lowering therapy, n (%)	27 (27.84)	27 (27.84)	13 (13.40)	p = 0.027	p = 0.048
ACE inhibitor or ARB therapy, n (%)	27 (27.84)	31 (31.96)	19 (19.59)	p = 0.179	p = 0.242
Peripheral artery disease history					
Previous peripheral revascularization (endovascular or surgical), n (%)		24 (24.74)	17 (17.52)	p = 0.796	
Previous amputation, n (%)		9 (9.28)	1 (1.03)	p = 0.034	
Fontaine Classification at randomization					
Stage I: Asymptomatic, n (%)		7 (11.48)	1 (1.64)	p = 0.081	
Stage II: Claudication, n (%)		21 (34.43)	22 (36.07)		
Stage III: Ischemia rest pain, n (%)		0	0		
Stage IV: Ulceration or gangrene, n (%)		7 (11.48)	3 (4.92)	p > 0.05	
SINBAD score		2.42	3		
Ankle Brachial index category/leg					
<0.9, n (%)	0	14 (11.20)	9 (7.20)	p = 0.958	p = 0.001
≥0.9, n (%)	63	24 (19.20)	15 (12.0)		
Transcutaneous oxygen pressure category/leg, mean, mm Hg	51.292 (11.150)	46.200 (10.774)	43.404 (9.828)	p = 0.082	p = 0.002

Table 2. The intergroup comparisons of patients with and without peripheral arterial disease at the beginning and at follow-up.

			No PAD + DAPAGLIFLOZIN (1)		PAD + DAPAGLIFLOZIN (2)		PAD + Other Antidiabetic Drugs (3)		p Value		
									1–2	1–3	2–3
			mean	sd	mean	sd	mean	sd			
BMI (kg/m^2)		baseline	30.387	5.717	28.999	7.453	26.571	3.742	0.354	0.002	0.0075
		follow up	29.936	5.508	28.604	7.671	26.573	3.603	0.326	0.010	0.029
HbA1c (%)		baseline	7.350	1.289	7.314	1.506	7.062	1.452	0.362	0.165	0.320
		follow up	6.747	0.586	6.786	0.938	6.815	1.215	0.286	0.246	0.430
Glucose fasting (mmol/L)		baseline	8.194	1.779	9.231	2.603	8.064	2.073	0.005	0.380	0.040
		follow up	8.036	1.620	7.777	1.478	8.588	2.811	0.276	0.473	0.318
Hematocrit (L/L)		baseline	0.431	0.003	0.419	0.035	0.424	0.037	0.077	0.414	0.401
		follow up	0.447	0.032	0.429	0.033	0.412	0.043	0.016	0.001	0.059
Hemoglobin (g/L)		baseline	145.944	13.852	135.614	27.125	140.038	12.032	0.009	0.018	0.471
		follow up	149.917	13.693	142.371	11.971	136.808	14.006	0.001	0.001	0.081
LDL cholesterol (mmol/L)		baseline	2.489	1.146	2.489	1.162	2.531	0.995	0.452	0.462	0.395
		follow up	2.178	0.933	1.817	0.936	1.900	0.940	0.024	0.061	0.392
Total cholesterol (mmol/L)		baseline	4.447	1.325	4.543	1.367	4.492	1.322	0.425	0.437	0.494
		follow up	4.175	1.250	3.760	1.041	4.504	3.749	0.054	0.108	0.403
Non-HDL cholesterol (mmol/L)		baseline	3.208	1.248	3.251	1.337	3.196	1.185	0.481	0.475	0.493
		follow up	2.883	1.064	2.474	0.983	2.504	0.967	0.020	0.05	0.406
HDL cholesterol (mmol/L)		baseline	1.239	0.289	1.291	0.342	1.298	0.409	0.237	0.396	0.346
		follow up	1.292	0.285	1.337	0.469	1.242	0.412	0.466	0.120	0.138
Triglycerides (mmol/L)		baseline	1.602	0.692	1.943	2.364	1.455	0.740	0.419	0.106	0.077
		follow up	1.567	0.770	1.440	0.591	1.315	0.683	0.31	0.031	0.081
C-reactive protein (mg/L)		baseline	2.923	3.113	5.397	9.204	3.212	4.160	0.198	0.36	0.335
		follow up	3.044	3.943	3.674	4.648	3.050	3.717	0.488	0.495	0.494
Creatinine (umol/L)		baseline	85.028	20.740	89.711	36.187	89.231	38.998	0.292	0.438	0.372
		follow up	85.417	21.089	100.829	43.377	91.500	38.801	0.055	0.431	0.097

PAD, peripheral arterial disease.

Significantly higher mean BMI values at baseline were observed in patients without PAD treated with dapagliflozin compared to patients with PAD treated with other antidiabetic drugs (30.387 kg/m^2 vs. 26.571 kg/m^2, $p = 0.002$) and between patients with PAD treated with dapagliflozin and patients with PAD treated with other antidiabetic drugs (28.999 kg/m^2 vs. 26.571 kg/m^2, $p = 0.0075$). During follow-up, there was a slight decrease in BMI values in the groups without PAD and with PAD treated with dapagliflozin and a minimal increase in BMI in patients with PAD treated with other antidiabetic drugs, but the differences between the previously mentioned groups remained statistically significant ($p = 0.010$, $p = 0.029$).

During follow-up, there was a slight increase in hematocrit in both groups treated with dapagliflozin (without PAD and with PAD), while there was a slight decrease in hematocrit in the group of patients with PAD treated with other antidiabetic drugs. Statistically significant differences were observed in mean hematocrit values at the end of follow-up between dapagliflozin-treated patients without PAD and dapagliflozin-treated patients with PAD (0.447 vs. 0.429, $p = 0.016$) and between patients without PAD treated with dapagliflozin and patients with PAD treated with other antidiabetic drugs (0.447 vs. 0.412, $p = 0.001$).

Hemoglobin levels were significantly higher in the dapagliflozin-treated patients without PAD than in the dapagliflozin-treated patients with PAD and the group of patients with PAD treated with other antidiabetic drugs, both at baseline (145.944 g/L vs. 135.614 g/L $p = 0.009$ and 145.944 g/L vs. 140.038 g/L, $p = 0.018$) and at follow-up (149.917 g/L vs. 142.371 g/L, $p = 0.001$ and 149.917 g/L vs. 136.808 g/L, $p = 0.001$), although hemoglobin levels increased in both groups treated with dapagliflozin and decreased in the group with PAD treated with other antidiabetic drugs.

LDL cholesterol levels were significantly higher in dapagliflozin-treated patients without PAD than in dapagliflozin-treated patients with PAD at follow-up (2.178 mmol/L, SD = 0.933 vs. 1.817, SD = 0.939, $p = 0.024$). In addition, non-HDL cholesterol levels at follow-up were also significantly higher in the group of dapagliflozin-treated patients without PAD than in the dapagliflozin-treated patients with PAD (2.883 mmol/L, SD = 1.064 vs. 2.474 mmol/L, SD = 0.983, $p = 0.020$).

Fasting glucose levels at baseline were significantly higher in the dapagliflozin-treated group with PAD than in the dapagliflozin-treated group without PAD and the group of patients with PAD treated with other antidiabetic drugs (9.231 vs. 8.194 mmol/L, $p = 0.005$ and 9.231 vs. 8.064 mmol/L $p = 0.040$), although the HbA1c values did not differ between the groups at baseline or follow-up.

None of the observed variables affected the statistical association (significance) between the outcome (tissue oxygenation) and explanatory variables, which means that, if a significant association was observed in the core analysis, it remained significant until the end of follow-up, and vice versa. Furthermore, regression factors were not changed considerably.

The mean value of TcPO2 in the subcutaneous tissue of the foot was highest at baseline in the group without PAD treated with dapagliflozin (51.292 mm Hg), and it was significantly higher than the TcPO2 value in the group with PAD treated with dapagliflozin (46.200 mm Hg, $p = 0.002$) and significantly higher than the TcPO2 value in the group with PAD treated with other antidiabetic medications (43.404 mm Hg, $p = 0.0002$, $p = 0.001$). No statistically significant difference was found between the mean TcPO2 values between the group with PAD treated with dapagliflozin and the group with PAD treated with other antidiabetic drugs (Table 3).

Table 3. Transcutaneous oxygen pressure at baseline.

GROUP	Mean/mm Hg	SD	n
1 No PAD + dapagliflozin	51.292	11.150	72
2 PAD + dapagliflozin	46.200	10.774	70
3 PAD + other antidiabetic drugs	43.404	9.828	52
Total	46.96	10.58	194

PAD, peripheral arterial disease; differences between groups: 1–2 $p = 0.002$; 1–3 $p = 0.001$; 2–3 $p = 0.082$.

At follow-up, there was a decrease in TcPO2 values in all groups, with the highest mean control TcPO2 value in the group without PAD treated with dapagliflozin of 47.83 mm Hg, which was significantly higher than the mean control TcPO2 value in the group with PAD treated with dapagliflozin (42.314 mm Hg, $p = 0.002$) and significantly higher than in the group with PAD treated with other antidiabetic drugs (41.923 mm Hg, $p = 0.001$). The observed differences in the mean values of the control TcPO2 were not statistically different between the group with PAD treated with dapagliflozin and the group with PAD treated with other antidiabetic drugs ($p = 0.495$) (Table 4).

At follow-up, the largest decrease in mean TcPO2 was observed in the group with PAD treated with dapagliflozin (3.88 mm Hg), followed by the group without PAD treated with dapagliflozin (1.76 mm Hg), and the smallest decrease was observed in the group with PAD treated with other antidiabetic drugs (1.48 mm Hg). No statistically significant differences were found in the decrease in mean TcPO2 between the groups (Table 5).

Table 4. Transcutaneous oxygen pressure at follow-up.

GROUP	Mean/mm Hg	SD	n
1 No PAD + dapagliflozin	49.528	12.238	72
2 PAD + dapagliflozin	42.314	13.304	70
3 PAD + other antidiabetic drugs	41.923	11.424	52
Total	44.58	12.322	194

PAD, peripheral arterial disease; differences between groups 1–2 $p = 0.002$; 1–3 $p = 0.001$; 2–3 $p = 0.495$.

Table 5. Decrease in mean transcutaneous oxygen pressure at follow-up.

GROUP	Mean/mm Hg	SD	n
1 No PAD + dapagliflozin	1.76	12.93	72
2 PAD + dapagliflozin	3.88	15.13	70
3 PAD + other antidiabetic drugs	1.48	11.55	52
Total	2.4536082	13.41372	194

PAD, peripheral arterial disease; differences between groups: $p = 0.379$; 1–2 $p = 0.130$; 1–3 $p = 0.414$; 2–3 $p = 0.106$.

When analyzing the distribution of patients (feet) with mean TcPO2 control values of less than 40 mm Hg and of 40 mm Hg or more by group, the largest proportion of patients (feet) had mean TcPO2 control values of ≥ 40 mm Hg (145 feet or 74.74% of the total number), while a smaller proportion of patients (feet) had mean TcPO2 control values < 40 mm Hg (49 feet or 25.25% of the total number). A slightly higher proportion of patients with mean control TcPO2 less than 40 mm Hg was found in the group with PAD treated with other antidiabetic drugs (17 feet; 32.69%) than in the group with PAD treated with dapagliflozin (21 feet, 30%), and the lowest proportion was found in the group without PAD treated with dapagliflozin (11 feet, 15.28%). The observed differences in the frequency of patients with TcPO2 < 40 mm Hg differed significantly between the three patient groups ($p = 0.046$), but not significantly between the two groups with PAD (2 vs. 3 $p = 0.751$). (Table 6, Figure 1).

Table 6. Distribution of patients with a mean transcutaneous oxygen pressure at follow-up of 40 mm Hg or more and with less than 40 mm Hg.

GROUP	Mean TcPO2 < 40 mm Hg n (%)	Mean TcPO2 ≥ 40 Mm Hg n (%)	Total
1 No PAD + dapagliflozin	11 (15.28)	61 (84.75)	72
2 PAD + dapagliflozin	21 (30)	49 (70)	70
3 PAD + other antidiabetic drugs	17 (32.69)	35 (67.31)	52
Total	49	145	194

PAD, peripheral arterial disease; differences between groups $p = 0.046$; 2–3 $p = 0.751$.

Among patients with a control transcutaneous oxygen pressure of less than 40 mm Hg, the highest control TcPO2 was observed in the group without PAD treated with dapagliflozin (31.36 mm Hg, SD 7.47) than in the group with PAD treated with other antidiabetic drugs (28.70 mm Hg SD 9.10), and the lowest control TcPO2 level was in the group with PAD treated with dapagliflozin (27.71 mm Hg, SD 11.87). There was no significant difference in the control TcPO2 between groups (Table 7).

Among patients with a control TcPO2 of less than 40 mm Hg, the mean TcPO2 decreased from the beginning of the study to the control in all groups. The largest decrease in mean TcPO2 was observed in the group of patients with PAD treated with dapagliflozin (15.80 mm Hg) than in the group of patients without PAD treated with dapagliflozin (12.90 mm Hg), and the smallest decrease in TcPO2 was observed in the group of patients with PAD treated with other antidiabetic drugs (9.35 mm Hg). No significant differences in the degree of TcPO2 reduction were found between the observed groups (Table 8, Figure 2).

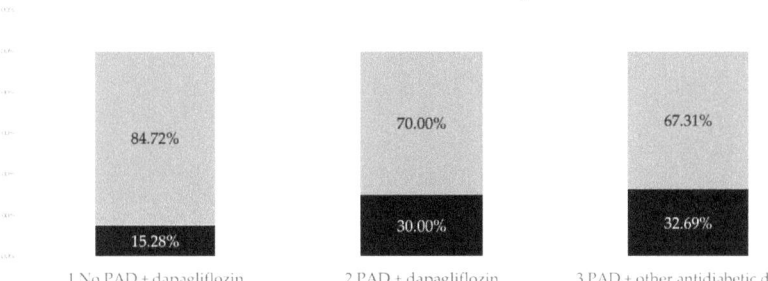

Figure 1. PAD, peripheral arterial disease.

Table 7. Transcutaneous oxygen pressure at follow-up in patients with a control transcutaneous oxygen pressure of less than 40 mm Hg.

GROUP	Mean	SD	n
1 No PAD + dapagliflozin	31.36	7.47	11
2 PAD + dapagliflozin	27.71	11.87	21
3 PAD + other antidiabetic drugs	28.70	9.10	17
Total	28.87	10.00	49

PAD, peripheral arterial disease; differences between groups: 1–2 $p = 0.224$; 1–3 $p = 0.164$; 2–3 $p = 0.384$.

Table 8. Decrease in mean transcutaneous oxygen pressure at follow-up in patients with a control transcutaneous oxygen pressure of less than 40 mm Hg.

GROUP	Mean	SD	n
1 No PAD + dapagliflozin	12.909091	14.046028	11
2 PAD + dapagliflozin	15.809524	16.064305	21
3 PAD + other antidiabetic drugs	9.3529412	11.999694	17
Total	12.918367	14.310306	49

PAD, peripheral arterial disease; differences between groups are not statistically significant $p > 0.05$; 1–2 $p = 0.355$; 1–3 $p = 0.226$; 2–3 $p = 0.094$.

Among patients who had a control TcPO2 of 40 mm Hg or higher, mean TcPO2 during follow-up increased slightly in all groups; in the group without PAD treated with dapagliflozin, by 0.24 mm Hg; in the group with PAD treated with dapagliflozin, by 1.22 mm Hg; and in the group with PAD treated with other antidiabetic drugs, by 2.34 mm Hg. No significant differences in the degree of TcPO2 elevation were found between the observed groups (Table 9, Figure 2).

A certain number of ABI index measurements failed because the device was unable to measure the ABI value due to the low systolic pressure at the ankle level. Since the number of failed measurements was significant in the group with PAD treated with dapagliflozin (45.71%) and in the group with PAD treated with other antidiabetic drugs (53.85%), we did not include these groups in the ABI analysis. In the group of patients without PAD treated with dapagliflozin, there were 12.50% unsuccessful measurements, which is, in our opinion, an insignificant and acceptable proportion, and these measurements were included in the analysis. According to the guidelines of the Society for Vascular Surgery and the American Venous Forum, a change in ABI of 0.15 is required to be considered clinically relevant, or more than 0.10 if accompanied by a change in clinical status. In the group of patients without PAD who were treated with dapagliflozin (a total of 63 legs were analyzed), there was a minimal and non-significant increase in ABI value after follow-up compared to baseline (0.04609375, SD 0.3567072; i.e., from 1.202, SD 0.136 at baseline to 1.212, SD 0.087 at follow-up) [23].

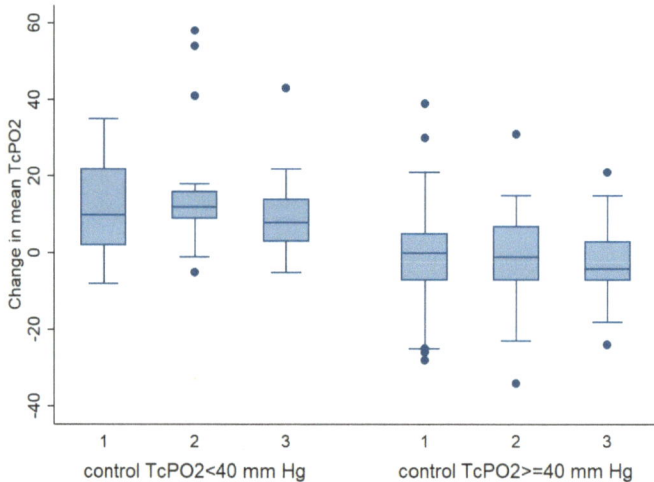

Figure 2. Change in mean TcPO2 at follow-up in patients with a control TcPO2 < 40 mm Hg and ≥40 mm Hg.

Table 9. Change in mean transcutaneous oxygen pressure at follow-up in patients with a control transcutaneous oxygen pressure ≥40 mm Hg.

GROUP	Mean/mm Hg	SD	n
1 No PAD + dapagliflozin	−0.24590164	11.75677	61
2 PAD + dapagliflozin	−1.2244898	11.53304	49
3 PAD + other antidiabetic drugs	−2.3428571	9.298775	35
Total	−1.0827586	11.09210	145

PAD, peripheral arterial disease; differences between groups: 1–2 $p = 0.451$; 1–3 $p = 0.128$; 2–3 $p = 0.162$.

In summary, the mean value of TcPO2 in the subcutaneous tissue of the foot was highest in the group without PAD (group 1) both at the beginning and at the end of the follow-up period, with a significant difference compared to both groups with PAD, while the TcPO2 finding in the foot was not significantly different between the group with PAD treated with dapagliflozin (group 2) and the group with PAD treated with other antidiabetic drugs (group 3), either at baseline or at the end of the follow-up period (Tables 3 and 4). During the follow-up period, there was a slight decrease in TcPO2 values in all groups (1.76, 3.88, 1.48 mm Hg), with the decrease in TcPO2 values being more pronounced in the group with PAD treated with dapagliflozin, although it did not differ significantly between the groups (1−2 $p = 0.130$; 1–3 $p = 0.414$; 2–3 $p = 0.106$). In the group of patients in whom TcPO2 decreased below 40 mm Hg after follow-up, which we consider a pathologic finding reflecting hypoxia of the feet, the lowest TcPO2 was in the PAD group treated with dapagliflozin (group 2) (31.36, 27.71, 28.70 mm Hg) but with no significant difference compared to the other groups (1–2 $p = 0.224$; 1–3 $p = 0.164$; 2–3 $p = 0.384$). In the group of patients with pathologic TcPO2 control findings (<40 mm Hg), the extent of TcPO2 reduction in the feet was much more pronounced in all groups and most pronounced in the PAD group treated with dapagliflozin (12.9, 15.80, 9.35 mm Hg), but again with no significant difference between groups (1–2 p = 0.224; 1–3 p = 0.164; 2–3 p = 0.384). When we analyzed the proportion of patients with pathological TcPO2 control findings (TcPO2 < 40 mm Hg) by group, we found no significant differences between the groups with PAD (32.69% vs. 30%, p = 0.751).

3. Discussion

In patients with diabetes, PAD develops early, progresses rapidly, and is often asymptomatic. The atherosclerotic process is more pronounced in the distal arteries of the legs (tibial artery, dorsalis pedis artery) with impaired microcirculation of the feet compared to patients with PAD without diabetes [2,24]. Together with diabetic neuropathy, this could at least partially explain the significantly higher incidence of non-traumatic leg amputations in patients with PAD and diabetes compared to patients with PAD without diabetes.

SGLT-2 inhibitors are now considered one of the most important antidiabetic agents in T2DM and are particularly recommended for the treatment of patients with diabetes with cardiovascular and renal comorbidities or risks for these conditions due to their proven cardio–renal protective effect, regardless of the degree of diabetes control [24]. Due to the significant and twofold increased risk of amputations in patients treated with the SGLT2 inhibitor canagliflozin compared with placebo, as shown in the CANVAS study, and considering the potential class effect, caution is recommended when introducing SGLT2 inhibitors into the treatment of patients at increased risk of lower limb amputations, which includes patients with PAD [17,25,26]. This puts the clinician in an unenviable position where, guided by the Latin maxim "Primum non nocere" (First do no harm), they often decide not to use an SGLT2 inhibitor in patients with diabetes with PAD and/or diabetic foot. Considering this context, this study aimed to determine whether therapy with the SGLT2 inhibitor dapagliflozin leads to changes in peripheral vascular oxygenation, i.e., feet microcirculation, in patients with diabetes without PAD compared to patients with diabetes with PAD and compared to patients with diabetes with PAD treated with other antidiabetic agents, with the exception of thiazolidinediones, GLP-1 receptor agonists, and the combination of GLP-1 and GIP receptor agonists. We also wanted to see the potential connection between metabolic parameters and possible changes in foot microcirculation.

We used TcPO2 to measure oxygen pressure on the skin surface of the foot to assess peripheral vascular oxygenation and obtain information on microcirculatory function. The TcPO2 is used in clinical practice to quantify the severity of peripheral arterial disease, determine the optimal level of amputation, assess revascularization procedures, select candidates for hyperbaric oxygen therapy, and predict its efficacy. TcPO2 is a better predictor for ulcer healing than the toe-brachial index in patients with diabetes with chronic foot ulcers [27]. According to the consensus of professional societies, TcPO2 values below 40 mm Hg represent hypoxia of the skin and subcutaneous tissues of the foot, which slows and prevents wounds or residual limbs from healing, while values of 40 (50) mm Hg or higher are considered normal findings [10–13,28]. In patients with TcPO2 values below 40 mm Hg, it is recommended to measure the TcPO2 additionally after elevating the legs. A drop in TcPO2 values of 10 mm Hg or more is considered clinically significant in terms of increased risk of a non-healing wound or amputation [29].

There are numerous positive effects of SGLT-2 inhibitors on microcirculation. The strongest effect is their nephroprotective effect, with a proven relative risk reduction of 39% in composite renal outcomes ($\geq$50% sustained decrease in eGFR, ESKD, renal, or CV death) [19]. The beneficial effects of dapagliflozin on the microvasculature in terms of reducing retinal capillary hyperperfusion and minimizing arteriole remodeling are evident after only 6 weeks of therapy with dapagliflozin compared to placebo [30]. Another study found that treatment with dapagliflozin improves endothelial function after 12 weeks of treatment compared to glibenclamide [31]. Compared to placebo, dapagliflozin as an add-on therapy to metformin for 16 weeks improved endothelial function in patients with inadequately controlled early-stage T2DM [32]. Even an acute treatment with dapagliflozin for two days significantly improves systemic endothelial function and arterial stiffness independently of blood pressure changes, suggesting a direct beneficial effect on the vasculature [33]. In patients with T2DM with underlying ischemic heart disease who were receiving metformin and insulin therapy, endothelial function in the group of patients with T2DM with an HbA1c > 7.0% at baseline was improved by additional therapy with dapagliflozin for 12 weeks compared to placebo [34]. As far as we know, the effects of SGLT2 inhibitor therapy on microcirculation in the feet have not yet been investigated using transcutaneous oximetry. As expected, TcPO2 at baseline showed a significantly better TcPO2 value in the group without PAD compared to the two groups with PAD ($p = 0.002$, $p = 0.001$). After an average follow-up period of 119.35 days, we observed a decrease in TcPO2 findings, i.e., a slight worsening of the microcirculation in all three patient groups. In the group with PAD treated with dapagliflozin, the decrease was more pronounced than in the group without PAD treated with dapagliflozin (3.88 mm Hg; 1.76 mm Hg), while the decrease in TcPO2 was the least pronounced in the group with PAD treated with other antidiabetic drugs (1.48 mm Hg).

Although we found no statistically significant difference in the magnitude of the decrease in TcPO2 levels between the groups, the question arises as to why there was no improvement in microcirculation in the dapagliflozin-treated groups, considering that numerous studies indicate a positive effect of dapagliflozin on microcirculation very soon after starting the medication, i.e., in a period of 42–112 days [30–32]. Simultaneously with the slight decrease in TcPO2 findings in the dapagliflozin-treated group without PAD, we noted a slight and insignificant improvement in ABI values, which neutralizes the effect of macrocirculation on the decrease in TcPO2 values. Unfortunately, we could not establish the aforementioned correlation between ABI and TcPO2 change in the groups with PAD due to technical limitations. A significant decrease in BMI and an increase in hemoglobin and hematocrit in the dapagliflozin-treated groups compared with the PAD group treated with other antidiabetic drugs could only have a positive effect on microcirculation. Slightly elevated levels of LDL and non-HDL cholesterol above target in all groups despite prescribed lipid-lowering therapy are the result of the small number of patients who did not take regular lipid-lowering therapy, and the significantly higher level of LDL and non-HDL cholesterol in the group without PAD treated with dapagliflozin compared to the groups with PAD is the result of less stringent lipid-lowering therapy according to their moderate to high degree of cardiovascular risk compared to the groups with PAD. However, their significant effect on the TcPO2 control values was excluded using stepwise regression to check for possible confounding variables. In the same way, possible confounding by other parameters that were statistically different between groups was also excluded (Tables 1 and 2) [35–37].

When analyzing the distribution of patients (feet) with mean TcPO2 control values of less than 40 mm and 40 mm Hg or more by group, the largest proportion had mean TcPO2 control values of 40 mm Hg or more (145 feet or 74.74% of the total number), while a smaller proportion had mean TcPO2 control values of less than 40 mm Hg (49 feet or 25.25% of the total number). The observed differences in the frequency of patients with TcPO2 < 40 mm Hg differed significantly between the three patient groups ($p = 0.046$) but not significantly between the two groups with PAD (2 vs. 3 $p = 0.751$).

As mentioned previously, in patients with a TcPO2 of less than 40 mm Hg, an additional decrease in TcPO2 after elevation of the leg of 10 mm Hg or more is considered clinically significant in terms of an increased risk of a non-healing wound or amputation. When we separately analyzed patients with a pathological TcPO2 control value indicating hypoxia of the foot (TcPO2 < 40 mm Hg), the degree of decrease in TcPO2 was even more pronounced. There was a clinically significant decrease in control TcPO2 in the dapagliflozin-treated groups (15.80 mm Hg in the group with PAD and 12.9 mm Hg in the group without PAD), whereas in the group with PAD treated with other antidiabetic medications, this decrease was below clinical significance and amounted to 9.35 mm Hg (Tables 7 and 8) [29]. Despite the clinically significant differences mentioned above, no significant differences between the groups were found in the statistical analysis. When we separately analyze patients whose control TcPO2 is normal (≥40 mm Hg), we find that there was a slight increase in TcPO2 values in all groups, i.e., an improvement in microcirculation, although the increase was insignificantly lower in the groups treated with dapagliflozin (without PAD 0.24 mm Hg, PAD 1.22 mm Hg) compared to the group with PAD treated with other antidiabetic drugs (2.34 mm Hg).

It can be further investigated with a larger number of patients whether and according to which metabolic, anamnestic, and therapeutic parameters the groups of patients with pathological and normal control TcPO2 values differ or whether there are additional factors that can be associated with such different dynamics of TcPO2 findings; i.e., the opposite effects on the microcirculation in all three groups of patients (Figure 1). For example, one possible explanation could be the nature of the other antidiabetic medications used to treat some of the patients in all three groups, such as DPP4 inhibitors, which have shown an antiatherosclerotic and vasculoprotective mechanism in clinical and experimental studies, or metformin, which has been shown to improve muscle microvascular insulin sensitivity in insulin-resistant humans and retinal capillary perfusion in diabetic mice [38–40].

In summary, patients with control TcPO2 findings suggestive of hypoxia (TcPO2 < 40 mm Hg) who were treated with dapagliflozin had a decrease in mean TcPO2 of 10 mm Hg or more, which is considered clinically significant. However, the aforementioned decrease in TcPO2 was not statistically significantly different from the decrease in TcPO2 in the group with PAD treated with other diabetic medications ($p = 0.226$, $p = 0.094$).

The present study has several potential limitations. First, this was a single hospital-based study with a limited number of study participants who most probably are not representative of the population with PAD. Therefore, the data must be confirmed in prospective studies with more patients. Second, the short follow-up time is an obvious limitation. Third, since our study only included patients from a white European population, there was no racial/ethnic diversity. Fourth, there was only one control measurement of the TcPO2 findings, which could be a limitation, especially if the changes in the TcPO2 findings are small. From the authors' point of view, the main limitation of the presented research is the relatively small sample size, which can result in low statistical power, making it difficult to detect meaningful effects or relationships, large variability, and outliers in some measurements.

4. Materials and Methods

4.1. Study Design and Subjects

This was a single-center prospective randomized pilot study conducted at the Polyclinic of the Vuk Vrhovac University Clinic for Diabetes, Endocrinology and Metabolic Diseases. This pilot study is part of a larger study that will last at least one year and include a larger number of patients. During this time, it is planned to repeat the TcPO2 measurements at 3- to 6-month intervals to monitor the dynamics of possible changes in the microcirculation of the feet over time and after therapeutic interventions.

This study initially included 97 T2DM patients who had previously been treated with standard antidiabetic therapy (metformin and/or DPP4 inhibitors and/or sulfonylureas and/or insulin), of whom 61 patients had PAD and 36 patients had no PAD. The patients with PAD were mostly recruited from the Department and Clinic of Vascular Surgery at the Merkur University Hospital and invited by telephone to the diabetology outpatient clinic for examination. A significantly smaller number of patients were discovered during examinations in the diabetology outpatient clinic, where they were referred by their GP for diabetes treatment. Patients without PAD were referred by their GP to the diabetology outpatient clinic for diabetes treatment and to investigate chronic diabetic complications.

The inclusion criteria were patients aged 40–85 years with T2DM with or without PAD. Acute and severely ill patients, patients with T2DM with or without PAD treated with thiazolidinediones, GLP-1 receptor agonists, or a combination of GLP-1 and GIP receptor agonists, patients with acute lower limb ischemia, patients who had undergone peripheral revascularization or angiography within the last 4 weeks, patients with advanced chronic kidney disease with an eGFR < 15 mL/min/1.73 m^2 and/or chronic dialysis therapy, patients with active malignant disease, and patients with scleroderma were not included in this study. Patients who underwent additional peripheral revascularization and/or hyperbaric oxygen therapy during the study (n = 1) or who were diagnosed with malignant disease and had started antineoplastic treatment, as well as patients who had started treatment with thiazolidinedione, GLP-1 receptor agonists, or a combination of GLP-1 and GIP receptor agonists, were excluded from the study. Patients with PAD were randomized into two groups according to the order of arrival at the diabetes outpatient clinic. In the first 22 patients with PAD, dapagliflozin was added to the therapy; in the next 22 patients with PAD, other antidiabetic drugs were added to the therapy, and then the procedure was repeated (Figure 3). Dapagliflozin was added to therapy in all patients without PAD.

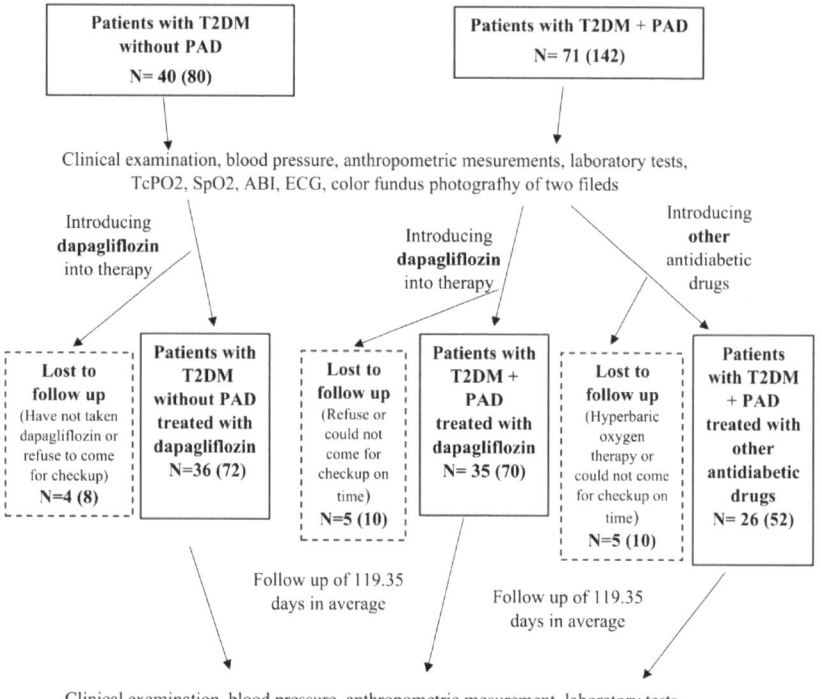

Figure 3. Study design diagram.

At the beginning of the study, before the therapeutic intervention, TcPO2, ankle-brachial index (ABI), peripheral oxygen saturation (SpO2), color fundus photography of two fields (macula-centered, optic disc-centered) of both eyes with a standard VISUCAM Zeiss camera, blood pressure, anthropometric parameters, and laboratory tests were measured in all patients, which were repeated after a follow-up period. In the data analysis, TcPO2 was evaluated separately for each leg; i.e., a total of 194 legs were analyzed, 122 in the group of patients with PAD and 72 in the group of patients without PAD (Figure 3).

In clinical practice, a single TcPO2 measurement on the foot is a sufficient finding to provide an indication for hyperbaric oxygen therapy. In recent studies, microcirculation at the foot was assessed by a single TcPO2 measurement [28,41]. Therefore, we believe that, for a pilot study and considering the number of legs included in the analysis (194), one TcPO2 measurement is sufficient for the assessment of microcirculation at the foot.

ABI measurements were performed on all 194 legs. A certain number of ABI index measurements failed because the device could not measure the ABI value due to the low systolic pressure at ankle level. Therefore, the results of the successful measurements on a total of 63 legs were analyzed.

All patients in the group with PAD were at very high cardiovascular risk, including patients in the group without PAD who had documented cerebrovascular and/or ischemic heart disease. Other patients in the group without PAD had moderate or high cardiovascular risk. Lipid-lowering therapy (statins with or without ezetimibe) was recommended to all patients according to estimated cardiovascular risk, and therapy with ACE inhibitors or ARBs according to the 2024 ADA guidelines and antiplatelet and/or anticoagulant therapy were recommended for all patients with PAD and/or cerebrovascular and/or cardiovascular disease.

Abbreviations used in the text are as follows: T2DM, diabetes mellitus type 2; PAD, peripheral arterial disease; TcPO2, transcutaneous oxygen pressure; SpO2, peripheral oxygen saturation; ABI, Ankle Brachial index; and ECG, electrocardiogram.

4.2. Data Collection

When taking the medical history, information on chronic diseases, smoking status, and medication intake was obtained. All data and results were stored via the hospital information system. Those who currently smoked or who had smoked at least 100 cigarettes in the past were defined as smokers. Hypertension was defined as a systolic blood pressure (SBP) $\geq$ 140 mm Hg and/or diastolic blood pressure (DBP) $\geq$ 90 mm Hg or antihypertensive treatment. Body mass index (BMI) was calculated by dividing weight in kilograms and height in meters squared. Waist circumference (WC) and hip circumference were measured with a tailor's tape measure at standard sites on bare skin to calculate the waist-to-hip ratio (WHR).

Complete blood count (CBC), white blood cell differential, prothrombin time (PT), prothrombin time—international normalized ratio (PT-INR), activated partial thromboplastin time (APTT), fibrinogen level test, D-dimer test, potassium, sodium, chloride, ionized calcium, total calcium, inorganic phosphate, iron, total iron binding capacity, unsaturated iron binding capacity, urea, creatinine, urate, total bilirubin, aspartate aminotransferase, alanine aminotransferase, gamma-glutamyl transferase, alkaline phosphatase, total cholesterol, high-density lipoprotein (HDL), low-density lipoprotein (LDL), very low-density lipoprotein (VLDL), triglycerides, non-HDL cholesterol, lipoprotein (a), proteins, albumin, C-reactive protein, proteins and creatinine in a single urine sample, protein/creatinine ratio, albumin/creatinine ratio, and urine biochemistry and sediment were measured on the day of hospital admission. All samples were collected after a 12-h fasting period.

Glycated hemoglobin (HbA1c) was measured using an automated turbidimetric inhibition immunoassay (HbA1c Gen 3, Cobas Integra 400 Plus, Roche Diagnostic, Basel, Switzerland) and expressed as a percentage according to the National Glycohemoglobin Standardization Program (NGSP).

4.3. Diagnosis of Peripheral Arterial Disease

Most patients with PAD were diagnosed by multislice spiral computed tomography (MSCT) and/or digital subtraction angiography, and most had previously undergone peripheral revascularization procedures. In a small number of patients, PAD was diagnosed by ultrasound Doppler, where the absence of a normal triphasic appearance of the signals, i.e., the presence of a biphasic and/or monophasic appearance of the signals in one of the lower limb arteries, was considered as PAD.

4.4. Microcirculation Assessment—Transcutaneous Oxygen Pressure Measurement

The sensor/electrode of the TCM5 series transcutaneous monitor (Radiometer, Denmark) was placed on the skin, where it heated the underlying tissue to produce local hyperemia (causing the artery to dilate), intensifying blood flow and increasing oxygen levels. The electrode was heated to 45 °C and delivered a temperature of about 43 °C to the skin, which improved the oxygenation of the capillary blood. The sensor received an electric current that corresponded to the oxygen concentration in the capillary blood. Depending on the pressure, the oxygen diffused from the capillary blood through the vascular epidermis to the electrode placed on the skin surface. After application, the sensor required a warm-up phase of 10 to 15 min and had to be calibrated every four to eight hours. An on-site measurement took an average of 35 min per leg. The sensor was placed on a homogeneous capillary bed without hair, skin defects, ulcers, or protruding veins. The electrode was not placed over bone, previous surgical sites, scar tissue, or severe edema, as this would have led to unreliable results. Patients were advised not to smoke or consume caffeine for at least one hour before the test. Patients were placed in the supine position during monitoring.

4.5. Ankle Brachial Index

This index is calculated by comparing the highest measured systolic pressure in the ankle area with the brachial systolic pressure, and their ratio gives the ABI. This index provides us with relative information about the severity of the peripheral arterial occlusive disease. Before starting the examination, the patient rests for at least 15 min in a supine position with the limbs parallel to the body. The MESI mTABLET device with blood pressure diagnostic module model/type TBPMD, serial number 1211-MDGO, manufactured in Ljubljana, Slovenia, consisting of four cuffs applied simultaneously to both upper arms and both ankles, was used to measure the ABI.

4.6. Fundus Photography

Most patients underwent fundus photography with a standard 45° VISUCAM Zeiss fundus camera (Carl Zeiss Meditec AG, Jena, Germany). Patients who underwent ophthalmologic monitoring and treatment at other institutions did not undergo fundus photography, but their medical records from another institution were reviewed. For all patients who underwent fundus photography, the images were analyzed by an ophthalmologist and a posterior segment subspecialist, and a report was written.

4.7. Statistical Analysis

We processed the numerical indicators statistically using the statistical program Stata 13.1 (Stata Corp., College Station, TX, USA). We checked the normality of the distribution of indicators expressed in continuous values using the Shapiro–Wilk test. We compared the values of each indicator between three groups of patients: (1). patients without PAD treated with dapagliflozin; (2). patients with PAD treated with dapagliflozin; and (3). patients with PAD treated with other antidiabetic drugs. In the statistical analysis of tissue oximetry, we considered each limb individually. We checked the statistical significance of differences in each indicator expressed in continuous values between patient groups using analysis of variance for indicators that met the criteria of normal distribution (Shapiro–Wilk test $p > 0.05$), and for indicators that did not meet the criteria of normal distribution (Shapiro–Wilk test $p < 0.05$), we used the non-parametric Kruskal–Wallis test. Tukey's post-hoc test and Dunn's test were used to compare groups after parametric and non-parametric statistical analysis. We tested the significance of differences in indicators expressed in binary values (yes/no) using the chi-square test and Fisher's exact test. Stepwise regression was performed to check possible confounding variables. Before regression analysis, variables were log transformed in order to normalize distribution.

5. Conclusions

Based on the available data, dapagliflozin appears to have an effect on tissue oxygenation in T2DM with PAD. The results show differences in TcPO2 control values between patients treated with dapagliflozin and those treated with other blood-glucose-lowering drugs. To better understand the specific effects of dapagliflozin on tissue oxygenation in the feet of T2DM patients with PAD, further studies and analyses are needed, in particular by extending the follow-up period and repeating the TcPO2 examinations and ideally by including a larger number of patients.

Author Contributions: Conceptualization, B.B. and T.B.; data curation, B.B. and B.B.L.; formal analysis, B.B., Ž.M. and M.B.; methodology, B.B. and T.B.; project administration, B.B. and B.B.L.; resources, B.B. and N.B.; supervision, T.B.; writing—original draft, B.B.; writing—review and editing, B.B. and T.B. All authors have read and agreed to the published version of the manuscript.

Funding: This research received no specific grant from any funding agency in the public, commercial, or not-for-profit sectors.

Institutional Review Board Statement: This research was conducted after consideration and approval by a Hospital's Ethics Committee (protocol number 2330/4, approval date: 19 March 2024) following the Declaration of Helsinki.

Informed Consent Statement: Before any procedures and inclusion in the study, patients received written and oral information about the study and signed informed consent.

Data Availability Statement: The data presented in this study are available on request from the corresponding author.

Conflicts of Interest: The authors declare no conflict of interest.

References

1. Gregg, E.W.; Sorlie, P.; Paulose-Ram, R.; Gu, Q.; Eberhardt, M.S.; Wolz, M.; Burt, V.; Curtin, L.; Engelgau, M.; Geiss, L. 1999–2000 national health and nutrition examination survey. Prevalence of lower-extremity disease in the US adult population ≥ 40 years of age with and without diabetes: 1999–2000 national health and nutrition examination survey. *Diabetes Care* **2004**, *27*, 1591–1597. [CrossRef] [PubMed]
2. Yang, S.L.; Zhu, L.Y.; Han, R.; Sun, L.L.; Li, J.X.; Dou, J.T. Pathophysiology of peripheral arterial disease in diabetes mellitus. *J. Diabetes* **2016**, *9*, 133–140. [CrossRef]
3. Sun, H.; Saeedi, P.; Karuranga, S.; Pinkepank, M.; Ogurtsova, K.; Duncan, B.B.; Stein, C.; Basit, A.; Chan, J.C.; Mbanya, J.C.; et al. IDF Diabetes Atlas: Global, regional and country-level diabetes prevalence estimates for 2021 and projections for 2045. *Diabetes Res. Clin. Pract.* **2022**, *183*, 109119, Erratum in *Diabetes Res. Clin. Pract.* **2023**, *204*, 110945. [CrossRef] [PubMed] [PubMed Central]
4. E Reiber, G.; Vileikyte, L.; Boyko, E.J.; del Aguila, M.; Smith, D.G.; A Lavery, L.; Boulton, A.J. Causal pathways for incident lower-extremity ulcers in patients with diabetes from two settings. *Diabetes Care* **1999**, *22*, 157–162. [CrossRef]
5. Hinchliffe, R.J.; Andros, G.; Apelqvist, J.; Bakker, K.; Fiedrichs, S.; Lammer, J.; Lepantalo, M.; Mills, J.L.; Reekers, J.; Shearman, C.P.; et al. A systematic review of the effectiveness of revascularization of the ulcerated foot in patients with diabetes and peripheral arterial disease. *Diabetes/Metab. Res. Rev.*. [CrossRef]
6. Boulton, A.J.M. The diabetic foot. Preface. *Med. Clin. N. Am.* **2013**, *97*, 775–992. [CrossRef]
7. Stoberock, K.; Kaschwich, M.; Nicolay, S.S.; Mahmoud, N.; Heidemann, F.; Rieß, H.C.; Debus, E.S.; Behrendt, C.-A. The interrelationship between diabetes mellitus and peripheral arterial disease. *Vasa* **2021**, *50*, 323–330. [CrossRef] [PubMed]
8. Mauricio, D.; Gratacòs, M.; Franch-Nadal, J. Diabetic microvascular disease in non-classical beds: The hidden impact beyond the retina, the kidney, and the peripheral nerves. *Cardiovasc. Diabetol.* **2023**, *22*, 314. [CrossRef] [PubMed] [PubMed Central]
9. Deng, H.; Li, B.; Shen, Q.; Zhang, C.; Kuang, L.; Chen, R.; Wang, S.; Ma, Z.; Li, G. Mechanisms of diabetic foot ulceration: A review. *J. Diabetes* **2023**, *15*, 299–312. [CrossRef] [PubMed] [PubMed Central]
10. Fife, C.E.; Smart, D.R.; Sheffield, P.J.; Hopf, H.W.; Hawkins, G.; Clarke, D. Transcutaneous oximetry in clinical practice: Consensus statements from an expert panel based on evidence. *Undersea Hyperb. Med.* **2009**, *36*, 43–53.
11. Larsen, J.F.; Christensen, K.S.; Egeblad, K. Transcutaneous oxygen tension exercise profile. A method for objectively assessing the results after reconstructive peripheral arterial surgery. *Eur. J. Vasc. Surg.* **1988**, *2*, 377–381. [CrossRef]
12. Rooke, T.W. The use of transcutaneous oximetry in the noninvasive vascular laboratory. *Int. Angiol.* **1992**, *11*, 36–40. [PubMed]
13. de Graaff, J.C.; Ubbink, D.T.; Legemate, D.A.; de Haan, R.J.; Jacobs, M.J. Interobserver and intraobserver reproducibility of peripheral blood and oxygen pressure measurements in the assessment of lower extremity arterial disease. *J. Vasc. Surg.* **2001**, *33*, 1033–1040. [CrossRef]

14. Sprenger, L.; Mader, A.; Larcher, B.; Mächler, M.; Vonbank, A.; Zanolin-Purin, D.; Leiherer, A.; Muendlein, A.; Drexel, H.; Saely, C.H. Type 2 diabetes and the risk of cardiovascular events in peripheral artery disease versus coronary artery disease. *BMJ Open Diabetes Res. Care* **2021**, *9*, e002407. [CrossRef] [PubMed] [PubMed Central]
15. Zinman, B.; Wanner, C.; Lachin, J.M.; Fitchett, D.; Bluhmki, E.; Hantel, S.; Mattheus, M.; Devins, T.; Johansen, O.E.; Woerle, H.J.; et al. Empagliflozin, cardiovascular outcomes, and mortality in type 2 diabetes. *N. Engl. J. Med.* **2015**, *373*, 2117–2128. [CrossRef] [PubMed]
16. American Diabetes Association Professional Practice Committee; ElSayed, N.A.; Aleppo, G.; Bannuru, R.R.; Bruemmer, D.; Collins, B.S.; Ekhlaspour, L.; Gaglia, J.L.; Hilliard, M.E.; Johnson, E.L.; et al. 9. Pharmacologic Approaches to Glycemic Treatment: Standards of Care in Diabetes—2024. *Diabetes Care* **2023**, *47* (Suppl. S1), S158–S178. [CrossRef]
17. Neal, B.; Perkovic, V.; Mahaffey, K.W.; de Zeeuw, D.; Fulcher, G.; Erondu, N.; Shaw, W.; Law, G.; Desai, M.; Matthews, D.R.; et al. Canagliflozin and cardiovascular and renal events in type 2 diabetes. *N. Engl. J. Med.* **2017**, *377*, 644–657. [CrossRef]
18. Bonaca, M.P.; Wiviott, S.D.; Zelniker, T.A.; Mosenzon, O.; Bhatt, D.L.; Leiter, L.A.; McGuire, D.K.; Goodrich, E.L.; Furtado, R.H.D.M.; Wilding, J.P.; et al. Dapagliflozin and Cardiac, Kidney, and Limb Outcomes in Patients with and Without Peripheral Artery Disease in DECLARE-TIMI 58. *Circulation* **2020**, *142*, 734–747. [CrossRef]
19. Chertow, G.M.; Correa-Rotter, R.; Vart, P.; Jongs, N.; McMurray, J.J.V.; Rossing, P.; Langkilde, A.M.; Sjöström, C.D.; Toto, R.D.; Wheeler, D.C.; et al. Effects of Dapagliflozin in Chronic Kidney Disease, With and Without Other Cardiovascular Medications: DAPA-CKD Trial. *J. Am. Hear. Assoc.* **2023**, *12*, e028739. [CrossRef]
20. The EMPA-KIDNEY Collaborative Group; Herrington, W.G.; Staplin, N.; Wanner, C.; Green, J.B.; Hauske, S.J.; Emberson, J.R.; Preiss, D.; Judge, P.; Mayne, K.J.; et al. Empagliflozin in Patients with Chronic Kidney Disease. *N. Engl. J. Med.* **2023**, *388*, 117–127. [CrossRef] [PubMed]
21. Packer, M.; Anker, S.D.; Butler, J.; Filippatos, G.; Pocock, S.J.; Carson, P.; Januzzi, J.; Verma, S.; Tsutsui, H.; Brueckmann, M.; et al. Cardiovascular and Renal Outcomes with Empagliflozin in Heart Failure. *N. Engl. J. Med.* **2020**, *383*, 1413–1424. [CrossRef] [PubMed]
22. McMurray, J.J.V.; DeMets, D.L.; Inzucchi, S.E.; Køber, L.; Kosiborod, M.N.; Langkilde, A.M.; Martinez, F.A.; Bengtsson, O.; Ponikowski, P.; Sabatine, M.S.; et al. A trial to evaluate the effect of the sodium–glucose co-transporter 2 inhibitor dapagliflozin on morbidity and mortality in patients with heart failure and reduced left ventricular ejection fraction (DAPA-HF). *Eur. J. Heart Fail.* **2019**, *21*, 665–675. [CrossRef] [PubMed] [PubMed Central]
23. Burke, D.T.; Armour, D.J.; McCargo, T.; Al-Adawi, S. Change over time of the ankle brachial index. *Wound Med.* **2019**, *28*, 100174. [CrossRef]
24. American Diabetes Association Professional Practice Committee; ElSayed, N.A.; Aleppo, G.; Bannuru, R.R.; Bruemmer, D.; Collins, B.S.; Das, S.R.; Ekhlaspour, L.; Hilliard, M.E.; Johnson, E.L.; et al. 10. Cardiovascular Disease and Risk Management: Standards of Care in Diabetes—2024. *Diabetes Care* **2023**, *47* (Suppl. S1), S179–S218. [CrossRef]
25. Forxiga 10 mg Film-Coated Tablets (07 March 2024). Available online: https://www.medicines.org.uk/emc/product/7607/smpc (accessed on 30 May 2024).
26. Jardiance 10 mg Film-Coated Tablets, (14 September 2023). Available online: https://www.medicines.org.uk/emc/product/5441/smpc#gref (accessed on 30 May 2024).
27. Kalani, M.; Brismar, K.; Fagrell, B.; Ostergren, J.; Jörneskog, G. Transcutaneous oxygen tension and toe blood pressure as predictors for outcome of diabetic foot ulcers. *Diabetes Care* **1999**, *22*, 147–151. [CrossRef]
28. Bulum, T.; Brkljačić, N.; Ivančić, A.T.; Čavlović, M.; Prkačin, I.; Tomić, M. In Association with Other Risk Factors, Smoking Is the Main Predictor for Lower Transcutaneous Oxygen Pressure in Type 2 Diabetes. *Biomedicines* **2024**, *12*, 381. [CrossRef]
29. Bacharach, J.M.; Rooke, T.W.; Osmundson, P.J.; Gloviczki, P. Predictive value of transcutaneous oxygen pressure and amputation success by use of supine and elevation measurements. *J. Vasc. Surg.* **1992**, *15*, 558–563. [CrossRef] [PubMed]
30. Ott, C.; Jumar, A.; Striepe, K.; Friedrich, S.; Karg, M.V.; Bramlage, P.; Schmieder, R.E. A randomised study of the impact of the SGLT2 inhibitor dapagliflozin on microvascular and macrovascular circulation. *Cardiovasc. Diabetol.* **2017**, *16*, 26. [CrossRef]
31. Sposito, A.C.; Breder, I.; Soares, A.A.S.; Kimura-Medorima, S.T.; Munhoz, D.B.; Cintra, R.M.R.; Bonilha, I.; Oliveira, D.C.; Breder, J.C.; Cavalcante, P.; et al. Dapagliflozin effect on endothelial dysfunction in diabetic patients with atherosclerotic disease: A randomized active-controlled trial. *Cardiovasc. Diabetol.* **2021**, *20*, 74. [CrossRef] [PubMed]
32. Shigiyama, F.; Kumashiro, N.; Miyagi, M.; Ikehara, K.; Kanda, E.; Uchino, H.; Hirose, T. Effectiveness of dapagliflozin on vascular endothelial function and glycemic control in patients with early-stage type 2 diabetes mellitus: DEFENCE study. *Cardiovasc. Diabetol.* **2017**, *16*, 84. [CrossRef]
33. Solini, A.; Giannini, L.; Seghieri, M.; Vitolo, E.; Taddei, S.; Ghiadoni, L.; Bruno, R.M. Dapagliflozin acutely improves endothelial dysfunction, reduces aortic stiffness and renal resistive index in type 2 diabetic patients: A pilot study. *Cardiovasc. Diabetol.* **2017**, *16*, 138. [CrossRef] [PubMed]
34. Zainordin, N.A.; Hatta, S.F.W.M.; Shah, F.Z.M.; Rahman, T.A.; Ismail, N.; Ismail, Z.; Ghani, R.A. Effects of Dapagliflozin on Endothelial Dysfunction in Type 2 Diabetes with Established Ischemic Heart Disease (EDIFIED). *J. Endocr. Soc.* **2019**, *4*, bvz017. [CrossRef] [PubMed]
35. Sorop, O.; Olver, T.D.; van de Wouw, J.; Heinonen, I.; van Duin, R.W.; Duncker, D.J.; Merkus, D. The microcirculation: A key player in obesity-associated cardiovascular disease. *Cardiovasc. Res.* **2017**, *113*, 1035–1045. [CrossRef] [PubMed]

36. Andreieva, I.O.; Riznyk, O.I.; Myrnyi, S.P.; Surmylo, N.N. State of Cutaneous Microcirculation in Patients with Obesity. *Wiad. Lek.* **2021**, *74*, 2039–2043. [CrossRef] [PubMed]
37. Sano, M.; Goto, S. Possible Mechanism of Hematocrit Elevation by Sodium Glucose Cotransporter 2 Inhibitors and Associated Beneficial Renal and Cardiovascular Effects. *Circulation* **2019**, *139*, 1985–1987. [CrossRef]
38. Avogaro, A.; Kreutzenberg, S.; Fadini, G. Dipeptidyl-peptidase 4 inhibition: Linking metabolic control to cardiovascular protection. *Curr. Pharm. Des.* **2014**, *20*, 2387–2394. [CrossRef] [PubMed] [PubMed Central]
39. Jahn, L.A.; Hartline, L.; Liu, Z.; Barrett, E.J. Metformin improves skeletal muscle microvascular insulin resistance in metabolic syndrome. *Am. J. Physiol. Endocrinol. Metab.* **2021**, *322*, E173–E180. [CrossRef] [PubMed] [PubMed Central]
40. Qiao, X.; Li, Y.; Zhou, T.; Edwards, P.A.; Gao, H. Metformin improves retinal capillary perfusion in diabetic mice. *Investig. Ophthalmol. Vis. Sci.* **2018**, *59*, 3591.
41. Schönborn, M.; Gregorczyk-Maga, I.; Batko, K.; Bogucka, K.; Maga, M.; Płotek, A.; Pasieka, P.; Słowińska-Solnica, K.; Maga, P. Circulating Angiogenic Factors and Ischemic Diabetic Foot Syndrome Advancement—A Pilot Study. *Biomedicines* **2023**, *11*, 1559. [CrossRef]

Disclaimer/Publisher's Note: The statements, opinions and data contained in all publications are solely those of the individual author(s) and contributor(s) and not of MDPI and/or the editor(s). MDPI and/or the editor(s) disclaim responsibility for any injury to people or property resulting from any ideas, methods, instructions or products referred to in the content.

 *pharmaceuticals*

Article

Potentially Inappropriate Medications Involved in Drug–Drug Interactions in a Polish Population over 80 Years Old: An Observational, Cross-Sectional Study

Emilia Błeszyńska-Marunowska [1,*], Kacper Jagiełło [2], Łukasz Wierucki [2], Marcin Renke [1], Tomasz Grodzicki [3], Zbigniew Kalarus [4] and Tomasz Zdrojewski [2]

[1] Department of Occupational, Metabolic and Internal Diseases, Medical University of Gdańsk, ul. Powstania Styczniowego 9B, 81-516 Gdynia, Poland; marcin.renke@gumed.edu.pl
[2] Department of Preventive Medicine and Education, Medical University of Gdańsk, 80-211 Gdańsk, Poland; kacper.jagiello@gumed.edu.pl (K.J.); lukasz.wierucki@gumed.edu.pl (Ł.W.); tz@gumed.edu.pl (T.Z.)
[3] Department of Internal Medicine and Gerontology, Jagiellonian University Medical College, 31-107 Kraków, Poland; tomasz.grodzicki@uj.edu.pl
[4] Department of Cardiology, Congenital Heart Disease and Electrotherapy, Silesian Center for Heart Disease, 41-800 Zabrze, Poland; zbigniew.kalarus@kalmet.com.pl
* Correspondence: e.bleszynska@gumed.edu.pl; Tel.: +48-605-881-185

Abstract: The clinical context of drug interactions detected by automated analysis systems is particularly important in older patients with multimorbidities. We aimed to provide unique, up-to-date data on the prevalence of potentially inappropriate medications (PIMs) and drug–drug interactions (DDIs) in the Polish geriatric population over 80 years old and determine the frequency and the most common PIMs involved in DDIs. We analyzed all non-prescription and prescription drugs in a representative national group of 178 home-dwelling adults over 80 years old with excessive polypharmacy ($\geq$10 drugs). The FORTA List was used to assess PIMs, and the Lexicomp® Drug Interactions database was used for DDIs. DDIs were detected in 66.9% of the study group, whereas PIMs were detected in 94.4%. Verification of clinical indications for the use of substances involved in DDIs resulted in a reduction in the total number of DDIs by more than 1.5 times, as well as in a nearly 3-fold decrease in the number of interactions requiring therapy modification and drug combinations that should be strictly avoided. The most common PIMs involved in DDIs were painkillers, and drugs used in psychiatry and neurology. Special attention should be paid to DDIs with PIMs since they could increase their inappropriate character. The use of automated interaction analysis systems, while maintaining appropriate clinical criticism, can increase both chances for a good therapeutic effect and the safety of the elderly during treatment processes.

Keywords: geriatrics; multimorbidity; polypharmacy; drug interactions; potentially inappropriate medications

1. Introduction

The increase in life expectancy in the 21st century alongside the declining fertility rates has led to a progressive aging of the population. According to the World Health Organization's report, the number of people over 65 years old is expected to increase from 524 million in 2010 (8% of the world's population) to 1.5 billion (16% of the world's population) by 2050 [1].

The main challenging factors in conducting safe and effective pharmacological therapy in older adults are changes in drug metabolism and elimination, as well as multimorbidity [2]. An additional disturbing aspect is the growing consumption of over-the-counter drugs (OTC), which are aggressively advertised and easily accessible [3].

Drug–drug interactions (DDIs) are reactions of at least two drugs that can lead to a quantitative and/or qualitative change in the action of one of them [4]. Potentially inappropriate medications (PIMs) can be defined as drugs for which use among older adults should be avoided due to the high risk of adverse reactions for this population and/or insufficient evidence of their benefits when safer and equally or more effective therapeutic alternatives are available [5]. The clinical context of drug–drug interactions detected by automated interaction analysis systems is exceptionally significant in seniors with multiple chronic diseases, who are in need of multidrug therapies according to guidelines of evidence-based medicine.

A reduction in inappropriate polypharmacy has been identified by the World Health Organization Third Global Patient Safety Challenge: Medication Without Harm as a major public health goal [6]. Properly conducted pharmacotherapy delivers multiple benefits such as a decreased risk of rehospitalization [7] and death [8] and cost reduction for the healthcare system [9], as well as an increased chance of achieving satisfactory therapeutic effects and improvement in life quality.

The aim of this study was to provide unique, up-to-date data on the prevalence of PIMs and DDIs in the Polish geriatric population over 80 years old. Furthermore, we aimed to determine the frequency and the most common PIMs involved in DDIs in our study group.

2. Results

In order to reflect the general population of Poland, the results were stratified according to the age structure of the Polish population over 80 years old in 2017. A detailed description of sampling and subsequent weighing can be found in the methodological publication [10].

2.1. Drug–Drug Interactions (DDIs)

DDIs were detected in 66.9% of the study group (n = 119) in a total number of 240 interactions. Detailed analysis showed that 60.7% of all respondents (n = 108) used drug combinations requiring treatment modification (Lexicomp category D; n = 197), while 17.4% of the entire group (n = 31) were treated with drug combinations that should be avoided (Lexicomp category X; n = 43). The mean number (95% CI) of DDIs per person was 1.35 (1.13–1.57), 1.11 (0.92–1.29) for category D interactions, and 0.24 (0.15–0.33) for category X interactions.

2.2. Potentially Inappropriate Medications (PIMs)

PIMs were detected in 94.4% of the study group (n = 168) in a total number of 473 substances. Detailed analysis showed that 84.8% of all respondents (n = 151) used substances with a questionable safety profile (FORTA class C; n = 317), while 62.9% of the entire group (n = 112) were treated with preparations that should be avoided in seniors (FORTA class D; n = 156). The mean number (95% CI) of PPIs per person was 2.66 (2.43–2.88), 1.78 (1.6–1.97) for class C substances, and 0.88 (0.74–1.01) for class D substances.

2.3. DDIs with PIMs

Verification of clinical indications for the use of substances involved in DDIs resulted in a reduction in the total number of DDIs by more than 1.5 times, as well as in a nearly 3-fold decrease in the number of interactions requiring therapy modification and drug combinations that should be strictly avoided.

However, despite that intervention, DDIs with PIMs were still found in 67 respondents (37.6% of all), with a mean number (95% CI) of 0.83 (0.56–1.10). Moreover, 40 respondents (22.5% of all) presented drug interactions requiring therapy monitoring (Lexicomp category D) due to the involvement of substances with a questionable safety profile (FORTA class C), while 4 people (2.5% of all) were treated with drug combinations that should be avoided (Lexicomp category X) with the involvement of substances contraindicated in seniors (FORTA class D). Detailed data are presented in Table 1. The qualitative analysis showed that the most common PIMs involved in DDIs were painkillers, and drugs used in psychiatry and neurology. Detailed data are presented in Tables 2 and 3.

Table 1. PIMs involved in DDIs.

	Lexicomp D + X	Classification of DDIs	
		Lexicomp D	Lexicomp X
Mean number of DDIs (95% CI)	1.35 (1.13–1.57)	1.11 (0.92–1.29)	0.24 (0.15–0.33)
Number of DDIs (n)	240	197	43
Number of patients with DDIs (n)	119	108	31
Mean number of DDIs with PIMs (95% CI)			
Forta Class C + D	0083 (0.56–1.1)	0.69 (0.49–0.9)	0.14 (0.03–0.25)
Forta Class C	0.43 (0.28–0.59)	0.38 (0.24–0.52)	0.06 (0.01–0.1)
Forta Class D	0.40 (0.17–0.63)	0.32 (0.16–0.46)	0.08 (0–0.18)
Number of DDIs with PIMs (n)			
Forta Class C + D	148	123	25
Forta Class C	77	67	10
Forta Class D	71	56	15
Number of patients with DDIs with PIMs (n)			
Forta Class C + D	67	62	12
Forta Class C	44	40	8
Forta Class D	33	31	4

DDIs—drug–drug interactions; PIMs—potentially inappropriate medications.

Table 2. Most common PIMs involved in DDIs from Lexicomp category D.

PIMs Involved in DDIs	Description	Solution
Painkillers		
NSAID–ASA	Increased risk of bleeding. Diminished cardioprotective effect of ASA. Decreased serum concentration of NSAIDs.	Monitor for increased risk of bleeding. Use alternative analgesics (e.g., acetaminophen).
NSAID–LD	Diminished diuretic effect of LD. Enhanced nephrotoxic effect of NSAIDs.	Monitor for decreased therapeutic effects of LD or evidence of acute kidney injury. Consider using an NSAID with lesser potential for interacting (e.g., diflunisal, flurbiprofen, ketoprofen, and ketorolac).
NSAID–VKA	Enhanced anticoagulant effect VKA. Increased risk of bleeding.	Monitor for increased risk of bleeding. Use alternative analgesics (e.g., acetaminophen).
NSAID–NSAID, e.g., Diclofenac–Aceclofenac; Diclofenac–Meloxicam; Diclofenac–Ketoprofen	Increased risk of some adverse effects of NSAIDs, including gastrointestinal adverse effects.	Monitor for increased risk of bleeding. Use alternative analgesics (e.g., acetaminophen).
Tramadol–H1-AH	Enhanced CNS depressant effect.	Avoid combination. If combined, limit the dosages and duration of each drug to the minimum possible while achieving the desired clinical effect.
Tramadol–BZD	Enhanced CNS depressant effect.	Avoid combination. If combined, limit the dosages and duration of each drug to the minimum possible while achieving the desired clinical effect.
Tramadol–Clonidine	Enhanced CNS depressant effect.	Avoid combination. If combined, limit the dosages and duration of each drug to the minimum possible while achieving the desired clinical effect.
Tramadol–Hydroxyzine	Enhanced CNS depressant effect.	Avoid combination. If combined, limit the dosages and duration of each drug to the minimum possible while achieving the desired clinical effect.
Tramadol–Tizanidine	Enhanced CNS depressant effect.	Avoid combination. If combined, limit the dosages and duration of each drug to the minimum possible while achieving the desired clinical effect.
Tramadol–Quetiapine	Enhanced CNS depressant effect.	Avoid combination. If combined, limit the dosages and duration of each drug to the minimum possible while achieving the desired clinical effect.
Tramadol–Mianserin	Enhanced CNS depressant effect.	Avoid combination. If combined, limit the dosages and duration of each drug to the minimum possible while achieving the desired clinical effect.
Tramadol–Z-drugs	Enhanced CNS depressant effect.	Avoid combination. If combined, limit the dosages and duration of each drug to the minimum possible while achieving the desired clinical effect.

Table 2. Cont.

PIMs Involved in DDIs	Description	Solution
Drugs in neurology and psychiatry		
Quetiapine–Ropinirole	Diminished therapeutic effect of dopamine agonist.	Consider using an alternative antipsychotic agent.
Quetiapine–Levodopa	Diminished therapeutic effect of dopamine agonist.	Consider using an alternative antipsychotic agent.
Carbamazepine–BB	Diminished therapeutic effect of beta-blockers.	Consider an alternative for one of the interacting drugs in order to avoid therapeutic failure.
Carbamazepine–HMGCRI	Diminished therapeutic effect of statins.	Consider an alternative for one of the interacting drugs in order to avoid therapeutic failure.
Carbamazepine–VKA	Decreased serum concentration of VKA	Monitor INR. Adjust the dose of VKA.
Carbamazepine–Quetiapine	Decreased serum concentration of Quetiapine. Increased serum concentration of Carbamazepine.	Adjust doses of both drugs. Monitor response.
Carbamazepine–DHP-CCB	Increased metabolism of DHP-CCBs.	Monitor for reduced therapeutic effects of DHP-CCBs, and adjust the dose. Consider alternatives to DHP-CCBs.
Ergotamine–Nebivolol	Enhanced vasoconstricting effect.	Monitor for evidence of excessive peripheral vasoconstriction. Consider alternative drugs.
Phenobarbital–Hydroxyzine	Enhanced CNS depressant effect.	Consider a decrease in barbiturate dose.
Phenobarbital–Z-drugs	Enhanced CNS depressant effect.	Avoid combination. If combined, limit the dosages and duration of each drug to the minimum possible while achieving the desired clinical effect.
Phenobarbital–HMGCRI	Diminished therapeutic effect of HMGCRIs.	Consider an alternative for one of the interacting drugs in order to avoid therapeutic failure.
Phenobarbital–Isosorbide mononitrate	Diminished therapeutic effect of Isosorbide.	Consider an alternative for one of the interacting drugs in order to avoid therapeutic failure.
Z-drugs–Hydroxyzine	Enhanced CNS depressant effect.	Avoid combination. If combined, limit the dosages and duration of each drug to the minimum possible while achieving the desired clinical effect.
Selegiline–Levodopa	Risk of hypertensive reaction.	Avoid combination.
Risperidone–Levodopa	Diminished therapeutic effect of dopamine agonist.	Consider using an alternative antipsychotic agent.
Others		
Clonidine–BB	Enhanced AV-blocking effect, sinus node dysfunction, rebound hypertensive effect.	Monitor heart rate and blood pressure.
SU–DPP-4I	Enhance hypoglycemic effect.	Consider a decrease in SU dose when initiating therapy with a DPP-4I and monitor patients for hypoglycemia.
MRA–Potassium	Risk of hyperkalemia.	Monitor serum potassium concentrations and for other evidence of hyperkalemia (e.g., muscular weakness, fatigue, arrhythmias, bradycardia).
NDHP-CCB–HMGCRI	Increased serum concentration of HMGCRI.	Consider using lower doses of statin, and monitor closely for signs of HMGCRI toxicity (e.g., myositis, rhabdomyolysis, hepatotoxicity). Fluvastatin, pravastatin, and rosuvastatin may be less affected.
Verapamil–Eplerenone	Increased serum concentration of Eplerenone	Adjust the dose of Eplerenone
Amiodarone–VKA	Increased serum concentration of VKA.	Monitor INR. Adjust the dose of VKA.
Ciprofibrate–HMGCRI	Enhance adverse/toxic effect of HMGCRI.	Monitor for signs/symptoms of muscle toxicity. Consider using alternative drugs.
Theophylline–BZD	Diminished therapeutic effect of BZD.	Monitor the effect of BZD.
Theophylline–Non-selective BB	Diminished bronchodilatory effect.	Monitor for symptoms of reduced theophylline efficacy.
Dietary supplements		
Gingko biloba–ASA	Increased risk of bleeding.	Monitor for signs and symptoms of bleeding (especially intracranial bleeding). Consider using alternative drugs.
Gingko biloba–Turmeric	Increased risk of bleeding.	Monitor for adverse effects (e.g., bleeding, bruising, altered mental status due to CNS bleeds).
Gingko biloba–Piracetam	Increased risk of bleeding.	Monitor for adverse effects (e.g., bleeding, bruising, altered mental status due to CNS bleeds).
Gingko biloba–NSAID	Increased risk of bleeding.	Monitor for adverse effects (e.g., bleeding, bruising, altered mental status due to CNS bleeds).

ASA—acetylsalicylic acid; BB—beta-blocker; BZD—benzodiazepine; DDIs—drug–drug interactions; DHP-CCB—dihydropyridine calcium channel blocker; DPP-4I—DPP4 inhibitor; H1-AH—H1-antihistamines; HMGCRI—HMG-CoA Reductase Inhibitor; LD—loop diuretic; MRA—aldosterone receptor antagonist; NDHP-CCB—non-dihydropyridine calcium channel blocker; NSAID—non-steroidal anti-inflammatory drug; PIMs—potentially inappropriate medications; SU—sulfonylurea; VKA—vitamin K antagonist.

Table 3. Most common PIMs involved in DDIs from Lexicomp® category X.

PIMs Involved in DDIs	Description	Solution
Painkillers		
Nimesulide–NSAID	Enhanced adverse/toxic effect of NSAIDs.	Avoid combination.
Dexketoprofen–NSAID	Dexketoprofen may enhance the adverse/toxic effect of NSAIDs.	Avoid combination.
Drugs in neurology and psychiatry		
Quetiapine–Potassium	Enhanced ulcerogenic effect.	Avoid solid oral dosage forms of potassium chloride. Liquid or effervescent potassium preparations are possible alternatives.
Quetiapine–Sotalol	Enhanced QTc-prolonging effect.	Avoid combination.
Carbamazepine–Apixaban	Decrease serum concentration of Apixaban.	Avoid combination.
Clomipramine–Potassium	Enhanced ulcerogenic effect.	Avoid solid oral dosage forms of potassium chloride. Liquid or effervescent potassium preparations are possible alternatives.
Risperidone–Ipratropium	Enhanced anticholinergic effect.	Avoid combination. If not possible, monitor for evidence of anticholinergic-related toxicities.
Others		
Spironolactone–Amiloride	Risk of hyperkalemia.	Avoid combination.
Tolterodine–Potassium	Enhanced ulcerogenic effect.	Avoid solid oral dosage forms of potassium chloride. Liquid or effervescent potassium preparations are possible alternatives.
Doxazosin–Tamsulosin	Risk of hypotension and syncope.	Avoid combination.

DDIs—drug–drug interactions; NSAID—non-steroidal anti-inflammatory drug; PIMs—potentially inappropriate medications.

3. Discussion

Due to the significant increase in life expectancy that long-term care carries over to older age [11], a growing number of chronic comorbidities requires the implementation of multidrug regimens according to evidence-based medicine, including drugs with a narrow safety profile especially in seniors [12]. Reducing exposure to PIMs is associated with a lower risk of adverse drug reactions and hospitalization in older individuals; however, this has no influence on mortality [13]. Physician-led interventions are based on standardized tools, such as STOPP criteria (Screening Tool of Older persons' Potentially Inappropriate Prescriptions) and START (Screening Tool to Alert doctors to the Right Treatment) [14], Beers' criteria [15], the PRISCUS list [16], the MAI (Medication Appropriateness Index) [17], the Good Palliative–Geriatric Practice Algorithm [18] or the FORTA (Fit For The Aged) List [19]. The FORTA List stands out from other methods and is classified as a patient-in-focus listing approach, which requires complex medical knowledge about individuals. The clinical usefulness of the FORTA List has been validated in randomized controlled clinical trials [19].

The incidence of PIMs in the available literature varies from 1.2% in Norway reported by Fog et al. to 93.9% in Finland stated by Toivo et al. We observed a higher consumption of PIMs in the Polish geriatric population over 80 years old than in most studies [19,20]. Differences in PIM prevalence can be explained by the application of various measurement tools, as well as different study settings (home-dwelling vs. primary care vs. hospital) [21].

The reported incidence of potential DDIs in other studies also varies depending on the settings and patient sample [22–24]. According to the available literature, the highest DDI prevalence of 88.4% was reported in the Slovak Republic by Kolar et al. while the lowest prevalence at a level of 4.4% in Norway was reported by Fog et al. The prevalence of DDIs in our study was relatively high, which can be explained by the fact that our study group included the oldest seniors over 80 years old with excessive polypharmacy ($\geq$10 drugs) who are at the highest risk of adverse drug reactions. In view of the gradual computerization of healthcare systems, clinical decision support programs are becoming more accessible. However, due to high sensitivity and low specificity, they can generate numerous alerts [25]. The available literature proves the effectiveness of decision support programs in deprescribing; however, in general, the interventions have little effect on hospital admissions or mortality [26].

Surprisingly, only a few studies involving a rather limited number of patients focused on the issue of DDIs occurring specifically with PIMs [21]. Furthermore, data concerning the oldest seniors are particularly sparse.

We have observed that the verification of clinical indications for the use of substances involved in DDIs resulted in a significant reduction in potentially clinically important DDIs. Nevertheless, DDIs with PIMs were still found in 37.6% of all respondents, with painkillers, and drugs used in psychiatry/neurology were the most common PIMs involved in DDIs. The few researchers who analyzed the topic reported similar benefits from such interventions [22,26–29].

To our knowledge, this is the first study in Poland and one of several studies in the world to present the important problem of DDIs with PIMs, with particular emphasis on the oldest people with excessive polypharmacy. The main limitation is the theoretical attempt to objectify the quality of pharmacotherapy with methods unable to fully reflect the complexity of clinical situations.

4. Materials and Methods

4.1. Study Design

The study group consisted of participants from the nationwide, cross-sectional observational study NOMED-AF (NOninvasive Monitoring for Early Detection of Atrial Fibrillation), which included ECG monitoring, the completion of a detailed questionnaire, a follow-up survey, blood pressure measurements, and blood/urine sample collection. A detailed description of the methodology and sampling of the NOMED-AF study was presented in a separate publication [10]. All participants provided written informed consent prior to participation. The study was approved by the Independent Bioethics Committee for Scientific Research at the Medical University of Gdańsk (13/2020; 2020-04-21) and by the Bioethics Commission at the Silesian Medical Chamber in Katowice (26/2015; 2015-07-01). All research procedures were conducted in keeping with the Declaration of Helsinki and Good Clinical Practice.

4.2. Setting

The study was conducted from 2017 to 2018. Respondents were randomly selected by the Ministry of Digitization of the Republic of Poland based on a social security number database; therefore, they constituted a representative sample of the Polish population in terms of sex, age, and place of residence. Based on a detailed questionnaire, the data were obtained by a trained nurse directly from the respondent, their family, or caregivers, followed by the presentation of the packaging of all drugs. The interview covered all preparations (prescription drugs, over-the-counter drugs, vitamins, nutritional preparations, and dietary supplements) taken at least once in the two weeks preceding the study (including drug name, form, single dose, and dosing frequency). Respondents provided information on diagnosed chronic diseases and were asked to present discharge cards from previous hospitalizations. Based on these data, individuals were assigned codes from the International Classification of Diseases, Tenth Revision (ICD-10).

4.3. Participants and Sample Size

The specific inclusion criteria for this study were an agreement to provide information on taken drugs, using at least ten active substances (excessive polypharmacy, EPP), and age over 80 years. The study group comprised 178 respondents, including 79 women and 99 men. The mean (SD) age of the entire sample was 85.8 (4.2) years (85.5 [4.2] years for women and 86.1 [4.2] years for men). The calculated maximum error in the study group was 2%. Detailed analyses of comorbidities and sociodemographic factors of the study group were presented in a separate publication [30].

4.4. Variables

The analysis of drug interactions between active substances was performed using Lexicomp® Drug Interactions by Wolters Kluwer Clinical Drug Information (www.wolterskluwer.com/en/solutions/lexicomp/, accessed on 10 March 2021), which enables a simultaneous analysis of 50 active substances. Detected interactions were classified into one of five categories: A = "no known interaction"; B = "no action required"; C = "monitor therapy"; D = "consider modifying therapy"; and X = "avoid combination". Category D and X interactions were considered for further analysis as drug–drug interactions (DDIs) of potential clinical significance.

The analysis of potentially inappropriate medications was performed using the FORTA List (Fit For The Aged) [19]. For each respondent, active substances were assigned to one of four classes: A = "indicated"; B = "favorable"; C = "carefully"; and D = "do not apply". Class C and D substances were considered for further analysis as potentially inappropriate medications (PIMs).

4.5. Statistical Methods

Post-stratification was used to adjust the sample structure against the Polish population in 2017. The results are presented as percentages, median values with first and third quartiles, and mean values with 95% confidence intervals (CIs). The analysis was performed using the statistical package R version 3.6.3 (R Foundation for Statistical Computing, Vienna, Austria) and SAS 9.4 TS Level 1M5 (SAS Institute, Inc., Cary, NC, USA).

5. Conclusions

Although it is difficult to avoid the limitations of theoretical considerations that are not able to reflect complex clinical situations, ideally PIMs and DDIs should always be reviewed in older patients. However, special attention should be paid to those DDIs occurring with PIMs since they could increase their inappropriate character. The use of automated interaction analysis systems, while maintaining appropriate clinical criticism, can increase the chances of achieving a therapeutic effect while increasing the safety of the elderly during the treatment process. Further studies would be necessary to investigate the potential negative outcomes of DDIs in PIMs on adverse drug reactions and their long-term consequences, such as hospitalization, morbidity, and mortality.

Author Contributions: Conceptualization, E.B.-M. and T.G.; methodology, E.B.-M. and T.G.; software, K.J.; validation, E.B.-M. and Ł.W.; formal analysis, K.J.; investigation, Ł.W. and T.Z.; data curation, E.B.-M., Ł.W., and K.J.; writing—original draft preparation, E.B.-M.; writing—review and editing, Ł.W., M.R., and T.G.; supervision, T.G., Z.K., and T.Z.; project administration, Z.K. and T.Z. All authors have read and agreed to the published version of the manuscript.

Funding: The study was funded by the National Centre for Research and Development under a grant agreement (No. STRATEGMED2/269343/18/NCBR/2016).

Institutional Review Board Statement: The study was conducted in accordance with the Declaration of Helsinki and approved by the Institutional Ethics Committee of the Medical University of Gdańsk (13/2020; 2020-04-21) and the Silesian Medical Chamber in Katowice (26/2015; 2015-07-01) for studies involving humans.

Informed Consent Statement: Informed consent was obtained from all subjects involved in the study.

Data Availability Statement: The data are contained within the article.

Conflicts of Interest: The authors declare no conflicts of interest.

References

1. World Health Organization. Global Health and Aging. 2011. Available online: https://www.nia.nih.gov/sites/default/files/d7/nia-who_report_booklet_oct-2011_a4__1-12-12_5.pdf (accessed on 22 March 2022).
2. Hilmer, S.N. Bridging geriatric medicine, clinical pharmacology and ageing biology to understand and improve outcomes of medicines in old age and frailty. *Ageing Res Rev.* **2021**, *71*, 101457. [CrossRef] [PubMed]

3. Mielke, N.; Huscher, D.; Douros, A.; Ebert, N.; Gaedeke, J.; van der Giet, M.; Kuhlmann, M.K.; Martus, P.; Schaeffner, E. Self-reported medication in community-dwelling older adults in Germany: Results from the Berlin Initiative Study. *BMC Geriatr.* **2020**, *20*, 22. [CrossRef] [PubMed]
4. Pazan, F.; Wehling, M. Polypharmacy in older adults: A narrative review of definitions, epidemiology and consequences. *Eur. Geriatr. Med.* **2021**, *12*, 443–452. [CrossRef] [PubMed]
5. Zhang, X.; Zhou, S.; Pan, K.; Li, X.; Zhao, X.; Zhou, Y.; Cui, Y.; Liu, X. Potentially inappropriate medications in hospitalized older patients: A cross-sectional study using the Beers 2015 criteria versus the 2012 criteria. *Clin. Interv. Aging* **2017**, *12*, 1697–1703. [CrossRef] [PubMed]
6. Donaldson, L.J.; Kelley, E.T.; Dhingra-Kumar, N.; Kieny, M.P.; Sheikh, A. Medication Without Harm: WHO's Third Global Patient Safety Challenge. *Lancet* **2017**, *389*, 1680–1681. [CrossRef] [PubMed]
7. Patel, N.S.; Patel, T.K.; Patel, P.B.; Naik, V.N.; Tripathi, C.B. Hospitalizations due to preventable adverse reactions-a systematic review. *Eur. J. Clin. Pharmacol.* **2017**, *73*, 385–398. [CrossRef] [PubMed]
8. Muhlack, D.C.; Hoppe, L.K.; Weberpals, J.; Brenner, H.; Schöttker, B. The Association of Potentially Inappropriate Medication at Older Age with Cardiovascular Events and Overall Mortality: A Systematic Review and Meta-Analysis of Cohort Studies. *J. Am. Med. Dir. Assoc.* **2017**, *18*, 211–220. [CrossRef] [PubMed]
9. Kim, J.; Parish, A.L. Polypharmacy and Medication Management in Older Adults. *Nurs. Clin. N. Am.* **2017**, *52*, 457–468. [CrossRef] [PubMed]
10. Kalarus, Z.; Balsam, P.; Bandosz, P.; Grodzicki, T.; Kaźmierczak, J.; Kiedrowicz, R.; Mitręga, K.; Noczyński, M.; Opolski, G.; Rewiuk, K.; et al. NOninvasive Monitoring for Early Detection of Atrial Fibrillation: Rationale and design of the NOMED-AF study. *Kardiol. Pol.* **2018**, *76*, 1482–1485. [CrossRef] [PubMed]
11. Li, F.; Otani, J. Financing elderly people's long-term care needs: Evidence from China. *Int. J. Health Plann. Manag.* **2018**, *33*, 479–488. [CrossRef] [PubMed]
12. Jungo, K.T.; Streit, S.; Lauffenburger, J.C. Patient factors associated with new prescribing of potentially inappropriate medications in multimorbid US older adults using multiple medications. *BMC Geriatr.* **2021**, *21*, 163. [CrossRef] [PubMed]
13. Xing, X.X.; Zhu, C.; Liang, H.Y.; Wang, K.; Chu, Y.Q.; Zhao, L.B.; Jiang, C.; Wang, Y.Q.; Yan, S.Y. Associations Between Potentially Inappropriate Medications and Adverse Health Outcomes in the Elderly: A Systematic Review and Meta-analysis. *Ann. Pharmacother.* **2019**, *53*, 1005–1019. [CrossRef] [PubMed]
14. O'Mahony, D. STOPP/START criteria for potentially inappropriate medications/potential prescribing omissions in older people: Origin and progress. *Expert Rev. Clin. Pharmacol.* **2020**, *13*, 15–22. [CrossRef]
15. American Geriatrics Society. Updated AGS Beers Criteria® for Potentially Inappropriate Medication Use in Older Adults. *J. Am. Geriatr. Soc.* **2019**, *67*, 674–694. [CrossRef] [PubMed]
16. Mühlbauer, B. The New PRISCUS List. *Dtsch. Arztebl. Int.* **2023**, *120*, 1–2. [CrossRef] [PubMed]
17. Hanlon, J.T.; Schmader, K.E. The Medication Appropriateness Index: A Clinimetric Measure. *Psychother. Psychosom.* **2022**, *91*, 78–83. [CrossRef] [PubMed]
18. Bilek, A.J.; Levy, Y.; Kab, H.; Andreev, P.; Garfinkel, D. Teaching physicians the GPGP method promotes deprescribing in both inpatient and outpatient settings. *Ther. Adv. Drug Saf.* **2019**, *10*, 204209861989591. [CrossRef] [PubMed]
19. Pazan, F.; Weiss, C.; Wehling, M.; FORTA. The FORTA (Fit fOR The Aged) List 2021: Fourth Version of a Validated Clinical Aid for Improved Pharmacotherapy in Older Adults. *Drugs Aging* **2022**, *39*, 245–247. [CrossRef] [PubMed]
20. Bhagavathula, A.S.; Vidyasagar, K.; Chhabra, M.; Rashid, M.; Sharma, R.; Bandari, D.K.; Fialova, D. Prevalence of Polypharmacy, Hyperpolypharmacy and Potentially Inappropriate Medication Use in Older Adults in India: A Systematic Review and Meta-Analysis. *Front Pharmacol.* **2021**, *12*, 685518. [CrossRef] [PubMed]
21. Tian, F.; Chen, Z.; Wu, J. Prevalence of Polypharmacy and Potentially Inappropriate Medications Use in Elderly Chinese Patients: A Systematic Review and Meta-Analysis. *Front. Pharmacol.* **2022**, *13*, 862561. [CrossRef] [PubMed]
22. Bories, M.; Bouzillé, G.; Cuggia, M.; Le Corre, P. Drug-Drug Interactions in Elderly Patients with Potentially Inappropriate Medications in Primary Care, Nursing Home and Hospital Settings: A Systematic Review and a Preliminary Study. *Pharmaceutics* **2021**, *13*, 266. [CrossRef] [PubMed]
23. Marinović, I.; Bačić Vrca, V.; Samardžić, I.; Marušić, S.; Grgurević, I. Potentially inappropriate medications involved in drug–drug interactions at hospital discharge in Croatia. *Int. J. Clin. Pharm.* **2021**, *43*, 566–576. [CrossRef] [PubMed]
24. Georgiev, K.D.; Hvarchanova, N.; Georgieva, M.; Kanazirev, B. The role of the clinical pharmacist in the prevention of potential drug interactions in geriatric heart failure patients. *Int. J. Clin. Pharm.* **2019**, *41*, 1555–1561. [CrossRef] [PubMed]
25. Stuhec, M.; Flegar, I.; Zelko, E.; Kovačič, A.; Zabavnik, V. Clinical pharmacist interventions in cardiovascular disease pharmacotherapy in elderly patients on excessive polypharmacy. *Wien Klin Wochenschr.* **2021**, *133*, 770–779. [CrossRef] [PubMed]
26. Rieckert, A.; Reeves, D.; Altiner, A.; Drewelow, E.; Esmail, A.; Flamm, M.; Hann, M.; Johansson, T.; Klaassen-Mielke, R.; Kunnamo, I.; et al. Use of an electronic decision support tool to reduce polypharmacy in elderly people with chronic diseases: Cluster randomised controlled trial. *BMJ* **2020**, *369*, m1822. [CrossRef] [PubMed]
27. Novaes, P.H.; da Cruz, D.T.; Lucchetti, A.L.G.; Leite, I.C.G.; Lucchetti, G. The "iatrogenic triad": Polypharmacy, drug-drug interactions, and potentially inappropriate medications in older adults. *Int. J. Clin. Pharm.* **2017**, *39*, 818–825. [CrossRef] [PubMed]
28. Hosia-Randell, H.M.; Muurinen, S.M.; Pitkälä, K.H. Exposure to potentially inappropriate drugs and drug-drug interactions in elderly nursing home residents in Helsinki, Finland: A cross-sectional study. *Drugs Aging* **2008**, *25*, 683–692. [CrossRef] [PubMed]

29. Varallo, F.R.; Ambiel, I.S.S.; Nanci, L.O.; Galduróz, J.C.F.; Mastroianni, P.D.C. Assessment of pharmacotherapeutic safety of medical prescriptions for elderly residents in a long-term care facility. *Braz. J. Pharm. Sci.* **2012**, *48*, 477–485. [CrossRef]
30. Błeszyńska-Marunowska, E.; Jagiełło, K.; Grodzicki, T.; Wierucki, Ł.; Sznitowska, M.; Kalarus, Z.; Renke, M.; Mitręga, K.; Rewiuk, K.; Zdrojewski, T. Polypharmacy among elderly patients in Poland: Prevalence, predisposing factors, and management strategies. *Pol. Arch. Intern. Med.* **2022**, *132*, 16347. [CrossRef] [PubMed]

Disclaimer/Publisher's Note: The statements, opinions and data contained in all publications are solely those of the individual author(s) and contributor(s) and not of MDPI and/or the editor(s). MDPI and/or the editor(s) disclaim responsibility for any injury to people or property resulting from any ideas, methods, instructions or products referred to in the content.

Review

Advances in Nanomedicine for Precision Insulin Delivery

Alfredo Caturano [1,2,*,†], Roberto Nilo [3,†], Davide Nilo [1,†], Vincenzo Russo [4,5], Erica Santonastaso [6], Raffaele Galiero [1], Luca Rinaldi [7], Marcellino Monda [2], Celestino Sardu [1], Raffaele Marfella [1] and Ferdinando Carlo Sasso [1]

1. Department of Advanced Medical and Surgical Sciences, University of Campania Luigi Vanvitelli, 80138 Naples, Italy
2. Department of Experimental Medicine, University of Campania Luigi Vanvitelli, 80138 Naples, Italy
3. Data Collection G-STeP Research Core Facility, Fondazione Policlinico Universitario A. Gemelli IRCCS, 00168 Roma, Italy
4. Department of Biology, College of Science and Technology, Sbarro Institute for Cancer Research and Molecular Medicine, Temple University, Philadelphia, PA 19122, USA
5. Division of Cardiology, Department of Medical Translational Sciences, University of Campania Luigi Vanvitelli, 80138 Naples, Italy
6. Independent Researcher, 81024 Maddaloni, Italy
7. Department of Medicine and Health Sciences "Vincenzo Tiberio", Università degli Studi del Molise, 86100 Campobasso, Italy
* Correspondence: alfredo.caturano@unicampania.it; Tel.: +39-3338616985
† These authors contribute equally to this work.

Abstract: Diabetes mellitus, which comprises a group of metabolic disorders affecting carbohydrate metabolism, is characterized by improper glucose utilization and excessive production, leading to hyperglycemia. The global prevalence of diabetes is rising, with projections indicating it will affect 783.2 million people by 2045. Insulin treatment is crucial, especially for type 1 diabetes, due to the lack of β-cell function. Intensive insulin therapy, involving multiple daily injections or continuous subcutaneous insulin infusion, has proven effective in reducing microvascular complications but poses a higher risk of severe hypoglycemia. Recent advancements in insulin formulations and delivery methods, such as ultra-rapid-acting analogs and inhaled insulin, offer potential benefits in terms of reducing hypoglycemia and improving glycemic control. However, the traditional subcutaneous injection method has drawbacks, including patient compliance issues and associated complications. Nanomedicine presents innovative solutions to these challenges, offering promising avenues for overcoming current drug limitations, enhancing cellular uptake, and improving pharmacokinetics and pharmacodynamics. Various nanocarriers, including liposomes, chitosan, and PLGA, provide protection against enzymatic degradation, improving drug stability and controlled release. These nanocarriers offer unique advantages, ranging from enhanced bioavailability and sustained release to specific targeting capabilities. While oral insulin delivery is being explored for better patient adherence and cost-effectiveness, other nanomedicine-based methods also show promise in improving delivery efficiency and patient outcomes. Safety concerns, including potential toxicity and immunogenicity issues, must be addressed, with the FDA providing guidance for the safe development of nanotechnology-based products. Future directions in nanomedicine will focus on creating next-generation nanocarriers with precise targeting, real-time monitoring, and stimuli-responsive features to optimize diabetes treatment outcomes and patient safety. This review delves into the current state of nanomedicine for insulin delivery, examining various types of nanocarriers and their mechanisms of action, and discussing the challenges and future directions in developing safe and effective nanomedicine-based therapies for diabetes management.

Keywords: nanomedicine; insulin; precision insulin delivery; next-generation nanocarriers; targeted drug delivery

Citation: Caturano, A.; Nilo, R.; Nilo, D.; Russo, V.; Santonastaso, E.; Galiero, R.; Rinaldi, L.; Monda, M.; Sardu, C.; Marfella, R.; et al. Advances in Nanomedicine for Precision Insulin Delivery. *Pharmaceuticals* **2024**, *17*, 945. https://doi.org/10.3390/ph17070945

Academic Editor: Kelong Fan

Received: 4 June 2024
Revised: 7 July 2024
Accepted: 12 July 2024
Published: 15 July 2024

Copyright: © 2024 by the authors. Licensee MDPI, Basel, Switzerland. This article is an open access article distributed under the terms and conditions of the Creative Commons Attribution (CC BY) license (https:// creativecommons.org/licenses/by/ 4.0/).

1. Introduction

Diabetes mellitus comprises a group of metabolic disorders affecting carbohydrate metabolism. In this condition, glucose is both inadequately utilized as an energy source and excessively produced due to inappropriate gluconeogenesis and glycogenolysis, leading to hyperglycemia [1].

The prevalence of diabetes has surged to alarming levels, with global estimates indicating that 10.5% of individuals aged 20–79 (536.6 million people) are affected, a figure projected to rise to 12.2% (783.2 million) by 2045 [2]. Over the next two decades, this prevalence is expected to double, impacting over half a billion people, with more than 75% residing in low- and middle-income countries [3]. The increase seen in developing nations is attributed to the adoption of Western lifestyle habits, including sedentary behavior, physical inactivity, and a high-energy diet [4,5].

The diagnosis of diabetes involves demonstrating elevated concentrations of glucose in the venous plasma or an increased glycated hemoglobin (A1C) level in the blood. Conventionally, diabetes is classified into various clinical categories, including type 1 or type 2 diabetes, gestational diabetes mellitus, and other specific types arising from various causes, such as genetic factors, exocrine pancreatic disorders, and certain medications [1,6]. Type 1 diabetes mellitus (T1DM) is characterized by an absence or near-absence of β-cell function, necessitating insulin therapy for survival. Insulin treatment stands as a cornerstone for individuals grappling with type 1 diabetes, given the characteristic absence or near-absence of β-cell function in this population. Apart from addressing hyperglycemia, insulinopenia, a hallmark of type 1 diabetes, can give rise to additional metabolic disturbances such as hypertriglyceridemia, ketoacidosis, and tissue catabolism, posing life-threatening risks [7]. Beyond its traditional metabolic effects, insulin has also been proven to have positive cardiovascular effects, which are diminished during insulin resistance [8,9]. For many decades after the groundbreaking discovery of insulin, severe metabolic decompensation was predominantly averted through once- or twice-daily injections. However, in the last four decades, mounting evidence supports the adoption of more intensive insulin replacement strategies for individuals with type 1 diabetes [10].

The Diabetes Control and Complications Trial demonstrated the efficacy of intensive therapy involving multiple daily injections or continuous subcutaneous insulin infusion using short-acting (regular) and intermediate-acting (NPH) human insulins [11–13]. Intensive control, achieving lower A1C levels (7%), correlated with an approximately 50% reduction in microvascular complications over a 6-year treatment period. Despite its benefits, intensive therapy was associated with a higher incidence of severe hypoglycemia compared to conventional treatment [11,14,15]. Follow-up studies revealed fewer macrovascular and microvascular complications in the group that underwent intensive treatment, and the positive impact persisted over two decades beyond the active treatment phase of the study [11–13,16]. Expanding our focus to type 2 diabetes (T2DM), insulin therapy also plays a crucial role in its management. While, initially, lifestyle modifications and oral medications are commonly employed, the progressive nature of type 2 diabetes often necessitates the inclusion of insulin therapy. Basal insulin, mealtime insulin, and correction insulin strategies are similarly applied in type 2 diabetes, offering personalized approaches to optimize glycemic control. The choice of insulin regimen depends on individual patient characteristics, preferences, and treatment goals [17–19]. Similarly to T1DM, T2DM is linked to an increased risk of both vascular and non-vascular complications. In fact, T2DM elevates the risk of a range of cardiovascular disorders approximately twofold [20–22]. Furthermore, T2DM is associated with diverse non-vascular conditions such as cancer, infections, liver disease, and sensorial, mental and nervous system disorders [7,23–30].

Insulin replacement plans typically encompass basal insulin, mealtime insulin, and correction insulin [17]. Basal insulin options include NPH insulin, long-acting insulin analogs, and continuous delivery of rapid-acting insulin through an insulin pump. Basal insulin analogs exhibit a prolonged duration of action with more consistent plasma concentrations compared to NPH insulin, while rapid-acting analogs (RAA) offer a quicker onset and peak with a shorter duration of action than regular human insulin. In individuals with type 1 diabetes, treatment with analog insulins is associated with reduced hypoglycemia, weight gain, and lower A1C compared to injectable human insulins [31–33].

Recent advancements have introduced two injectable ultra-rapid-acting analog (URAA) insulin formulations designed to expedite absorption and provide heightened activity in the initial phase of their profile compared to other RAA [34,35]. Inhaled human insulin, featuring a rapid peak and shortened duration of action, presents an alternative option. These newer formulations exhibit potential advantages, including a reduced hypoglycemia risk, improved postprandial glucose control, and enhanced administration flexibility in relation to prandial intake compared to traditional RAA [36–38].

Subcutaneous insulin injections are effective. Despite being the current gold standard for diabetes management, these injections can lead to poor adherence to treatment plans, thereby compromising effective diabetes control, due to the high risk of severe hypoglycemia and the burdensome need for multiple daily injections [18,19,39]. This issue persists even with newer devices like microneedle delivery systems, which are simpler, more adaptable, and tend to be more acceptable to patients compared to traditional needle injections [40]. The necessity for novel approaches like nanomedicine becomes evident in addressing these challenges. Nanomedicine offers a promising alternative by enhancing medication targeting and stability, potentially reducing the frequency of administration and minimizing the risk of hypoglycemia. By improving drug delivery mechanisms, nanomedicine can significantly increase patient compliance and overall treatment efficacy, revolutionizing diabetes care and offering a more convenient and safer therapeutic option for patients. While inhaled insulin provides a non-invasive alternative, it may pose challenges related to pulmonary deposition and long-term safety [41]. Among these, oral insulin delivery holds significant promise due to its potential to improve patient adherence and quality of life. Despite its potential, oral insulin delivery faces substantial barriers, primarily due to the harsh gastrointestinal (GI) environment, which degrades insulin and impedes its absorption. Strategies to overcome these barriers include the development of advanced delivery systems, such as nanoparticles, which enhance insulin bioavailability and permeability across the intestinal mucosa [42]. Nanocarriers, such as liposomes, chitosan, and PLGA (Poly(lactic-co-glycolic) Acid), provide protection against enzymatic degradation, improving drug stability and controlled release. These nanocarriers offer unique advantages, ranging from enhanced bioavailability and sustained release to specific targeting capabilities [19,42].

Nanomedicine presents innovative solutions to the challenges of insulin delivery, offering promising avenues for overcoming current drug limitations, enhancing cellular uptake, and improving pharmacokinetics and pharmacodynamics [43]. This review delves into the current state of nanomedicine for insulin delivery, examining various types of nanocarriers and their mechanisms of action, and discussing the challenges and future directions in developing safe and effective nanomedicine-based therapies for diabetes management.

2. Challenges to Oral Insulin Administration and Nanomedicines

Insulin, a globular protein with a molecular weight of 5808 Daltons, consists of two chains, A (21 residues) and B (30 residues), linked together by disulfide bonds [44]. Currently, insulin is primarily administered through subcutaneous injections [45]. However, this delivery method is associated with several drawbacks, including lipohypertrophy, obesity, retinopathy, hypoglycemia, neuropathy, lipoatrophy, allergic reactions, and peripheral hyperinsulinemia [46,47]. The need for multiple insulin injections throughout the day can be burdensome for patients, leading to non-compliance. To address these challenges, there is a growing demand for controlled and prolonged release systems that reduce the injection frequency and enhance patient compliance. The Health Care Costs Institute reports a doubling of the cost of insulin for patients, emphasizing the urgent need for more patient-friendly alternatives. While subcutaneous injections are the current norm, oral delivery of insulin presents an attractive alternative due to the improved patient adherence and cost-effective manufacturing processes compared to injections [48].

2.1. Physiological Barriers in the Gastrointestinal Tract (GIT)

The harsh conditions of the gastrointestinal tract pose significant challenges to the oral absorption of peptide drugs. Factors such as the low pH, the presence of peptidases and proteases, and poor absorption through the intestinal epithelial layer hinder effective drug delivery [49,50].

2.1.1. Gastric Acid Barrier

The stomach serves as one of the most formidable chemical barriers to insulin bioavailability. Gastric acid, which includes hydrochloric acid, is a fluid produced by parietal cells within the gastric glands of the stomach and used for digestion and protection from external potential infections. This secretion creates a highly acidic environment (pH 1–2) [51]. In such acidic conditions, free insulin is prone to degradation and proteolysis by the enzyme pepsin. Therefore, protecting insulin from the acidic and enzymatic environment is crucial for the effective oral delivery of insulin [52].

2.1.2. Intestinal Mucosal Barrier

The intestine, the longest segment of the digestive tract, plays a critical role in both digestion and nutrient absorption, as well as in immune defense. The intestinal mucosal lining, also known as the intestinal barrier, functions to prevent pathogens from entering the bloodstream and other tissues [53]. The surface of the intestine is coated with mucus, which acts as a barrier between the luminal contents and the intestinal epithelium. This mucus is a gel-like network composed of highly glycosylated mucin molecules, which ensures the structural integrity and protective function of the intestinal barrier. Goblet cells are responsible for maintaining and renewing the mucus layer approximately every 4–5 h by secreting adhesion proteins and endophytes. The mucus selectively allows small molecules to permeate, while larger particles and pathogens are blocked, providing an anti-infective effect [54,55]. Hence, an effective oral insulin delivery system must traverse the mucus barrier to reach the intestinal epithelium.

Below the mucus layer, the intestine is lined with epithelial cells, which are the primary components of the intestinal mucosal barrier (Figure 1). For oral insulin to be effective, it must cross the intestinal epithelium via transcellular or paracellular pathways before it can enter the bloodstream and regulate blood glucose levels (Figure 2) [56]. The paracellular pathway, regulated by tight junctions between adjacent epithelial cells, is the preferred route for hydrophilic molecule translocation. However, most oral drugs are absorbed through the transcellular pathway, which involves processes such as intermembrane transport, adsorption, fusion, and endocytosis. The paracellular pathway is much more restrictive due to the tight junctions that create a sealing space of about 8–13 Å, which limits transport to hydrophilic molecules with molecular weights under 200 Da. In fact, the hydrophilic nature and large molecular size of peptide drugs further limit their oral absorption. One major obstacle to oral insulin delivery is the low bioavailability, often falling below 1%, attributed to enzymatic degradation in the gastrointestinal tract and inefficient absorption through intestinal epithelial cells [57]. Overcoming these barriers is crucial for the successful development of an effective and reliable oral insulin delivery system.

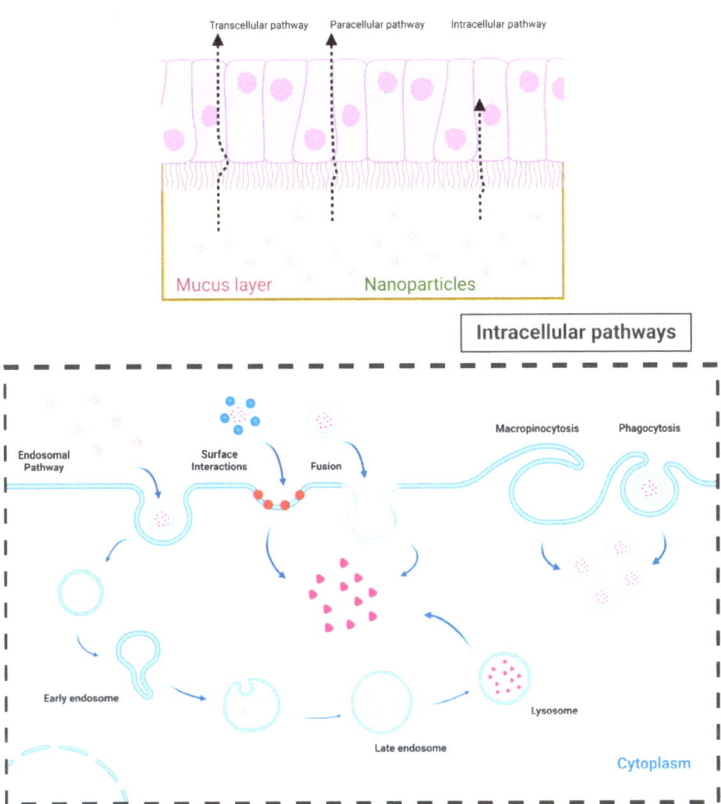

Figure 1. Barriers in the gastrointestinal tract to oral insulin delivery and potential pathways.

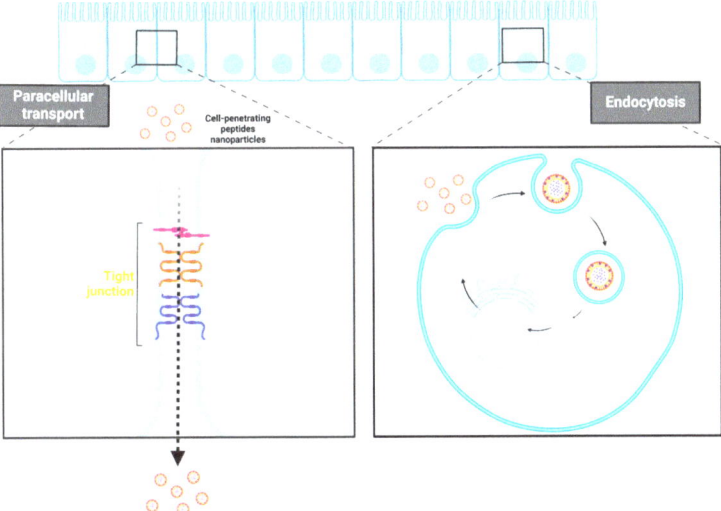

Figure 2. Mechanisms of the gastrointestinal barrier: gap junctions and intracellular pathways for oral insulin delivery.

2.1.3. Intestinal Wall Receptors and Delivery Systems

The intestinal wall hosts a variety of receptors, including vitamin B12 and folate receptors, and neonatal Fc receptors (FcRn). These receptors, located primarily on the apical surface of enterocytes, are crucial for facilitating the uptake of nutrients and can be strategically targeted to enhance the delivery and absorption of various therapeutic agents [58].

Vitamins are frequently used as ligands to enhance functional systems due to their safety, stability, and ease of modification. Among these, B12 and folic acid are the most extensively studied for oral targeted insulin delivery [59].

Vitamin B12 is a complex water-soluble molecule that includes a "corrin ring" and a nucleotide. It is absorbed orally after binding to haptocorrin, a salivary enzyme that protects and transports B12 to the small intestine [60]. In the intestine, B12 binds to an intrinsic factor to form a complex that interacts with receptors on enterocytes in the ileum. This mechanism has been used to improve the oral bioavailability of various poorly absorbed drugs, proteins, and peptides, including insulin [60]. In fact, B12-decorated dextran nanoparticles containing insulin significantly increased pharmacological availability (2.6-fold higher) compared to non-targeted nanoparticles [61]. By adjusting the parameters of the crosslinking and using an amino alkyl B12 derivative, researchers achieved a significant hypoglycemic effect and extended the antidiabetic effects, increasing insulin's pharmacological availability up to 29.4% [62]. However, the potential applications of B12 are limited by its relatively slower uptake compared to other vitamins and the restricted availability of absorption sites, primarily in the distal ileum [63,64].

Folic acid (vitamin B9) is absorbed through a saturable, pH- and sodium ion-dependent pathway that is sensitive to metabolic inhibitors [65]. Unlike B12, folic acid receptors are abundant and enhance the uptake and transport of bioactive molecules across the gastrointestinal tract. Moreover, folate receptors exhibit a high affinity for folate and folate-conjugated compounds, making them effective targets for drug delivery via receptor-mediated endocytosis. Folate-decorated nanoparticles show great promise in enhancing the bioavailability of insulin. When insulin is encapsulated within these nanoparticles, they can specifically bind to folate receptors on the surface of intestinal epithelial cells. This binding facilitates receptor-mediated endocytosis, allowing the nanoparticles to cross the intestinal barrier and release insulin into the bloodstream. This targeted delivery not only protects insulin from enzymatic degradation and acidic conditions in the gastrointestinal tract but also enhances its absorption and bioavailability. By leveraging specific interactions with folate receptors, folate-decorated nanoparticles offer significant advancements in the oral delivery of insulin, potentially leading to better therapeutic outcomes and improved patient compliance [66–70].

The use of neonatal Fc receptors (FcRn) to develop intestinal targeted delivery systems has gained momentum [71]. These receptors, available from the neonatal stage into adulthood in the human intestine, bind immunoglobulin G (IgG) and albumin, protecting them from intracellular enzymatic breakdown [72]. By using the Fc portion of IgG or albumin as a targeting ligand, drugs/nanocarriers can be transcytosed by endosomes and released in the extracellular space at a physiological pH [73,74]. FcRn-targeted poly(lactic acid)–poly(ethylene glycol) (PEG-PLA) block copolymer-based nanoparticles, with the Fc portion of IgG linked to their surface and loaded with insulin, demonstrating a hypoglycemic effect in vivo [75]. Finally, when insulin was loaded into nanoparticles functionalized with the Fc fragment of IgG as a targeting ligand to the FcRn, its permeation across an in vitro cell culture model was significantly enhanced [76].

2.1.4. Nanomedicines: A Potential Solution to Insulin Administration Challenges

Innovative approaches, such as advanced drug delivery technologies and formulation strategies, are being explored to enhance the bioavailability and therapeutic efficacy of orally administered insulin. These advancements aim to revolutionize diabetes management by providing a more patient-friendly and cost-effective alternative to traditional injection methods [77].

Over the years, the field of nanoscience has witnessed exponential growth, driving promising advancements in the management of diabetes conditions [78]. Nanotechnology, serving as a transformative force in the medical realm, equips researchers with tools to design efficient nano-systems for delivering therapeutic molecules with enhanced benefits. These principles are applied in creating nanomedicines or nanotherapeutics, allowing for the loading of therapeutic agents, thereby improving physicochemical properties and achieving enhanced therapeutic outcomes with precise targeting [79]. Nanotechnology has not only played a pivotal role in the development of cutting-edge glucose monitoring devices but also empowered scientists to create various delivery systems aimed at improving the efficacy of insulin and other antidiabetic molecules in the systemic circulation. These nano-systems offer a superior approach compared to conventional methods, circumventing the harsh metabolic pathways that often reduce the effectiveness of these molecules [80].

Diabetes, especially type 2 with its challenges of insulin resistance or deficiency, presents unique complexities. In a sub-category of type 2 diabetes, a significant number of patients exhibit varying blood glucose levels due to obesity, independent of insulin, while others suffer from insulin deficiency or resistance [81]. Scientists have directed their attention to these diverse diabetic conditions, recognizing the potential of nanomedicines in managing these categories. Recent advancements in nanotechnological approaches have led to innovative delivery systems capable of enhancing the potential of anti-diabetic molecules [82]. Studies emphasize the vital role of various nano-formulations, especially those designed with novel smart polymers, in shielding drug molecules from harsh metabolic pathways and facilitating a controlled release pattern, ensuring sustained levels of insulin in patients [83].

Nanoparticle (NP)-based drug delivery systems (DDSs) have garnered significant attention for their efficacy in transporting therapeutic agents to target cells. Nanocarriers, including liposomes, polymeric NPs, solid lipid NPs (SLNs), chitosan, exosomes, micelles, nanogels, and dendrimers, play a crucial role in entrapping drugs, limiting side effects, and improving bioavailability. These nanocarriers are characterized by biodegradability, biocompatibility, non-toxicity, and the ability to escape the reticuloendothelial system [84].

Insulin encapsulation onto NPs provides protection against enzymatic degradation in the gastrointestinal (GI) tract, thereby improving bioavailability when compared to previous formulations. NPs also enhance insulin permeability across the intestinal mucosa by opening the tight junctions between epithelial cells [84]. Recent developments include long-acting NP formulations containing insulin to reduce the injection frequency and extensive exploration of nanocarriers for oral insulin delivery [85]. Smart nanocarrier-based drug delivery systems, such as glucose-responsive NPs synthesized from dextran, have demonstrated rapid and extended self-regulated insulin delivery, effectively reducing elevated blood glucose levels in mice and minimizing the risk of hypoglycemia [86]. Glucose-responsive self-assembled polyamine nanocarriers have also proven effective in regulating blood glucose concentrations [87]. In the context of oral insulin delivery, nanocarriers enhance transport through both paracellular and transcellular pathways. Chitosan, for instance, enhances intestinal permeability by opening the tight junctions between cells, thereby facilitating paracellular transport. This occurs through interactions of positively charged polymers with the negatively charged cell membrane. The transcellular pathway involves processes such as fusion, endocytosis, and adsorption, with receptor-mediated endocytosis being a major route for insulin-loaded nanocarriers to enter cells [52,88].

3. Nanomedicine Technological Advancements

Traditional injectable insulin has been superseded by novel delivery strategies, which are capable of overcoming well-known delivery limitations. These strategies involve releasing the drug at the target site through diverse carriers employing various mechanisms, following more or less complex kinetics based on the composition of the delivery system.

Nanocarriers (Figure 3) have garnered significant attention as the preferred formulations due to their potential to exploit dimensional characteristics for enhanced cellular uptake. Depending on their structural attributes, nanocarriers offer diverse advantages and disadvantages, resulting in a wide array of preparations (Table 1). In the following sections, we provide an overview of different nano-systems and their applications in insulin delivery.

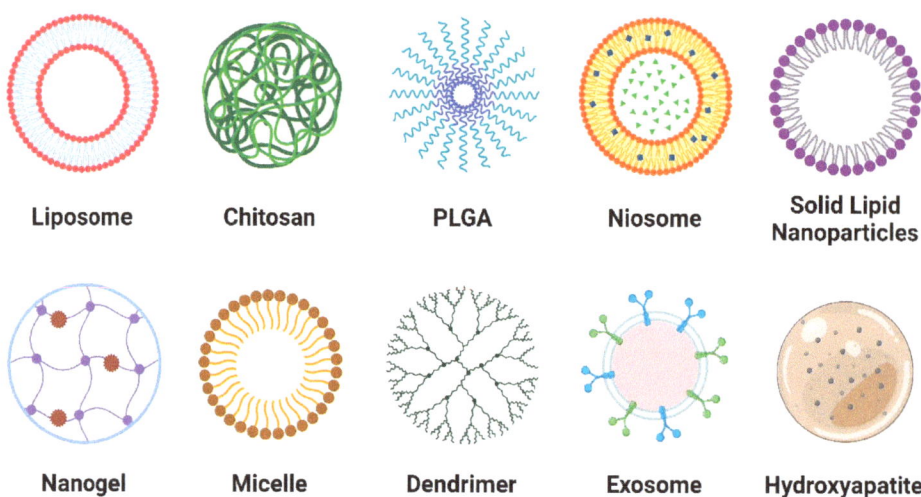

Figure 3. Structural overview of nanocarriers used for insulin delivery.

Table 1. Systems for delivering insulin nanoparticulates to treat diabetes mellitus.

Nanocarriers	Administration Route	Effects In Vivo	Challenges
Liposomes	Oral	- Increased hypoglycemic effect [89] - Improved insulin absorption and oral bioavailability [90] - Conquer mucus and epithelium barriers [91] - Increased retention time in lungs, reducing extra pulmonary side effects [91] - Improved proteolytic stability in oral administration [66] - Chemical responsive release [92] - Sustained release and transmucosal delivery [93]	- Non-uniform coating [66] - Lower entrapment efficiency compared to polymeric carriers [94] - Leakage due to instability, especially autoxidation of cholesterol [94] - Possible allergic reactions due to lipid presence [95,96]
Chitosan	Oral/Nasal/Transdermal	- Overcome mucus and epithelium barriers [97] - Increased bioavailability [98] - 54.19% reduction in blood sugar level after 4 h [99] - Normal histological findings, no signs of inflammation or ulceration [99]	- Insulin-loading efficiency in chitosan nano-formulations [100] - Ensuring stability and controlled release of insulin [100] - Better understanding of chitosan nanoparticles' interactions with biological barriers [101]

Table 1. *Cont.*

Nanocarriers	Administration Route	Effects In Vivo	Challenges
PLGA	Oral	- Reduction of blood sugar level [102] - Prolonged hypoglycemic effect [102]	- Potential lack selectivity in interacting with mucosal surfaces [103]
SLNs	Oral	- Reduction of blood sugar level [43]	- Low oral bioavailability due to stomach degradation [43] - Inactivation and digestion by proteases in the luminal cavity [43] - High molecular weight and lack of lipophilicity, leading to reduced intestinal absorption [104]
Nanogels	Oral	- Improved hypoglycemic effects [105] - Increased bioavailability [105]	- Specific size and stability [105] - Precise controlled release [106] - Tissue penetration [107] - Biocompatibility and immunogenicity [108]
Micelles	Oral	- Prevention of insulin aggregation [109] - Increased bioavailability [109]	- Limited payload capacity [110] - Stability issues [111] - Biocompatibility concerns [112] - Controlled release challenges [113]
HAP	Oral	- Regulation of blood glucose levels in rats by continuous insulin release [114]	- Low insulin-loading capacity [115] - Limited bioadhesiveness [116]
Dextran nanoparticles	Subcutaneous	- Prolonged hypoglycemic effect [117]	- Biodegradation and clearance [118] - Possible immunogenicity [119]
Polyethylene glycol (PEG) nanoparticles	Oral	- Enhanced hypoglycemic effects [120] - Improved bioavailability [120]	- Possible allergic reactions [121] - Controlled release and targeting efficiency [120]
Hydrogels	Oral/Subcutaneous	- Targeted delivery [122] - Enhanced bioavailability [122]	- Possible adverse immune responses or tissue reactions [123] - Insulin stability [123] - Precise control over insulin release kinetics to match physiological requirement [123]
Niosomes	Oral/Mucosal	- Stabilized against enzymatic degradation in oral and vaginal delivery [124] - Prolonged bioactivity for 6 h [124]	- Low entrapment efficiency [94] - Instability due to alteration in molecular arrangement of surfactants [94]

3.1. Liposomes

Liposomes, spherical vesicles with lipid bilayers containing an aqueous phase, serve as effective carriers for hydrophilic drugs such as insulin due to their internal structure. This characteristic facilitates the encapsulation of therapeutic agents, thereby offering protection during transportation. The degree of biocompatibility and biodegradability of liposomes varies depending on their composition and synthesis methods [125]. While liposomes typically exhibit particle sizes ranging from a few micrometers to 30 nanometers, polar lipids can also self-assemble into various colloidal particle forms beyond the conventional bilayer shapes, influenced by factors such as the temperature, molecule shape, and environmental conditions. Recent studies have demonstrated the potential of liposomes with specific characteristics, such as a size range of 150–210 nm and negative surface charge, in decreasing blood glucose levels and increasing insulin concentrations in vivo [50]. To overcome challenges like instability in the gastrointestinal tract, modifications such as functionalization with cell-penetrating peptides (CPPs) show promise in improving stability and uptake through endocytosis [50,126]. Additionally, folate-targeted PEGylated liposomes, designed for insulin delivery, demonstrated enhanced bioavailability and cellular uptake through folate receptor-mediated endocytosis [50]. Liposomal chitosan gel, developed for wound healing with encapsulated insulin, exhibited stability and sustained release, showcasing potential clinical benefits [127]. Moreover, sodium-glycodeoxycholate (SGDC)-incorporated elastic liposomes significantly enhanced insulin permeation across buccal tissues [128]. Innovative approaches, such as the development of glucose-sensitive multivesicular liposomes using the double emulsion technique by Liu et al., show promise for self-regulated insulin administration. Liu et al. demonstrated both in vitro and in vivo that these systems would be able to control blood glucose levels within a normal range [129]. Shafiq et al. have explored the potential of using liposomes derived from the fat globule membrane (MFGM) of camel milk as carriers for insulin. In an in vivo study, diabetic rats induced by STZ were treated with these camel milk-derived liposomes. The results revealed substantial reductions in the blood glucose levels, as well as significant decreases in the bilirubin, alkaline phosphatase, albumin, and alanine aminotransferase levels, highlighting the effectiveness of this novel insulin delivery method [130]. Wu et al. tested liposomes loaded with arginine–insulin complexes (AINS) incorporated into a hydrogel prepared from cysteine-modified alginate (Cys-Alg) to form liposome-in-alginate hydrogels (AINS-Lip-Gel). Their ex vivo study demonstrated that the intestinal permeation of AINS and AINS-Lip was approximately 2.0- and 6.0-fold higher, respectively, compared to free insulin. Moreover, their in vivo evaluation showed that the hydrogel effectively retarded the early release of insulin (~30%) from the liposomes and enhanced the intestinal mucosal retention. In vivo experiments further revealed that the AINS-Lip-Gel released insulin in a controlled manner and exhibited strong hypoglycemic effects [131].

3.2. Chitosan

Chitosan, a natural polysaccharide renowned for its mucoadhesive properties and facile encapsulation capabilities, is widely recognized as a pivotal material in the formulation of nanocarriers for oral insulin delivery. Chitosan NPs offer further advantages in overcoming the limitations associated with administering insulin alone. These nanocarriers possess a positive surface charge which, in acidic environments like the gastric tract, leads to enhanced protonation, rendering them more cationic [132]. This promotes greater bioavailability as they interact with negatively charged mucus, thereby prolonging the residence time and improving the drug absorption by creating space between the tight junctions among epithelial cells. However, the acidic environment is only characteristic of the stomach. Starting from the small intestine (pH about 6.8), the GI environment becomes closer to alkaline (up to 7.4 in the ileum), which leads to deprotonation of chitosan amino groups and destruction of polyelectrolyte complexes [51,133]. Moreover, the preparation of chitosan NPs is relatively straightforward, involving electrostatic interactions to complex with anionic insulin, minimizing the risk of affecting its structure and integrity. Despite chitosan's known drawbacks, such as its poor mechanical properties and low antibacterial activity, these challenges can be mitigated by complexing it with other natural or synthetic polymers [134]. Innovation lies in researchers' ability to combine chitosan with other materials to enhance its desirable properties. For example, folate-conjugated chitosan nanoparticles (NPs) increased insulin stability in the harsh GI tract and improved cellular uptake, resulting in enhanced hypoglycemic activity in vivo [135]. Modification with poly(sodium 4-styrenesulfonate, PSS) improved stability in acidic conditions, while a carboxymethyl-β-cyclodextrin-grafted chitosan NPs formulation exhibited excellent encapsulation efficiency and sustained drug release, significantly reducing the blood glucose levels in mice [136,137], or a thermosensitive copolymer incorporating the chitosan–zinc–insulin complex showcased reduced burst release and improved stability during storage [138].

Studies conducted both in vitro and in vivo have demonstrated that chitosans and their derivatives possess anti-diabetic properties by affecting insulin resistance, glucose uptake, and lipid metabolism [139]. Ju et al. showed that in insulin-resistant rats induced by a high-energy diet combined with streptozocin (STZ), chitosan administration led to reductions in the fasting insulin and blood glucose levels, improved insulin sensitivity, and enhanced oral glucose tolerance [140]. Furthermore, in STZ-induced diabetic mice, chitosan supplementation attenuated damage to pancreatic islets and prevented nuclear pyknosis and atrophy of pancreatic cells [141]. Pang et al. developed an oral insulin delivery system using chitosan (Chi) and alginate (Alg) dual-coated double-layered hydroxide (LDH) nanocomposites (Alg-Chi-LDH@INS). These nanocomposites, with a size of ~350.8 nm and a charge of ~−13.0 mV, effectively mitigated burst insulin release in acidic conditions. Flow cytometry showed enhanced uptake of LDH@INS by Caco-2 cells due to the chitosan coating. The in vivo study on diabetic mice demonstrated significant hypoglycemic effects, reducing the blood glucose levels by ~50% at 4 h, highlighting the system's potential for diabetes treatment [142]. Abd-Alhussain et al. compared insulin-loaded nanoparticles (NPs) to subcutaneous insulin in diabetic rats. They used biodegradable chitosan-capped NPs with soluble human insulin (20 IU/kg). While the glucose reduction at 6 h was not significant, the insulin levels were significantly higher at 12 and 24 h in the NP-treated rats compared to those administered subcutaneous insulin. The results suggest that chitosan-based NPs can maintain good glycemic control for up to 24 h and are a potential carrier for oral insulin delivery [143]. Maurya et al. developed mannose ligand-conjugated nanoparticles using a quality-by-design approach to enhance oral insulin delivery. They identified critical formulation attributes and process parameters affecting nanoparticle quality. Mannosylated chitosan nanoparticles, prepared by the inotropic gelation method and encapsulated to protect from acidic environments, showed optimal size, charge, and drug entrapment. Cell studies confirmed the safety and selective uptake by M-cells, enhancing insulin bioavailability [144]. Additional recent preclinical studies have highlighted the potential of chitosan-based NPs in enhancing oral insulin delivery. Thiolated chitosan NPs were incorporated into fast-disintegrating dosage forms, showing promising mucoadhesive properties and hypoglycemic effects [145]. Mercaptonicotinic acid-activated thiolated chitosan NPs demonstrated improved mucoadhesivity and insulin permeability [146]. Virus-capsid mimicking NPs with biotin-grafted chitosan exhibited superior transmucosal penetration and hypoglycemic response [147]. Dinitro salicylic acid-functionalized chitosan NPs effectively reduced the serum glucose levels in diabetic rats [148]. A "ternary mutual-assist" system with vitamin B12–chitosan improved insulin stability and absorption [149]. Chitosan derivative-based NPs loaded with fibroblast growth factor-21 (FGF-21) showed durable drug release and biocompatibility [150]. Net-neutral particles formed with chitosan improved mucosal transport and bioavailability [151]. Acetylated cashew gum and chitosan NPs reduced blood glucose without toxicity [152]. Silica-coated chitosan–dextran sulfate NPs provided controlled insulin release [153]. Lastly, chitosan/PEG and albumin-coated NPs maintained insulin bioactivity and stability in simulated gastrointestinal conditions, collectively indicating chitosan's pivotal role in optimizing oral insulin delivery systems [154].

In a clinical study involving Korean patients aged 20 to 75, daily chitosan supplementation of 1.5 g for three months effectively lowered the postprandial serum glucose levels [155]. Finally, it has been reported that chitosan oligosaccharides may contribute to the protection against oxidative stress [156]. Given the importance of managing and preventing complications associated with diabetes, this property could present another valuable tool [26,103].

3.3. PLGA (Poly(lactic-co-glycolic) Acid)

PLGA, renowned for its exceptional biocompatibility and biodegradability, stands out as a widely utilized organic material in the development of nanocarriers tailored for insulin delivery. Lactic and glycolic acid monomers undergo copolymerization to generate PLGA, an aliphatic polymer with a polyester backbone. This copolymerization process allows for the creation of a wide array of formulations by adjusting the ratio of lactic and glycolic acid monomers. By varying this composition, the crystallinity, hydrophilicity, and glass transition temperature of the copolymer can all be controlled. Since lactic acid is more hydrophobic than glycolic acid, a higher ratio of lactic acid results in a more hydrophobic copolymer. Consequently, modifying the composition in this way can decrease the rate at which water penetrates a device [157]. Additionally, PLGA is renowned for its non-toxic nature and is duly approved by both the Food and Drug Administration and the European Medicines Agency for medical applications. Within the body, it is metabolized into lactide and glycolate monomers, which are further transformed into CO_2 and H_2O through the Krebs cycle. The rate of drug release from PLGA is determined by its degradation, which can be regulated by the ratio of lactide to glycolide monomers and the molecular weight. Despite the previously demonstrated in vitro cellular uptake of insulin trapped in PLGA NPs through the early and late endosomes pathway, in vivo the effects of the administration of unmodified PLGA alone on blood glucose levels are limited [158]. This limitation is primarily attributed to its poor muco-permeabilization and muco-adhesion capacity, stemming to its hydrophobic and anionic nature, as well as the inability to completely protect insulin in the gastric environment. Consequently, there is a need to coat and/or modify the PLGA. The most prevalent modification involves conjugation to PEG, although the literature is continually expanding with various conjugation strategies. For instance, PLGA, a widely used polymeric NP system, enhances insulin absorption through the transcellular pathway. PLGA/chitosan NPs demonstrated significant hypoglycemic effects [159], while montmorillonite PLGA nanocomposites exhibited increased insulin stability in low-pH environments and sustained release in simulated gastrointestinal fluid [160]. The studies on oral insulin delivery using PLGA-based systems show promising advancements in improving the bioavailability and efficacy. Zhou et al. demonstrated that a "ternary mutual-assist" nano-delivery system using PLGA, ionic liquids, and vitamin B12–chitosan (VB12-CS) enhanced insulin protection and absorption, leading to a significant blood glucose reduction in diabetic mice [149]. Liu et al. reported that zwitterionic materials, although challenging to coat on hydrophobic nanoparticles, were successfully applied using zwitterionic Pluronic analogs, resulting in stable PLGA@PPP4K nanoparticles that effectively lowered the blood glucose levels in diabetic rats [161]. Asal et al. investigated chitosan-based nanoparticles; particularly, they showed that chitosan gold nanoparticles functionalized with PLGA achieved high insulin entrapment and controlled release, significantly reducing the blood glucose levels in diabetic rats [162]. Furthermore, Li et al. studied a novel PLGA-Hyd-PEG nanoplatform that exhibited pH-responsive properties, enhancing mucosal penetration and cellular uptake, and effectively lowering the blood glucose levels in diabetic rats [163].

3.4. Niosomes

Niosomes, synthetic microscopic vesicles mainly comprised of non-ionic surfactants and cholesterol as excipients, represent a versatile and promising class of drug delivery vehicles. These nanocarriers exhibit excellent potential in drug delivery systems due to their ability to serve as reservoirs for drugs, leading to sustained and prolonged release profiles while accommodating substances with varying solubilities [124]. Additionally, their inherent biocompatibility and low toxicity make them attractive candidates for therapeutic applications [164]. Niosomes can be categorized into different types based on their sizes and bilayers, including large unilamellar vesicles (LUVs), small unilamellar vesicles (SUVs), and multilamellar vesicles (MLVs). They offer advantages such as enhanced drug entrapment efficiency and controlled release kinetics [165]. Ning et al. investigated niosomes as vaginal delivery systems for insulin, demonstrating promising results. Specifically, they showed significant reductions in the blood glucose levels with sustained hypoglycemic effects [124]. In a study, niosomes made from polyoxyethylene alkyl ethers (Brij™) were developed for insulin encapsulation. The niosomes, particularly those with Brij 92 and cholesterol in a 7:3 molar ratio, protected insulin against enzymatic degradation, showing potential as sustained-release oral dosage forms for insulin delivery [94].

3.5. Solid Lipid Nanoparticles

Solid lipid nanoparticles (SLNs) are composed of solid lipids, such as purified triacyl-glycerols, waxes, or complex mixtures of acylglycerols, along with surfactants forming an aqueous dispersion around a crystalline lipid core [166]. These submicron (50–1000 nm) colloidal carriers offer biodegradability and are advantageous for large-scale production. Their degradation rate is composition-dependent, being inversely proportional to the length of the fatty acid chains of the acylglycerols, thus enabling predictable and regulated drug release. Various SLN formulations for loading insulin have been developed, typically including lecithin and poly-oxyethylene esters of 12-hydroxystearic acid as surfactants, along with solid lipids like cetyl palmitate, glyceryl monostearate, or glyceryl palmitostearate [167]. VB12-coated gel-core SLNs demonstrated prolonged hypoglycemic effects in diabetic rats by preventing burst release in low-pH environments [108]. Additionally, cationic SLNs protected insulin from enzymatic activity, ensuring stability and sustained release in the media [168]. Muntoni et al. developed SLNs loaded with glargine insulin, which were formulated into solid oral dosage forms to enhance stability and prolong shelf-life. Evaluation in both healthy and diabetic rat models demonstrated significant hypoglycemic effects, particularly with the capsule formulation, suggesting their promise as effective agents for lowering blood glucose levels. However, further optimization is required to refine these SLNs for practical use as orally active insulin preparations [169]. Meanwhile, Zheng et al. focused on SLNs loaded with insulin (INS-SLNs), prepared using a methanol–chloroform mixed solvent system. These nanoparticles exhibited desirable properties such as a small size, excellent stability, and sustained release in simulated intestinal conditions. Cellular studies indicated enhanced uptake and improved transcytosis across intestinal epithelial cells compared to free insulin, highlighting their potential to enhance the delivery of peptide and protein drugs via the oral route [170].

3.6. Nanogels

Nanogels are hydrogels at the nanometer scale, ranging from a few nanometers up to 300 nm. They consist of a network of polymer chains containing basic or acidic groups, allowing them to absorb and retain large amounts of aqueous solution. The primary characteristic of nanogels is their swelling capacity, maintained through crosslinks between polymers, which can be either homopolymers or copolymers. Nanogels come in various types: synthetic, natural, and hybrid, as well as neutral, anionic, or cationic, depending on their surface charge. They can internally load biological cargo, facilitated by electrostatic interactions, and externally conjugate different molecules via functional group bonds [171]. Unlike other nanoparticles, nanogels can be structured to respond to specific changes in their environment. Dextran-crosslinked glucose-responsive nanogels demonstrated high encapsulation efficiency and glucose-dependent insulin release [172], while pH- and temperature-sensitive HPMC nanogels provided controlled release, and a multi-responsive nanogel adjusted the release rate based on the glucose concentration [172–174].

Baloch et al. explored a novel approach for oral insulin delivery using an insulin-intercalated graphene oxide (In@GO NgC) nanogel composite. Their study investigated the release profile of insulin from In@GO NgC in simulated gastric and intestinal fluids, demonstrating enhanced release in intestinal conditions and stability against enzymatic degradation over 6 h. Comparative analysis with non-intercalated GO nanogels and nanogels without GO underscored the superior release profile and enzymatic stability of In@GO NgC, highlighting its potential for effective oral insulin delivery [174]. Mudassir et al. evaluated pH-sensitive methyl methacrylate/itaconic acid (MMA/IA) nanogels as carriers for oral insulin delivery. Their optimized formulations exhibited a favorable particle size, zeta potential, and high entrapment efficiency crucial for insulin stability and release characteristics in simulated gastrointestinal fluids. The study demonstrated that lyophilization with trehalose preserved the stability of these nanogels, maintaining the insulin integrity. In vivo studies in diabetic rats showed significant reductions in the blood glucose levels following oral administration of the optimized nanogel formulation, indicating their potential as effective carriers for enhancing oral insulin absorption and therapeutic efficacy [172]. Wang et al. synthesized hydroxyethyl methacrylate (HEMA) nanogels via emulsion polymerization and investigated their potential as oral insulin delivery systems. Their characterization studies highlighted the morphology, stability, and enhanced circulation half-life of HEMA nanogels, with minimal uptake by macrophage cells. In vivo studies demonstrated that HEMA nanogels facilitated efficient intestinal absorption of encapsulated insulin, sustaining blood glucose control for up to 12 h with improved bioavailability compared to free insulin. These findings suggest that HEMA nanogels could serve as promising alternatives for oral insulin delivery, offering potential benefits in managing diabetes without the need for injections [175]. Chou et al. developed an injectable insulin-loaded gel composed of self-assembled nanoparticles from carboxymethyl-hexanoyl chitosan (CHC) integrated with lysozyme for sustained basal insulin delivery. In vitro evaluations confirmed the controlled biodegradation and insulin release kinetics of the CHC-lysozyme gel, demonstrating the cytocompatibility of the degradation products. In vivo studies in diabetic mouse models revealed that a single injection of the gel maintained the fasting blood glucose levels within normal ranges for up to 10 days. This injectable system shows promise as a novel long-acting insulin delivery method, potentially improving treatment adherence and reducing the complications associated with conventional insulin therapies [176].

3.7. Micelles

Polymeric micelles serve as widely used nanocarriers owing to their inherent properties. In an aqueous milieu, amphiphilic polymers spontaneously organize into a distinct conformation. The resultant structure comprises a hydrophilic outer region and a hydrophobic core, offering potential as a reservoir for sustained drug release. While encapsulating hydrophilic insulin presents complexities, numerous techniques have been developed over time to enable efficient loading. Generally, polymeric complex micelles have demonstrated enhanced insulin-loading efficiency and stability under physiological conditions [177–179]. Moreover, innovative delivery strategies have significantly improved insulin delivery. Notably, Liu et al. showcased that glucose and H_2O_2 dual-responsive polymeric micelles exhibited considerable hypoglycemic effects in vivo while maintaining good biocompatibility [180]. Bahman et al. designed a poly-(styrene-co-maleic acid) (SMA) micellar system for oral insulin delivery, addressing challenges such as rapid insulin degradation in the stomach and enhancing intestinal absorption. Insulin was encapsulated into SMA micelles in a pH-dependent manner, characterized using dynamic light scattering and HPLC. In vitro studies with Caco-2 cells and ex vivo rat intestinal sections, along with in vivo experiments in diabetic mice, demonstrated effective stimulation of glucose uptake by SMA-insulin. The negatively charged micelles, with a mean diameter of 179.7 nm, showed promising hypoglycemic effects lasting up to 3 h post-administration.

Overall, SMA micelles offer a promising strategy for oral insulin delivery, potentially improving diabetes management [181]. Han et al. focused on overcoming the barriers to oral protein drug delivery by developing a zwitterionic micelle platform. These micelles feature a virus-mimetic zwitterionic surface and a betaine side chain, with an ultralow critical micelle concentration, enabling drug penetration through intestinal mucus and enhancing epithelial absorption via transporters. An oral insulin prototype, encapsulated in enteric-coated capsules, exhibited high oral bioavailability (>40%) and customizable insulin action profiles. This biocompatible formulation promises long-term safety and represents a significant advance in oral protein drug delivery, potentially transforming treatment for patients requiring regular injections [182]. Italiya et al. described the development of an orally active nanoformulation for lisofylline (LSF). LSF was encapsulated as its ester prodrug (LSF-linoleic acid (LA) prodrug) into biodegradable polymeric micelles (LSF-LA PLMs), enhancing its pharmacokinetic profile, with significantly improved oral bioavailability (74.86%) compared to free LSF. In vitro studies confirmed the formulation's stability and efficacy in insulin-secreting cells. In vivo experiments in STZ-induced diabetic rats demonstrated reduced fasting glucose levels and increased insulin levels via both oral and intraperitoneal routes. Additionally, the LSF-LA PLM formulation mitigated inflammation and preserved pancreatic integrity in diabetic animals, suggesting its potential as a therapeutic strategy in type 1 diabetes [183]. Hu et al. explored amphiphilic pH-sensitive block copolymer poly(methyl methacrylate-co-methacrylic acid)-b-poly(2-amino ethyl methacrylate) (P(MMA-co-MAA)-b-PAEMA) as carriers for oral insulin delivery. Synthesized via ARGET ATRP, these copolymers self-assembled into pH-responsive cationic polymeric micelles (PCPMs), characterized by 1H-NMR, FT-IR, and GPC. The PCPMs exhibited pH-sensitive behavior, maintaining stability in acidic environments and increasing in size at a neutral pH. Insulin was efficiently loaded into the PCPMs with low toxicity and a controlled pH-triggered release profile, demonstrating their potential as effective carriers for oral insulin delivery [184].

3.8. Dendrimers

Dendrimers are synthetic polymers synthesized by reacting a diamine with methyl acrylate. Their generation can proceed in both divergent and convergent manners, resulting in a macromolecule composed of a core, branching units, and surface functional groups that dictate their properties. Their highly regular architecture facilitates effective surface functionalization with molecules tailored to achieve specific properties [185,186]. For instance, modification with PEG enhances stabilization, preventing macrophage attack [187]. Another example is the caproyl-modified G2 PAMAM dendrimer, which efficiently increased pulmonary insulin absorption through the paracellular and transcellular pathways [188]. Amphiphilic dendrimers based on multi-armed poly(ethylene glycol) (PEG) showed stability under an acidic pH and reduced blood glucose levels [189]. Xian et al. developed a nanoscale complex for autonomous insulin therapy responsive to glucose levels, aiming to enhance diabetes management. Their approach combined a synthetic dendrimer carrier with an insulin analog modified using a high-affinity glucose-binding motif, employing electrostatic and dynamic-covalent interactions. This resulted in an injectable insulin depot capable of providing both glucose-directed and long-lasting insulin availability. The nanocomplex, administered via a single injection, maintained controlled blood glucose levels for at least one week in diabetic swine subjected to daily oral glucose challenges. The serum insulin concentrations increased correspondingly with the elevated blood glucose levels, a notable achievement in glucose-responsive insulin therapy [190].

3.9. Exosomes

Exosomes are membrane vesicles composed of specific proteins with vesicular fusion and fission functions, along with lipids. Ranging in size from 30 to 150 nm, exosomes play a physiological role in interacting with target cells and serving as cargo delivery agents within them. Notably, exosomes derived from the pancreas of patients with type 2 diabetes have been found to influence the survival and apoptosis of pancreatic β cells through the miRNAs they transport [191]. Exosomes hold potential as insulin delivery systems. Cell-derived exosomes encapsulated with insulin demonstrated enhanced transport and metabolism of glucose in cells, suggesting their value as nanocarriers for diabetic treatment [192]. Morales et al. achieved successful encapsulation of insulin into exosomes using an electroporation technique. Specifically, they mixed insulin with exosomes and subjected them to electroporation with parameters set at 200 V and 50 μF, followed by an incubation period of 1 h at 37 °C. They noted that the loading efficiency varied among exosomes derived from different cell sources, with optimal efficiency observed under these specified electroporation conditions. Importantly, this method exhibited significantly higher loading efficiency compared to conventional room temperature incubation methods for loading exosomes [189]. Treatment with exosomes isolated from plasma improved the glucose tolerance, insulin sensitivity, and reduced plasma triglyceride levels in mice [193]. Wu et al. investigated the potential of milk-derived exosomes as oral drug delivery vehicles, focusing on insulin encapsulation (EXO@INS) and their effects in type I diabetic rats. They found that EXO@INS exhibited a significantly enhanced and sustained hypoglycemic effect compared to subcutaneously injected insulin. Mechanistic studies revealed that milk-derived exosomes possess active multi-targeting uptake mechanisms, adapt to pH changes during gastrointestinal transit, activate ERK1/2 and p38 MAPK signaling pathways related to nutrient assimilation, and penetrate intestinal mucus effectively [194].

3.10. Hydroxyapatite (HAP)

Hydroxyapatite is a highly porous material, making it an excellent candidate material for drug transport, particularly in the development of nanocarriers for insulin delivery. Additionally, it has been shown to be biocompatible and bioactive. Consequently, hydroxyapatite nanoparticles ($Ca_{10}(PO_4)_6(OH)_2$) serve as a foundation for delivery systems aimed at insulin release [84,114]. PEG-functionalized HAP demonstrated effectiveness in this system [195]. Insulin encapsulated into the HAP crystal lattice showcased constant release, regulating the blood glucose levels in rats [77]. Moreover, Zhang et al. developed nano HAP nanoparticles for oral insulin delivery by surface-wrapping them with PEG and conjugating insulin (INS) and gallic acid (GA) with PEG. This innovative approach aimed to improve on traditional nanoparticle carriers by enhancing the hydrophilicity and stability in the gastrointestinal tract, addressing toxicity and biocompatibility concerns. In vivo studies in rat small intestines showed absorption of HAP-PEG-GA-INS by the epithelium, resulting in reduced blood glucose levels in type 1 diabetes rats after intragastric administration [195]. Scudeller et al. investigated insulin-loaded calcium phosphate nanoparticles (HA, SrHA, ZnHA) for oral diabetes treatment and bone cell stimulation. They found that insulin adsorption was influenced by electrostatic forces, with ZnHA enhancing adsorption and SrHA inhibiting it due to surface changes. Circular dichroism revealed insulin conformational changes, particularly pronounced on ZnHA. SrHA-loaded insulin improved cell proliferation in vitro, while HA and ZnHA had minimal effects [196].

4. Safety Considerations

The safety concerns associated with nanomedicine, including the potential toxicity, immunogenicity, and long-term effects on physiological systems, must still be carefully addressed to ensure patient safety [197]. Nanoparticles (NPs) can elicit toxicity due to their small size and high surface area, which may lead to interactions with biological molecules and cellular structures. Additionally, certain nanomaterials may provoke an immune response, causing immunogenicity issues that could compromise treatment efficacy. The long-term effects on physiological systems remain a concern, as the biodistribution and accumulation of NPs in organs over time are not yet fully understood [197]. To mitigate these risks, the FDA has recently issued guidance aimed at promoting the safe development of nanotechnology-based products for clinical use [198]. These guidelines emphasize the importance of extensive characterization of nanomaterials, understanding their intended applications, including the stability and release mechanisms, and assessing how these attributes impact product quality, safety, and efficacy, especially concerning routes of administration like insulin delivery. Additionally, the breakdown, elimination, and biodegradation of nanomaterials, initially evaluated through animal studies, are critical aspects. Furthermore, rigorous characterization of excipients is essential, given their potential impact on nano-level drug absorbency [198]. Current nanomedicine research increasingly aligns with essential safety criteria by focusing on the rigorous characterization of nanomaterials and comprehensive assessment of their applications, and by conducting thorough preclinical and clinical assessments, potential adverse effects can be identified and managed effectively, ensuring the safety and efficacy of nanomedicine formulations for diabetes management and other therapeutic applications [199].

5. Future Directions

In the field of nanomedicine for precision insulin delivery, future directions hold promise for transformative advancements aimed at addressing current limitations and optimizing therapeutic outcomes. These include challenges such as achieving sufficient oral bioavailability due to enzymatic degradation in the gastrointestinal tract and poor absorption across the intestinal epithelium, which limits the effectiveness of oral insulin formulations as a non-invasive alternative to injections. The immunogenicity of nanomaterials used in insulin delivery systems is another significant concern, potentially compromising the therapeutic benefits. Additionally, scaling up production and ensuring consistency in nanoparticle synthesis pose challenges, with variability in the size, shape, and surface properties impacting performance and safety. Researchers are exploring solutions such as protective coatings or modifications to shield insulin from degradation, surface modifications to reduce immunogenicity, and advanced manufacturing techniques to ensure uniform nanoparticle production [200]. A primary focus involves the development of next-generation nanocarriers with enhanced targeting capabilities, allowing for precise delivery of insulin to specific tissues or cells associated with diabetes pathophysiology. This includes exploring stimuli-responsive nanomaterials that can selectively release insulin in response to physiological cues, such as glucose levels or pH variations [201]. Farokhzad et al. are working on a novel approach to overcoming the gastrointestinal barriers for oral delivery of biologics by targeting nanoparticles to the FcRn in the intestines. Using insulin as a model, they aim to develop FcRn-targeted, insulin-encapsulated nanoparticles that can efficiently cross the intestinal epithelium via transcytosis, enhancing bioavailability (project number: 5R01EB015419-02) [202]. On the other hand, Li et al. aim to develop a fast-acting oral insulin formulation using milk protein casein-coated nanoparticles (casNPs) to encapsulate insulin and the absorption enhancer sodium caprate (C10). This innovative approach targets the small intestine, protecting insulin from gastric degradation and enabling enzyme-triggered release for improved bioavailability. The study involves optimizing the casNP/insulin/C10 formulation, tracking delivery and release in diabetic mice, and evaluating its efficacy in controlling hyperglycemia (project number: 1R41DK131761-01) [203]. Furthermore, Majeti et al. aim at developing an effective oral insulin delivery system using ligand-directed nanoparticles (project number: 5R01DK125372-04). This project focuses on optimizing nanoparticle chemistry to enhance insulin stability and absorption in the gastrointestinal tract. The study will evaluate the drug disposition and therapeutic outcomes under various physiological and pathological conditions, establishing a robust method for early insulin therapy in type 2 diabetic patients, potentially overcoming the drawbacks of current insulin injections and improving patient compliance and management [204].

Integrating advanced imaging modalities and biosensors into nanomedicine platforms enables real-time monitoring of the insulin distribution and therapeutic efficacy, facilitating personalized treatment strategies. This real-time feedback can significantly improve the management of insulin levels, reducing the risk of hypoglycemia and other complications [205]. The convergence of nanotechnology with emerging fields such as gene editing and regenerative medicine also holds revolutionary potential for diabetes management.

This includes engineering insulin-producing cells or tissues for transplantation, offering potential long-term solutions for insulin dependence. Moreover, there is potential in developing hybrid systems that combine nanocarriers with smart devices or wearable technology, providing continuous monitoring and automated insulin delivery systems that adjust to the user's needs dynamically [205,206]. Furthermore, in recent years, glucose-responsive insulin release nano-systems have become a significant focus of research [103]. These systems leverage glucose-sensitive elements such as glucose oxidase, concanavalin A, and phenylboronic acid to release insulin in response to hyperglycemic conditions, while minimizing or preventing insulin release under normal glucose levels [207–209]. This "smart" system primarily targets subcutaneous injection, with limited exploration in oral insulin delivery systems [210]. The "smart" glucose-responsive mechanism prevents insulin overdosage and reduces the risk of hypoglycemia. If successfully adapted for oral administration, this system could better mimic endogenous insulin secretion and distribution, thereby improving blood glucose level regulation and control, especially if incorporated with a hybrid system for automated insulin delivery [211,212]. Gu Zhen et al. are working on the development of a glucose-responsive insulin delivery system using glucose derivative-modified insulin and glucose transporters on red blood cells (project number: 7R01DK112939-02). The system will include polymeric and liposomal nanoparticles, integrated into a painless microneedle-array patch for precise blood glucose control [213]. Similarly, Li et al. propose an innovative glucose-responsive insulin delivery system that mimics pancreatic β-cells to improve diabetes care (project number: 5R01DK112939-06). Their system uses Glu-insulin interacting with GLUTs on red blood cells to release insulin during hyperglycemia. Li et al. will develop two formulations: polymeric nanoparticles coated with red blood cell membranes and liposomal nanoparticles integrated with glucose transporters. These will be incorporated into a "smart insulin patch" for up to 48 h of regulation. Their study will optimize the glucose responsiveness, effectiveness, and biocompatibility in diabetic mouse and rat models, aiming to revolutionize insulin-dependent diabetes therapy [214]. Collaborative efforts between multidisciplinary research teams, alongside sustained investment in preclinical and clinical studies, will be crucial in driving these future directions toward clinical translation. Such efforts are essential to ultimately improving outcomes for individuals living with diabetes by delivering more effective, personalized, and less invasive treatments.

6. Conclusions

In today's pharmaceutical landscape, drug delivery systems based on nanoparticles (NPs) play a crucial role. Nanocarriers are utilized to load drugs, aiming to mitigate their adverse effects while enhancing their bioavailability and efficacy. Particularly, in recent years, there has been significant interest in developing insulin delivery systems for the treatment of diabetes. NPs have demonstrated the ability to enhance insulin permeability across the intestinal mucosa by opening the tight junctions between epithelial cells. Looking ahead, it is conceivable that nanocarrier-based insulin delivery systems may eventually replace conventional methods such as subcutaneous insulin injections in the treatment of diabetic patients.

Author Contributions: Conceptualization, A.C., D.N. and R.N.; writing—original draft preparation, A.C., D.N. and R.N.; writing—review and editing, V.R., E.S., R.G., L.R. and C.S.; supervision R.M., M.M. and F.C.S. All authors have read and agreed to the published version of the manuscript.

Funding: The authors received no financial support for the research, authorship, and/or publication of this article.

Institutional Review Board Statement: The authors have reviewed the literature data and have reported results coming from studies approved by local ethics committees.

Informed Consent Statement: Not applicable.

Data Availability Statement: No dataset was generated for the publication of this article.

Acknowledgments: We wish to thank Francesca Dello Iacovo for the English revision of the manuscript.

Conflicts of Interest: The authors declare no conflicts of interest.

References

1. American Diabetes Association Professional Practice Committee. Diagnosis and Classification of Diabetes: Standards of Care in Diabetes-2024. *Diabetes Care* **2024**, *47* (Suppl. 1), S20–S42. [CrossRef]
2. Zaccardi, F.; Webb, D.R.; Yates, T.; Davies, M.J. Pathophysiology of type 1 and type 2 diabetes mellitus: A 90-year perspective. *Postgrad. Med. J.* **2016**, *92*, 63–69. [CrossRef]
3. International Diabetes Federation. *IDF Diabetes Atlas*; International Diabetes Federation: Brussels, Belgium, 2021. Available online: https://diabetesatlas.org/atlas/tenth-edition/ (accessed on 30 May 2024).
4. Shaw, J.E.; Sicree, R.A.; Zimmet, P.Z. Global estimates of the prevalence of diabetes for 2010 and 2030. *Diabetes Res. Clin. Pract.* **2010**, *87*, 4–14. [CrossRef]
5. Chan, J.C.; Malik, V.; Jia, W.; Kadowaki, T.; Yajnik, C.S.; Yoon, K.H.; Hu, F.B. Diabetes in Asia: Epidemiology, risk factors, and pathophysiology. *JAMA* **2009**, *301*, 2129–2140. [CrossRef]
6. Sasso, F.C.; Simeon, V.; Galiero, R.; Caturano, A.; De Nicola, L.; Chiodini, P.; Rinaldi, L.; Salvatore, T.; Lettieri, M.; Nevola, R.; et al. The number of risk factors not at target is associated with cardiovascular risk in a type 2 diabetic population with albuminuria in primary cardiovascular prevention. Post-hoc analysis of the NID-2 trial. *Cardiovasc. Diabetol.* **2022**, *21*, 235. [CrossRef]
7. Tomic, D.; Shaw, J.E.; Magliano, D.J. The burden and risks of emerging complications of diabetes mellitus. *Nat. Rev. Endocrinol.* **2022**, *18*, 525–539. [CrossRef]
8. Sasso, F.C.; Carbonara, O.; Cozzolino, D.; Rambaldi, P.; Mansi, L.; Torella, D.; Gentile, S.; Turco, S.; Torella, R.; Salvatore, T. Effects of insulin-glucose infusion on left ventricular function at rest and during dynamic exercise in healthy subjects and noninsulin dependent diabetic patients: A radionuclide ventriculographic study. *J. Am. Coll. Cardiol.* **2000**, *36*, 219–226. [CrossRef]
9. Sasso, F.C.; Carbonara, O.; Nasti, R.; Marfella, R.; Esposito, K.; Rambaldi, P.; Mansi, L.; Salvatore, T.; Torella, R.; Cozzolino, D. Effects of insulin on left ventricular function during dynamic exercise in overweight and obese subjects. *Eur. Heart J.* **2005**, *26*, 1205–1212. [CrossRef]
10. American Diabetes Association Professional Practice Committee. Pharmacologic Approaches to Glycemic Treatment: Standards of Care in Diabetes-2024. *Diabetes Care* **2024**, *47* (Suppl. 1), S158–S178. [CrossRef]
11. Diabetes Control and Complications Trial (DCCT)/Epidemiology of Diabetes Interventions and Complications (EDIC) Study Research Group. Mortality in Type 1 Diabetes in the DCCT/EDIC Versus the General Population. *Diabetes Care* **2016**, *39*, 1378–1383. [CrossRef]
12. Lachin, J.M.; Bebu, I.; Nathan, D.M.; DCCT/EDIC Research Group. The Beneficial Effects of Earlier Versus Later Implementation of Intensive Therapy in Type 1 Diabetes. *Diabetes Care* **2021**, *44*, 2225–2230. [CrossRef]
13. Lachin, J.M.; Nathan, D.M.; DCCT/EDIC Research Group. Understanding Metabolic Memory: The Prolonged Influence of Glycemia During the Diabetes Control and Complications Trial (DCCT) on Future Risks of Complications During the Study of the Epidemiology of Diabetes Interventions and Complications (EDIC). *Diabetes Care* **2021**, *44*, 2216–2224. [CrossRef]
14. Caturano, A.; Galiero, R.; Pafundi, P.C.; Cesaro, A.; Vetrano, E.; Palmiero, G.; Sardu, C.; Marfella, R.; Rinaldi, L.; Sasso, F.C. Does a strict glycemic control during acute coronary syndrome play a cardioprotective effect? Pathophysiology and clinical evidence. *Diabetes Res. Clin. Pract.* **2021**, *178*, 108959. [CrossRef]
15. Marfella, R.; Sasso, F.C.; Cacciapuoti, F.; Portoghese, M.; Rizzo, M.R.; Siniscalchi, M.; Carbonara, O.; Ferraraccio, F.; Torella, M.; Petrella, A.; et al. Tight glycemic control may increase regenerative potential of myocardium during acute infarction. *J. Clin. Endocrinol. Metab.* **2012**, *97*, 933–942. [CrossRef]
16. Holman, R.R.; Paul, S.K.; Bethel, M.A.; Matthews, D.R.; Neil, H.A. 10-year follow-up of intensive glucose control in type 2 diabetes. *N. Engl. J. Med.* **2008**, *359*, 1577–1589. [CrossRef]
17. Holt, R.I.G.; DeVries, J.H.; Hess-Fischl, A.; Hirsch, I.B.; Kirkman, M.S.; Klupa, T.; Ludwig, B.; Nørgaard, K.; Pettus, J.; Renard, E.; et al. The Management of Type 1 Diabetes in Adults. A Consensus Report by the American Diabetes Association (ADA) and the European Association for the Study of Diabetes (EASD). *Diabetes Care* **2021**, *44*, 2589–2625. [CrossRef]
18. Gong, Y.; Wei, T.; Liu, Y.; Wang, J.; Yan, J.; Yang, D.; Luo, S.; Weng, J.; Zheng, X. Continuous subcutaneous insulin infusion versus multiple daily injection therapy in pregnant women with type 1 diabetes. *J. Diabetes* **2024**, *16*, e13558. [CrossRef]
19. Barfar, A.; Alizadeh, H.; Masoomzadeh, S.; Javadzadeh, Y. Oral Insulin Delivery: A Review on Recent Advancements and Novel Strategies. *Curr. Drug Deliv.* **2024**, *21*, 887–900. [CrossRef]
20. Salvatore, T.; Galiero, R.; Caturano, A.; Vetrano, E.; Loffredo, G.; Rinaldi, L.; Catalini, C.; Gjeloshi, K.; Albanese, G.; Di Martino, A.; et al. Coronary Microvascular Dysfunction in Diabetes Mellitus: Pathogenetic Mechanisms and Potential Therapeutic Options. *Biomedicines* **2022**, *10*, 2274. [CrossRef]
21. Emerging Risk Factors Collaboration; Sarwar, N.; Gao, P.; Seshasai, S.R.; Gobin, R.; Kaptoge, S.; Di Angelantonio, E.; Ingelsson, E.; Lawlor, D.A.; Selvin, E.; et al. Diabetes mellitus, fasting blood glucose concentration, and risk of vascular disease: A collaborative meta-analysis of 102 prospective studies. *Lancet* **2010**, *375*, 2215–2222. [CrossRef] [PubMed]
22. Salvatore, T.; Caturano, A.; Galiero, R.; Di Martino, A.; Albanese, G.; Vetrano, E.; Sardu, C.; Marfella, R.; Rinaldi, L.; Sasso, F.C. Cardiovascular Benefits from Gliflozins: Effects on Endothelial Function. *Biomedicines* **2021**, *9*, 1356. [CrossRef]

23. Vetrano, E.; Rinaldi, L.; Mormone, A.; Giorgione, C.; Galiero, R.; Caturano, A.; Nevola, R.; Marfella, R.; Sasso, F.C. Non-alcoholic Fatty Liver Disease (NAFLD), Type 2 Diabetes, and Non-viral Hepatocarcinoma: Pathophysiological Mechanisms and New Therapeutic Strategies. *Biomedicines* **2023**, *11*, 468. [CrossRef]
24. Rao Kondapally Seshasai, S.; Kaptoge, S.; Thompson, A.; Di Angelantonio, E.; Gao, P.; Sarwar, N.; Whincup, P.H.; Mukamal, K.J.; Gillum, R.F.; Holme, I.; et al. Diabetes mellitus, fasting glucose, and risk of cause-specific death. *N. Engl. J. Med.* **2011**, *364*, 829–841. [CrossRef]
25. Galiero, R.; Caturano, A.; Vetrano, E.; Beccia, D.; Brin, C.; Alfano, M.; Di Salvo, J.; Epifani, R.; Piacevole, A.; Tagliaferri, G.; et al. Peripheral Neuropathy in Diabetes Mellitus: Pathogenetic Mechanisms and Diagnostic Options. *Int. J. Mol. Sci.* **2023**, *24*, 3554. [CrossRef]
26. Caturano, A.; D'Angelo, M.; Mormone, A.; Russo, V.; Mollica, M.P.; Salvatore, T.; Galiero, R.; Rinaldi, L.; Vetrano, E.; Marfella, R.; et al. Oxidative Stress in Type 2 Diabetes: Impacts from Pathogenesis to Lifestyle Modifications. *Curr. Issues Mol. Biol.* **2023**, *45*, 6651–6666. [CrossRef]
27. Pafundi, P.C.; Garofalo, C.; Galiero, R.; Borrelli, S.; Caturano, A.; Rinaldi, L.; Provenzano, M.; Salvatore, T.; De Nicola, L.; Minutolo, R.; et al. Role of Albuminuria in Detecting Cardio-Renal Risk and Outcome in Diabetic Subjects. *Diagnostics* **2021**, *11*, 290. [CrossRef]
28. Sasso, F.C.; Salvatore, T.; Tranchino, G.; Cozzolino, D.; Caruso, A.A.; Persico, M.; Gentile, S.; Torella, D.; Torella, R. Cochlear dysfunction in type 2 diabetes: A complication independent of neuropathy and acute hyperglycemia. *Metabolism* **1999**, *48*, 1346–1350. [CrossRef]
29. Sasso, F.C.; Pafundi, P.C.; Gelso, A.; Bono, V.; Costagliola, C.; Marfella, R.; Sardu, C.; Rinaldi, L.; Galiero, R.; Acierno, C.; et al. High HDL cholesterol: A risk factor for diabetic retinopathy? Findings from NO BLIND study. *Diabetes Res. Clin. Pract.* **2019**, *150*, 236–244. [CrossRef]
30. Sasso, F.C.; Pafundi, P.C.; Gelso, A.; Bono, V.; Costagliola, C.; Marfella, R.; Sardu, C.; Rinaldi, L.; Galiero, R.; Acierno, C.; et al. Relationship between albuminuric CKD and diabetic retinopathy in a real-world setting of type 2 diabetes: Findings from No blind study. *Nutr. Metab. Cardiovasc. Dis.* **2019**, *29*, 923–930. [CrossRef]
31. Tricco, A.C.; Ashoor, H.M.; Antony, J.; Beyene, J.; Veroniki, A.A.; Isaranuwatchai, W.; Harrington, A.; Wilson, C.; Tsouros, S.; Soobiah, C.; et al. Safety, effectiveness, and cost effectiveness of long acting versus intermediate acting insulin for patients with type 1 diabetes: Systematic review and network meta-analysis. *BMJ* **2014**, *349*, g5459. [CrossRef]
32. Bartley, P.C.; Bogoev, M.; Larsen, J.; Philotheou, A. Long-term efficacy and safety of insulin detemir compared to Neutral Protamine Hagedorn insulin in patients with Type 1 diabetes using a treat-to-target basal-bolus regimen with insulin aspart at meals: A 2-year, randomized, controlled trial. *Diabet. Med.* **2008**, *25*, 442–449. [CrossRef]
33. DeWitt, D.E.; Hirsch, I.B. Outpatient insulin therapy in type 1 and type 2 diabetes mellitus: Scientific review. *JAMA* **2003**, *289*, 2254–2264. [CrossRef]
34. Heise, T.; Pieber, T.R.; Danne, T.; Erichsen, L.; Haahr, H. A Pooled Analysis of Clinical Pharmacology Trials Investigating the Pharmacokinetic and Pharmacodynamic Characteristics of Fast-Acting Insulin Aspart in Adults with Type 1 Diabetes. *Clin. Pharmacokinet.* **2017**, *56*, 551–559. [CrossRef]
35. Aronson, R.; Biester, T.; Leohr, J.; Pollom, R.; Linnebjerg, H.; LaBell, E.S.; Zhang, Q.; Coutant, D.E.; Danne, T. Ultra rapid lispro showed greater reduction in postprandial glucose versus Humalog in children, adolescents and adults with type 1 diabetes mellitus. *Diabetes Obes. Metab.* **2023**, *25*, 1964–1972. [CrossRef]
36. Klaff, L.; Cao, D.; Dellva, M.A.; Tobian, J.; Miura, J.; Dahl, D.; Lucas, J.; Bue-Valleskey, J. Ultra rapid lispro improves postprandial glucose control compared with lispro in patients with type 1 diabetes: Results from the 26-week PRONTO-T1D study. *Diabetes Obes. Metab.* **2020**, *22*, 1799–1807. [CrossRef]
37. Bode, B.W.; McGill, J.B.; Lorber, D.L.; Gross, J.L.; Chang, P.C.; Bregman, D.B.; Affinity 1 Study Group. Inhaled Technosphere Insulin Compared with Injected Prandial Insulin in Type 1 Diabetes: A Randomized 24-Week Trial. *Diabetes Care* **2015**, *38*, 2266–2273. [CrossRef]
38. Russell-Jones, D.; Bode, B.W.; De Block, C.; Franek, E.; Heller, S.R.; Mathieu, C.; Philis-Tsimikas, A.; Rose, L.; Woo, V.C.; Østerskov, A.B.; et al. Fast-Acting Insulin Aspart Improves Glycemic Control in Basal-Bolus Treatment for Type 1 Diabetes: Results of a 26-Week Multicenter, Active-Controlled, Treat-to-Target, Randomized, Parallel-Group Trial (onset 1). *Diabetes Care* **2017**, *40*, 943–950. [CrossRef]
39. Caturano, A.; Galiero, R.; Pafundi, P.C. Metformin for Type 2 Diabetes. *JAMA* **2019**, *322*, 1312. [CrossRef]
40. Starlin Chellathurai, M.; Mahmood, S.; Mohamed Sofian, Z.; Hee, C.W.; Sundarapandian, R.; Ahamed, H.N.; Kandasamy, C.S.; Hilles, A.R.; Hashim, N.M.; Janakiraman, A.K. Biodegradable polymeric insulin microneedles—A design and materials perspective review. *Drug Deliv.* **2024**, *31*, 2296350. [CrossRef]
41. Galiero, R.; Caturano, A.; Vetrano, E.; Monda, M.; Marfella, R.; Sardu, C.; Salvatore, T.; Rinaldi, L.; Sasso, F.C. Precision Medicine in Type 2 Diabetes Mellitus: Utility and Limitations. *Diabetes Metab. Syndr. Obes.* **2023**, *16*, 3669–3689. [CrossRef]
42. Tan, S.Y.; Mei Wong, J.L.; Sim, Y.J.; Wong, S.S.; Mohamed Elhassan, S.A.; Tan, S.H.; Ling Lim, G.P.; Rong Tay, N.W.; Annan, N.C.; Bhattamisra, S.K.; et al. Type 1 and 2 diabetes mellitus: A review on current treatment approach and gene therapy as potential intervention. *Diabetes Metab. Syndr.* **2019**, *13*, 364–372. [CrossRef]
43. Souto, E.B.; Souto, S.B.; Campos, J.R.; Severino, P.; Pashirova, T.N.; Zakharova, L.Y.; Silva, A.M.; Durazzo, A.; Lucarini, M.; Izzo, A.A.; et al. Nanoparticle Delivery Systems in the Treatment of Diabetes Complications. *Molecules* **2019**, *24*, 4209. [CrossRef]

44. Mayer, J.P.; Zhang, F.; Di Marchi, R.D. Insulin structure and function. *Biopolymers* **2007**, *88*, 687–713. [CrossRef]
45. Sims, E.K.; Carr, A.L.J.; Oram, R.A.; Di Meglio, L.A.; Evans-Molina, C. 100 years of insulin: Celebrating the past, present and future of diabetes therapy. *Nat. Med.* **2021**, *27*, 1154–1164. [CrossRef]
46. Home, P.; Itzhak, B. Is Insulin Therapy Safe? *Am. J. Ther.* **2020**, *27*, e106–e114. [CrossRef]
47. Sharma, G.; Sharma, A.R.; Nam, J.S.; Doss, G.P.; Lee, S.S.; Chakraborty, C. Nanoparticle based insulin delivery system: The next generation efficient therapy for Type 1 diabetes. *J. Nanobiotechnol.* **2015**, *13*, 74. [CrossRef]
48. Chatterjee, S.; Bhushan Sharma, C.; Lavie, C.J.; Adhikari, A.; Deedwania, P.; O'keefe, J.H. Oral insulin: An update. *Minerva Endocrinol.* **2020**, *45*, 49–60. [CrossRef]
49. Liu, J.; Hirschberg, C.; Fanø, M.; Mu, H.; Müllertz, A. Evaluation of self-emulsifying drug delivery systems for oral insulin delivery using an in vitro model simulating the intestinal proteolysis. *Eur. J. Pharm. Sci.* **2020**, *147*, 105272. [CrossRef]
50. Yazdi, J.R.; Tafaghodi, M.; Sadri, K.; Mashreghi, M.; Nikpoor, A.R.; Nikoofal-Sahlabadi, S.; Chamani, J.; Vakili, R.; Moosavian, S.A.; Jaafari, M.R. Folate targeted PEGylated liposomes for the oral delivery of insulin: In vitro and in vivo studies. *Colloids Surf. B Biointerfaces* **2020**, *194*, 111203. [CrossRef]
51. Yamamura, R.; Inoue, K.Y.; Nishino, K.; Yamasaki, S. Intestinal and fecal pH in human health. *Front. Microbiomes* **2023**, *2*, 1192316. [CrossRef]
52. Xiao, Y.; Tang, Z.; Wang, J.; Liu, C.; Kong, N.; Farokhzad, O.C.; Tao, W. Oral Insulin Delivery Platforms: Strategies To Address the Biological Barriers. *Angew. Chem. Int. Ed. Engl.* **2020**, *59*, 19787–19795. [CrossRef]
53. Schoultz, I.; Keita, Å.V. The Intestinal Barrier and Current Techniques for the Assessment of Gut Permeability. *Cells* **2020**, *9*, 1909. [CrossRef]
54. Lock, J.Y.; Carlson, T.L.; Carrier, R.L. Mucus models to evaluate the diffusion of drugs and particles. *Adv. Drug Deliv. Rev.* **2018**, *124*, 34–49. [CrossRef]
55. Murgia, X.; Loretz, B.; Hartwig, O.; Hittinger, M.; Lehr, C.M. The role of mucus on drug transport and its potential to affect therapeutic outcomes. *Adv. Drug Deliv. Rev.* **2018**, *124*, 82–97. [CrossRef]
56. Ensign, L.M.; Cone, R.; Hanes, J. Oral drug delivery with polymeric nanoparticles: The gastrointestinal mucus barriers. *Adv. Drug Deliv. Rev.* **2012**, *64*, 557–570. [CrossRef]
57. Kumar, V.; Choudhry, I.; Namdev, A.; Mishra, S.; Soni, S.; Hurkat, P.; Jain, A.; Jain, D. Oral Insulin: Myth or Reality. *Curr. Diabetes Rev.* **2018**, *14*, 497–508. [CrossRef]
58. Xu, Y.; Shrestha, N.; Préat, V.; Beloqui, A. Overcoming the intestinal barrier: A look into targeting approaches for improved oral drug delivery systems. *J. Control. Release* **2020**, *322*, 486–508. [CrossRef]
59. des Rieux, A.; Pourcelle, V.; Cani, P.D.; Marchand-Brynaert, J.; Préat, V. Targeted nanoparticles with novel non-peptidic ligands for oral delivery. *Adv. Drug Deliv. Rev.* **2013**, *65*, 833–844. [CrossRef]
60. Petrus, A.K.; Fairchild, T.J.; Doyle, R.P. Traveling the vitamin B12 pathway: Oral delivery of protein and peptide drugs. *Angew. Chem. Int. Ed. Engl.* **2009**, *48*, 1022–1028. [CrossRef]
61. Chalasani, K.B.; Russell-Jones, G.J.; Yandrapu, S.K.; Diwan, P.V.; Jain, S.K. A novel vitamin B12-nanosphere conjugate carrier system for peroral delivery of insulin. *J. Control. Release* **2007**, *117*, 421–429. [CrossRef]
62. Chalasani, K.B.; Russell-Jones, G.J.; Jain, A.K.; Diwan, P.V.; Jain, S.K. Effective oral delivery of insulin in animal models using vitamin B12-coated dextran nanoparticles. *J. Control. Release* **2007**, *122*, 141–150. [CrossRef]
63. Zhang, X.; Wu, W. Ligand-mediated active targeting for enhanced oral absorption. *Drug Discov. Today* **2014**, *19*, 898–904. [CrossRef]
64. Gedawy, A.; Martinez, J.; Al-Salami, H.; Dass, C.R. Oral insulin delivery: Existing barriers and current counter-strategies. *J. Pharm. Pharmacol.* **2018**, *70*, 197–213. [CrossRef]
65. Anderson, K.E.; Stevenson, B.R.; Rogers, J.A. Folic acid-PEO-labeled liposomes to improve gastrointestinal absorption of encapsulated agents. *J. Control. Release* **1999**, *60*, 189–198. [CrossRef]
66. Agrawal, A.K.; Harde, H.; Thanki, K.; Jain, S. Improved stability and antidiabetic potential of insulin containing folic acid functionalized polymer stabilized multilayered liposomes following oral administration. *Biomacromolecules* **2014**, *15*, 350–360. [CrossRef]
67. Anderson, K.E.; Eliot, L.A.; Stevenson, B.R.; Rogers, J.A. Formulation and evaluation of a folic acid receptor-targeted oral vancomycin liposomal dosage form. *Pharm. Res.* **2001**, *18*, 316–322. [CrossRef]
68. Guo, S.; Li, H. Chitosan-Derived Nanocarrier Polymers for Drug Delivery and pH-Controlled Release in Type 2 Diabetes Treatment. *J. Fluoresc.* **2024**. [CrossRef]
69. Heyns, I.M.; Ganugula, R.; Varma, T.; Allamreddy, S.; Kumar, N.; Garg, P.; Kumar, M.N.V.R.; Arora, M. Rationally Designed Naringenin-Conjugated Polyester Nanoparticles Enable Folate Receptor-Mediated Peroral Delivery of Insulin. *ACS Appl. Mater. Interfaces* **2023**, *15*, 45651–45657. [CrossRef]
70. Nabi-Afjadi, M.; Ostadhadi, S.; Liaghat, M.; Pasupulla, A.P.; Masoumi, S.; Aziziyan, F.; Zalpoor, H.; Abkhooie, L.; Tarhriz, V. Revolutionizing type 1 diabetes management: Exploring oral insulin and adjunctive treatments. *Biomed. Pharmacother.* **2024**, *176*, 116808. [CrossRef]
71. Al Qaraghuli, M.M.; Kubiak-Ossowska, K.; Ferro, V.A.; Mulheran, P.A. Exploiting the Fc base of IgG antibodies to create functional nanoparticle conjugates. *Sci. Rep.* **2024**, *14*, 14832. [CrossRef]

72. Martins, J.P.; Kennedy, P.J.; Santos, H.A.; Barrias, C.; Sarmento, B. A comprehensive review of the neonatal Fc receptor and its application in drug delivery. *Pharmacol. Ther.* **2016**, *161*, 22–39. [CrossRef]
73. Toh, W.H.; Louber, J.; Mahmoud, I.S.; Chia, J.; Bass, G.T.; Dower, S.K.; Verhagen, A.M.; Gleeson, P.A. FcRn mediates fast recycling of endocytosed albumin and IgG from early macropinosomes in primary macrophages. *J. Cell Sci.* **2019**, *133*, jcs235416. [CrossRef]
74. Sockolosky, J.T.; Szoka, F.C. The neonatal Fc receptor, FcRn, as a target for drug delivery and therapy. *Adv. Drug Deliv. Rev.* **2015**, *91*, 109–124. [CrossRef]
75. Pridgen, E.M.; Alexis, F.; Kuo, T.T.; Levy-Nissenbaum, E.; Karnik, R.; Blumberg, R.S.; Langer, R.; Farokhzad, O.C. Transepithelial transport of Fc-targeted nanoparticles by the neonatal fc receptor for oral delivery. *Sci. Transl. Med.* **2013**, *5*, 213ra167. [CrossRef]
76. Martins, J.P.; Figueiredo, P.; Wang, S.; Espo, E.; Celi, E.; Martins, B.; Kemell, M.; Moslova, K.; Mäkilä, E.; Salonen, J.; et al. Neonatal Fc receptor-targeted lignin-encapsulated porous silicon nanoparticles for enhanced cellular interactions and insulin permeation across the intestinal epithelium. *Bioact. Mater.* **2021**, *9*, 299–315. [CrossRef]
77. Wang, M.; Wang, C.; Ren, S.; Pan, J.; Wang, Y.; Shen, Y.; Zeng, Z.; Cui, H.; Zhao, X. Versatile Oral Insulin Delivery Nanosystems: From Materials to Nanostructures. *Int. J. Mol. Sci.* **2022**, *23*, 3362. [CrossRef]
78. Villena Gonzales, W.; Mobashsher, A.T.; Abbosh, A. The Progress of Glucose Monitoring-A Review of Invasive to Minimally and Non-Invasive Techniques, Devices and Sensors. *Sensors* **2019**, *19*, 800. [CrossRef]
79. Prasad, M.; Lambe, U.P.; Brar, B.; Shah, I.; Manimegalai, J.; Ranjan, K.; Rao, R.; Kumar, S.; Mahant, S.; Khurana, S.K.; et al. Nanotherapeutics: An insight into healthcare and multi-dimensional applications in medical sector of the modern world. *Biomed. Pharmacother.* **2018**, *97*, 1521–1537. [CrossRef]
80. Lemmerman, L.R.; Das, D.; Higuita-Castro, N.; Mirmira, R.G.; Gallego-Perez, D. Nanomedicine-Based Strategies for Diabetes: Diagnostics, Monitoring, and Treatment. *Trends Endocrinol. Metab.* **2020**, *31*, 448–458. [CrossRef]
81. Lee, S.H.; Park, S.Y.; Choi, C.S. Insulin Resistance: From Mechanisms to Therapeutic Strategies. *Diabetes Metab. J.* **2022**, *46*, 15–37. [CrossRef]
82. Kerry, R.G.; Mahapatra, G.P.; Maurya, G.K.; Patra, S.; Mahari, S.; Das, G.; Patra, J.K.; Sahoo, S. Molecular prospect of type-2 diabetes: Nanotechnology based diagnostics and therapeutic intervention. *Rev. Endocr. Metab. Disord.* **2021**, *22*, 421–451. [CrossRef] [PubMed]
83. He, Y.; Al-Mureish, A.; Wu, N. Nanotechnology in the Treatment of Diabetic Complications: A Comprehensive Narrative Review. *J. Diabetes Res.* **2021**, *2021*, 6612063. [CrossRef] [PubMed]
84. Norouzi, P.; Rastegari, A.A.; Mottaghitalab, F.; Farokhi, M.; Zarrintaj, P.; Saeb, M.R. Nanoemulsions for intravenous drug delivery. In *Nanoengineered Biomaterials for Advanced Drug Delivery*; Elsevier: Amsterdam, The Netherlands, 2020; Volume 1, pp. 581–601.
85. He, H.; Lu, Y.; Qi, J.; Zhao, W.; Dong, X.; Wu, W. Biomimetic thiamine- and niacin-decorated liposomes for enhanced oral delivery of insulin. *Acta Pharm. Sin. B* **2018**, *8*, 97–105. [CrossRef] [PubMed]
86. Volpatti, L.R.; Matranga, M.A.; Cortinas, A.B.; Delcassian, D.; Daniel, K.B.; Langer, R.; Anderson, D.G. Glucose-Responsive Nanoparticles for Rapid and Extended Self-Regulated Insulin Delivery. *ACS Nano* **2020**, *14*, 488–497. [CrossRef] [PubMed]
87. Agazzi, M.L.; Herrera, S.E.; Cortez, M.L.; Marmisollé, W.A.; Tagliazucchi, M.; Azzaroni, O. Insulin Delivery from Glucose-Responsive, Self-Assembled, Polyamine Nanoparticles: Smart 'Sense-and-Treat' Nanocarriers Made Easy. *Chemistry* **2020**, *26*, 2456–2463. [CrossRef] [PubMed]
88. Alai, M.S.; Lin, W.J.; Pingale, S.S. Application of polymeric nanoparticles and micelles in insulin oral delivery. *J. Food Drug Anal.* **2015**, *23*, 351–358. [CrossRef] [PubMed]
89. Zhang, X.; Qi, J.; Lu, Y.; Hu, X.; He, W.; Wu, W. Enhanced hypoglycemic effect of biotin-modified liposomes loading insulin: Effect of formulation variables, intracellular trafficking, and cytotoxicity. *Nanoscale Res. Lett.* **2014**, *9*, 185. [CrossRef] [PubMed]
90. Wang, A.H.; Yang, T.T.; Fan, W.W.; Yang, Y.W.; Zhu, Q.L.; Guo, S.; Zhu, C.; Yuan, Y.; Zhang, T.; Gan, Y. Protein corona liposomes achieve efficient oral insulin delivery by overcoming mucus and epithelial barriers. *Adv. Healthc. Mater.* **2019**, *8*, e1801123. [CrossRef]
91. Huang, Y.Y.; Wang, C.H. Pulmonary delivery of insulin by liposomal carriers. *J. Control. Release* **2007**, *113*, 9–14. [CrossRef]
92. Karathanasis, E.; Bhavane, R.; Annapragada, A.V. Triggered release of inhaled insulin from the agglomerated vesicles: Pharmacodynamic studies in rats. *J. Control. Release* **2006**, *113*, 117–127. [CrossRef]
93. Jain, A.K.; Chalasani, K.B.; Khar, R.K.; Ahmed, F.J.; Diwan, P.V. Muco-adhesive multivesicular liposomes as an effective carrier for transmucosal insulin delivery. *J. Drug Target.* **2007**, *15*, 417–427. [CrossRef] [PubMed]
94. Pardakhty, A.; Varshosaz, J.; Rouholamini, A. In vitro study of polyoxyethylene alkyl ether niosomes for delivery of insulin. *Int. J. Pharm.* **2007**, *328*, 130–141. [CrossRef] [PubMed]
95. Hopkins, G.V.; Cochrane, S.; Onion, D.; Fairclough, L.C. The role of lipids in allergic sensitization: A systematic review. *Front. Mol. Biosci.* **2022**, *9*, 832330–832419. [CrossRef] [PubMed]
96. Zhang, Z.-H.; Zhang, Y.-L.; Zhou, J.-P.; Lv, H.-X. Solid lipid nanoparticles modified with stearic acid–octaarginine for oral administration of insulin. *Int. J. Nanomed.* **2012**, *7*, 3333–3339.
97. Wong, C.Y.; Al-Salami, H.; Dass, C.R. Potential of insulin nanoparticle formulations for oral delivery and diabetes treatment. *J. Control. Release* **2017**, *264*, 247–275. [CrossRef] [PubMed]
98. He, H.; Lu, Y.; Qi, J.; Zhu, Q.; Chen, Z.; Wu, W. Adapting Liposomes for Oral Drug Delivery. *Acta Pharm. Sin. B* **2019**, *9*, 36–48. [CrossRef] [PubMed]

99. Kondiah, P.P.; Choonara, Y.E.; Tomar, L.K.; Tyagi, C.; Kumar, P.; du Toit, L.C.; Marimuthu, T.; Modi, G.; Pillay, V. Development of a gastric absorptive, immediate responsive, oral protein-loaded versatile polymeric delivery system. *AAPS PharmSciTech* **2017**, *1*, 2479–2493. [CrossRef] [PubMed]
100. Shalaby, T.I.; El-Refaie, W.M. Bioadhesive Chitosan-Coated Cationic Nanoliposomes with Improved Insulin Encapsulation and Prolonged Oral Hypoglycemic Effect in Diabetic Mice. *J. Pharm. Sci.* **2018**, *107*, 2136–2143. [CrossRef] [PubMed]
101. Barbosa, F.C.; Silva, M.C.d.; Silva, H.N.d.; Albuquerque, D.; Gomes, A.A.R.; Silva, S.M.d.L.; Fook, M.V.L. Progress in the Development of Chitosan Based Insulin Delivery Systems: A Systematic Literature Review. *Polymers* **2020**, *12*, 2499. [CrossRef]
102. Sheng, J.Y.; Han, L.M.; Qin, J.; Ru, G.; Li, R.X.; Wu, L.H.; Cui, D.; Yang, P.; He, Y.; Wang, J. N-trimethyl chitosan chloride-coated PLGA nanoparticles overcoming multiple barriers to oral insulin absorption. *ACS Appl. Mater. Interfaces* **2015**, *7*, 15430–15441. [CrossRef]
103. Pang, H.; Huang, X.; Xu, Z.P.; Chen, C.; Han, F.Y. Progress in oral insulin delivery by PLGA nanoparticles for the management of diabetes. *Drug Discov. Today* **2023**, *28*, 103393. [CrossRef]
104. Carino, G.P.; Mathiowitz, E. Oral insulin delivery. *Adv. Drug Deliv. Rev.* **1999**, *35*, 249–257. [CrossRef] [PubMed]
105. Hajebi, S.; Rabiee, N.; Bagherzadeh, M.; Ahmadi, S.; Rabiee, M.; Roghani-Mamaqani, H.; Tahriri, M.; Tayebi, L.; Hamblin, M.R. Stimulus-responsive polymeric nanogels as smart drug delivery systems. *Acta Biomater.* **2019**, *92*, 1–18. [CrossRef] [PubMed]
106. Wu, W.; Mitra, N.; Yan, E.C.; Zhou, S. Multifunctional hybrid nanogel for integration of optical glucose sensing and self-regulated insulin release at physiological pH. *ACS Nano* **2010**, *4*, 4831–4839. [CrossRef] [PubMed]
107. Shahid, N.; Erum, A.; Hanif, S.; Malik, N.S.; Tulain, U.R.; Syed, M.A. Nanocomposite Hydrogels-A Promising Approach towards Enhanced Bioavailability and Controlled Drug Delivery. *Curr. Pharm. Des.* **2024**, *30*, 48–62. [CrossRef]
108. Suhail, M.; Rosenholm, J.M.; Minhas, M.U.; Badshah, S.F.; Naeem, A.; Khan, K.U.; Fahad, M. Nanogels as drug-delivery systems: A comprehensive overview. *Ther. Deliv.* **2019**, *10*, 697–717. [CrossRef] [PubMed]
109. Johnson, O.L.; Jaworowicz, W.; Cleland, J.L.; Bailey, L.; Charnis, M.; Duenas, E.; Wu, C.C.; Shepard, D.; Magil, S.; Last, T.; et al. The stabilization and encapsulation of human growth hormone into biodegradable microspheres. *Pharm. Res.* **1997**, *14*, 730–735. [CrossRef]
110. Feng, S.S.; Mu, L.; Win, K.Y.; Huang, G. Nanoparticles of biodegradable polymers for clinical administration of paclitaxel. *Curr. Med. Chem.* **2004**, *11*, 413–424. [CrossRef]
111. Kotta, S.; Aldawsari, H.M.; Badr-Eldin, S.M.; Nair, A.B.; YT, K. Progress in Polymeric Micelles for Drug Delivery Applications. *Pharmaceutics* **2022**, *14*, 1636. [CrossRef]
112. Gaucher, G.; Satturwar, P.; Jones, M.C.; Furtos, A.; Leroux, J.C. Polymeric micelles for oral drug delivery. *Eur. J. Pharm. Biopharm.* **2010**, *76*, 147–158. [CrossRef]
113. Majumder, N.; G Das, N.; Das, S.K. Polymeric micelles for anticancer drug delivery. *Ther. Deliv.* **2020**, *11*, 613–635. [CrossRef] [PubMed]
114. Shyong, Y.J.; Tsai, C.C.; Lin, R.F.; Soung, H.S.; Hsieh, H.C.; Hsueh, Y.S.; Chang, K.C.; Lin, F.H. Insulin loaded hydroxyapatite combined with macrophage activity to deliver insulin for diabetes mellitus. *J. Mater. Chem. B* **2015**, *3*, 2331–2340. [CrossRef] [PubMed]
115. Lara-Ochoa, S.; Ortega-Lara, W.; Guerrero-Beltrán, C.E. Hydroxyapatite Nanoparticles in Drug Delivery: Physicochemistry and Applications. *Pharmaceutics* **2021**, *13*, 1642. [CrossRef] [PubMed]
116. Yang, Y.H.; Liu, C.H.; Liang, Y.H.; Lin, F.H.; Wu, K.C. Hollow mesoporous hydroxyapatite nanoparticles (hmHANPs) with enhanced drug loading and pH-responsive release properties for intracellular drug delivery. *J. Mater. Chem. B* **2013**, *1*, 2447–2450. [CrossRef] [PubMed]
117. Gu, Z.; Aimetti, A.A.; Wang, Q.; Dang, T.T.; Zhang, Y.L.; Veiseh, O.; Cheng, H.; Langer, R.S.; Anderson, D.G. Injectable nano-network for glucose-mediated insulin delivery. *ACS Nano* **2017**, *7*, 4194–4201. [CrossRef] [PubMed]
118. Yao, Y.; Zhou, Y.; Liu, L.; Xu, Y.; Chen, Q.; Wang, Y.; Wu, S.; Deng, Y.; Zhang, J.; Shao, A. Nanoparticle-Based Drug Delivery in Cancer Therapy and Its Role in Overcoming Drug Resistance. *Front. Mol. Biosci.* **2020**, *7*, 193. [CrossRef] [PubMed]
119. Petrovici, A.R.; Pinteala, M.; Simionescu, N. Dextran Formulations as Effective Delivery Systems of Therapeutic Agents. *Molecules* **2023**, *28*, 1086. [CrossRef] [PubMed]
120. Danhier, F.; Ansorena, E.; Silva, J.M.; Coco, R.; Le Breton, A.; Préat, V. PLGA-based nanoparticles: An overview of biomedical applications. *J. Control. Release* **2012**, *161*, 505–522. [CrossRef] [PubMed]
121. Perkins, G.B.; Tunbridge, M.J.; Hurtado, P.R.; Zuiani, J.; Mhatre, S.; Yip, K.H.; Le, T.T.A.; Yuson, C.; Kette, F.; Hissaria, P. PEGylated Liposomes for Diagnosis of Polyethylene Glycol Allergy. *J. Allergy Clin. Immunol.* **2024**. [CrossRef] [PubMed]
122. Wood, K.M.; Stone, G.M.; Peppas, N.A. Wheat germ agglutinin functionalized complexation hydrogels for oral insulin delivery. *Biomacromolecules* **2008**, *9*, 1293–1298. [CrossRef]
123. Mansoor, S.; Kondiah, P.P.D.; Choonara, Y.E. Advanced Hydrogels for the Controlled Delivery of Insulin. *Pharmaceutics* **2021**, *13*, 2113. [CrossRef]
124. Ning, M.; Guo, Y.; Pan, H.; Yu, H.; Gu, Z. Niosomes with Sorbitan Monoester as a carrier for vaginal delivery of insulin: Studies in rats. *Drug Deliv.* **2005**, *12*, 399–407. [CrossRef]
125. Zhao, R.; Lu, Z.; Yang, J.; Zhang, L.; Li, Y.; Zhang, X. Drug Delivery System in the Treatment of Diabetes Mellitus. *Front. Bioeng. Biotechnol.* **2020**, *8*, 880. [CrossRef]

126. Zuben, E.D.; Eloy, J.O.; Araújo, V.H.; Gremião, M.P.; Chorilli, M. Insulin-loaded liposomes functionalized with cell-penetrating peptides: Influence on drug release and permeation through porcine nasal mucosa. *Colloids Surf. A Physicochem. Eng. Asp.* **2021**, *622*, 126624. [CrossRef]
127. Dawoud, M.H.S.; Yassin, G.E.; Ghorab, D.M.; Morsi, N.M. Insulin Mucoadhesive Liposomal Gel for Wound Healing: A Formulation with Sustained Release and Extended Stability Using Quality by Design Approach. *AAPS PharmSciTech* **2019**, *20*, 158. [CrossRef]
128. Bashyal, S.; Seo, J.-E.; Keum, T.; Noh, G.; Lamichhane, S.; Lee, S. Development, Characterization, and Ex Vivo Assessment of Elastic Liposomes for Enhancing the Buccal Delivery of Insulin. *Pharmaceutics* **2021**, *13*, 565. [CrossRef]
129. Liu, G.; He, S.; Ding, Y.; Chen, C.; Cai, Q.; Zhou, W. Multivesicular Liposomes for Glucose-Responsive Insulin Delivery. *Pharmaceutics* **2021**, *14*, 21. [CrossRef]
130. Shafiq, S.; Siddiq Abduh, M.; Iqbal, F.; Kousar, K.; Anjum, S.; Ahmad, T. A novel approach to insulin delivery via oral route: Milk fat globule membrane derived liposomes as a delivery vehicle. *Saudi J. Biol. Sci.* **2024**, *31*, 103945. [CrossRef]
131. Wu, H.; Nan, J.; Yang, L.; Park, H.J.; Li, J. Insulin-loaded liposomes packaged in alginate hydrogels promote the oral bioavailability of insulin. *J. Control. Release* **2023**, *353*, 51–62. [CrossRef]
132. Skovstrup, S.; Hansen, S.; Skrydstrup, T.; Schiøtt, B. Conformational flexibility of chitosan: A molecular modeling study. *Biomacromolecules* **2010**, *11*, 3196–3207. [CrossRef]
133. Rinaudo, M. Chitin and chitosan: Properties and applications. *Prog. Polym. Sci.* **2006**, *31*, 603–632. [CrossRef]
134. El Leithy, E.S.; Abdel-Bar, H.M.; Ali, R.A. Folate-chitosan nanoparticles triggered insulin cellular uptake and improved in vivo hypoglycemic activity. *Int. J. Pharm.* **2019**, *571*, 118708. [CrossRef]
135. Song, M.; Wang, H.; Chen, K.; Zhang, S.; Yu, L.; Elshazly, E.H.; Ke, L.; Gong, R. Oral insulin delivery by carboxymethyl-β-cyclodextrin-grafted chitosan nanoparticles for improving diabetic treatment. *Artif. Cells Nanomed. Biotechnol.* **2018**, *46* (Suppl. 3), 774–782. [CrossRef]
136. Agrawal, A.K.; Urimi, D.; Harde, H.; Kushwah, V.; Jain, S. Folate appended chitosan nanoparticles augment the stability, bioavailability and efficacy of insulin in diabetic rats following oral administration. *RSC Adv.* **2015**, *5*, 105179–105193. [CrossRef]
137. Sharma, D.; Arora, S.; Singh, J. Smart Thermosensitive Copolymer Incorporating Chitosan-Zinc-Insulin Electrostatic Complexes for Controlled Delivery of Insulin: Effect of Chitosan Chain Length. *Int. J. Polym. Mater.* **2020**, *69*, 1054–1068. [CrossRef]
138. Tzeng, H.-P.; Liu, S.-H.; Chiang, M.-T. Antidiabetic Properties of Chitosan and Its Derivatives. *Mar. Drugs.* **2022**, *20*, 784. [CrossRef]
139. Yuan, W.P.; Liu, B.; Liu, C.H.; Wang, X.J.; Zhang, M.S.; Meng, X.M.; Xia, X.K. Antioxidant activity of chito-oligosaccharides on pancreatic islet cells in streptozotocin-induced diabetes in rats. *World J. Gastroenterol.* **2009**, *15*, 1339–1345. [CrossRef]
140. Ju, C.; Yue, W.; Yang, Z.; Zhang, Q.; Yang, X.; Liu, Z.; Zhang, F. Antidiabetic effect and mechanism of chitooligosaccharides. *Biol. Pharm. Bull.* **2010**, *33*, 1511–1516. [CrossRef]
141. Kim, H.J.; Ahn, H.Y.; Kwak, J.H.; Shin, D.Y.; Kwon, Y.I.; Oh, C.G.; Lee, J.H. The effects of chitosan oligosaccharide (GO2KA1) supplementation on glucose control in subjects with prediabetes. *Food Funct.* **2014**, *5*, 2662–2669. [CrossRef]
142. Pang, H.; Wu, Y.; Chen, Y.; Chen, C.; Nie, X.; Li, P.; Huang, G.; Xu, Z.P.; Han, F.Y. Development of polysaccharide-coated layered double hydroxide nanocomposites for enhanced oral insulin delivery. *Drug Deliv. Transl. Res.* **2024**. [CrossRef]
143. Abd-Alhussain, G.K.; Alatrakji, M.Q.Y.M.; Ahmed, S.J.; Fawzi, H.A. Efficacy of oral insulin nanoparticles for the management of hyperglycemia in a rat model of diabetes induced with streptozotocin. *J. Med. Life* **2024**, *17*, 217–225. [CrossRef]
144. Maurya, R.; Ramteke, S.; Jain, N.K. Quality by design (QbD) approach-based development of optimized nanocarrier to achieve quality target product profile (QTPP)-targeted lymphatic delivery. *Nanotechnology* **2024**, *35*, 265101. [CrossRef] [PubMed]
145. Chamsai, B.; Opanasopit, P.; Samprasit, W. Fast disintegrating dosage forms of mucoadhesive-based nanoparticles for oral insulin delivery: Optimization to in vivo evaluation. *Int. J. Pharm.* **2023**, *647*, 123513. [CrossRef] [PubMed]
146. Pratap-Singh, A.; Guo, Y.; Baldelli, A.; Singh, A. Mercaptonicotinic acid activated thiolated chitosan (MNA-TG-chitosan) to enable peptide oral delivery by opening cell tight junctions and enhancing transepithelial transport. *Sci. Rep.* **2023**, *13*, 17343. [CrossRef] [PubMed]
147. Cui, Z.; Cui, S.; Qin, L.; An, Y.; Zhang, X.; Guan, J.; Wong, T.W.; Mao, S. Comparison of virus-capsid mimicking biologic-shell based versus polymeric-shell nanoparticles for enhanced oral insulin delivery. *Asian J. Pharm. Sci.* **2023**, *18*, 100848. [CrossRef]
148. AlSalem, H.S.; Abdulsalam, N.M.; Khateeb, N.A.; Binkadem, M.S.; Alhadhrami, N.A.; Khedr, A.M.; Abdelmonem, R.; Shoueir, K.R.; Nadwa, E.H. Enhance the oral insulin delivery route using a modified chitosan-based formulation fabricated by microwave. *Int. J. Biol. Macromol.* **2023**, *247*, 125779. [CrossRef] [PubMed]
149. Zhou, J.; Zhang, J.; Sun, Y.; Luo, F.; Guan, M.; Ma, H.; Dong, X.; Feng, J. A nano-delivery system based on preventing degradation and promoting absorption to improve the oral bioavailability of insulin. *Int. J. Biol. Macromol.* **2023**, *244*, 125263. [CrossRef] [PubMed]
150. Wu, J.; Chen, Q.; Wang, W.; Lin, Y.; Kang, H.; Jin, Z.; Zhao, K. Chitosan Derivative-Based Microspheres Loaded with Fibroblast Growth Factor for the Treatment of Diabetes. *Polymers* **2023**, *15*, 3099. [CrossRef]
151. Lu, X.; Li, J.; Xue, M.; Wang, M.; Guo, R.; Wang, B.; Zhang, H. Net-Neutral Nanoparticles-Extruded Microcapsules for Oral Delivery of Insulin. *ACS Appl. Mater. Interfaces* **2023**, *15*, 33491–33503. [CrossRef]

152. Vasconcelos Silva, E.L.; Oliveira, A.C.J.; Moreira, L.M.C.C.; Silva-Filho, E.C.; Wanderley, A.G.; Soares, M.F.R.; Soares-Sobrinho, J.L. Insulin-loaded nanoparticles based on acetylated cashew gum/chitosan complexes for oral administration and diabetes treatment. *Int. J. Biol. Macromol.* **2023**, *242 Pt 1*, 124737. [CrossRef]
153. Fathy, M.M.; Hassan, A.A.; Elsayed, A.A.; Fahmy, H.M. Controlled release of silica-coated insulin-loaded chitosan nanoparticles as a promising oral administration system. *BMC Pharmacol. Toxicol.* **2023**, *24*, 21. [CrossRef] [PubMed]
154. Pessoa, B.; Collado-Gonzalez, M.; Sandri, G.; Ribeiro, A. Chitosan/Albumin Coating Factorial Optimization of Alginate/Dextran Sulfate Cores for Oral Delivery of Insulin. *Mar. Drugs* **2023**, *21*, 179. [CrossRef] [PubMed]
155. Liu, H.T.; Li, W.M.; Xu, G.; Li, X.Y.; Bai, X.F.; Wei, P.; Yu, C.; Du, Y.G. Chitosan oligosaccharides attenuate hydrogen peroxide-induced stress injury in human umbilical vein endothelial cells. *Pharmacol. Res.* **2009**, *59*, 167–175. [CrossRef] [PubMed]
156. Salvatore, T.; Pafundi, P.C.; Galiero, R.; Albanese, G.; Di Martino, A.; Caturano, A.; Vetrano, E.; Rinaldi, L.; Sasso, F.C. The Diabetic Cardiomyopathy: The Contributing Pathophysiological Mechanisms. *Front. Med.* **2021**, *8*, 695792. [CrossRef]
157. Mohammadpour, F.; Hadizadeh, F.; Tafaghodi, M.; Sadri, K.; Mohammadpour, A.H.; Kalani, M.R.; Gholami, L.; Mahmoudi, A.; Chamani, J. Preparation, in vitro and in vivo evaluation of PLGA/Chitosan based nano-complex as a novel insulin delivery formulation. *Int. J. Pharm.* **2019**, *572*, 118710. [CrossRef] [PubMed]
158. Lal, S.; Perwez, A.; Rizvi, M.M.; Datta, M.K. Design and development of a biocompatible montmorillonite PLGA nanocomposites to evaluate in vitro oral delivery of insulin. *Appl. Clay Sci.* **2017**, *147*, 69–79. [CrossRef]
159. Ag Seleci, D.; Seleci, M.; Walter, J.G.; Stahl, F.; Scheper, T. Niosomes as Nanoparticular Drug Carriers: Fundamentals and Recent Applications. *J. Nanomater.* **2016**, *2016*, 7372306. [CrossRef]
160. Kazi, K.M.; Mandal, A.S.; Biswas, N.; Guha, A.; Chatterjee, S.; Behera, M.; Kuotsu, K. Niosome: A future of targeted drug delivery systems. *J. Adv. Pharm. Technol. Res.* **2010**, *1*, 374–380. [CrossRef]
161. Liu, K.; Chen, Y.; Yang, Z.; Jin, J. zwitterionic Pluronic analog-coated PLGA nanoparticles for oral insulin delivery. *Int. J. Biol. Macromol.* **2023**, *236*, 123870. [CrossRef]
162. Asal, H.A.; Shoueir, K.R.; El-Hagrasy, M.A.; Toson, E.A. Controlled synthesis of in-situ gold nanoparticles onto chitosan functionalized PLGA nanoparticles for oral insulin delivery. *Int J Biol Macromol.* **2022**, *209 Pt B*, 2188–2196. [CrossRef]
163. Li, J.; Qiang, H.; Yang, W.; Xu, Y.; Feng, T.; Cai, H.; Wang, S.; Liu, Z.; Zhang, Z.; Zhang, J. Oral insulin delivery by epithelium microenvironment-adaptive nanoparticles. *J. Control. Release* **2022**, *341*, 31–43. [CrossRef] [PubMed]
164. Souto, E.B.; Doktorovova, S. Chapter 6—Solid lipid nanoparticle formulations pharmacokinetic and biopharmaceutical aspects in drug delivery. *Methods Enzym.* **2009**, *464*, 105–129.
165. Yang, R.; Gao, R.; Li, F.; He, H.; Tang, X. The influence of lipid characteristics on the formation, in vitro release, and in vivo absorption of protein-loaded SLN prepared by the double emulsion process. *Drug Dev. Ind. Pharm.* **2011**, *37*, 139–148. [CrossRef] [PubMed]
166. He, H.; Wang, P.; Cai, C.; Yang, R.; Tang, X. VB12-coated Gel-Core-SLN containing insulin: Another way to improve oral absorption. *Int. J. Pharm.* **2015**, *493*, 451–459. [CrossRef] [PubMed]
167. Hecq, J.; Amighi, K.; Goole, J. Development and evaluation of insulin-loaded cationic solid lipid nanoparticles for oral delivery. *J. Drug Deliv. Sci. Technol.* **2016**, *36*, 192–200. [CrossRef]
168. Elshaarani, T.; Yu, H.; Wang, L.; Lin, L.; Wang, N.; Zhang, L.; Han, Y.; Fahad, S.; Ni, Z. Dextran-crosslinked glucose responsive nanogels with a self-regulated insulin release at physiological conditions. *Eur. Polym. J.* **2020**, *125*, 109505. [CrossRef]
169. Muntoni, E.; Anfossi, L.; Milla, P.; Marini, E.; Ferraris, C.; Capucchio, M.T.; Colombino, E.; Segale, L.; Porta, M.; Battaglia, L. Glargine insulin loaded lipid nanoparticles: Oral delivery of liquid and solid oral dosage forms. *Nutr. Metab. Cardiovasc. Dis.* **2021**, *31*, 691–698. [CrossRef]
170. Zheng, Y.X.; He, Q.; Xu, M.; Huang, Y. Construction of Oral Insulin-Loaded Solid Lipid Nanoparticles and Their Intestinal Epithelial Cell Transcytosis Study. *J. Sichuan Univ. Med. Sci. Ed.* **2021**, *52*, 570–576. [CrossRef]
171. Zhao, D.; Shi, X.; Liu, T.; Lu, X.; Qiu, G.; Shea, K.J. Synthesis of surfactant-free hydroxypropyl methylcellulose nanogels for controlled release of insulin. *Carbohydr. Polym.* **2016**, *151*, 1006–1011. [CrossRef]
172. Mudassir, J.; Darwis, Y.; Muhamad, S.; Khan, A.A. Self-assembled insulin and nanogels polyelectrolyte complex (Ins/NGs-PEC) for oral insulin delivery: Characterization, lyophilization and in-vivo evaluation. *Int. J. Nanomed.* **2019**, *14*, 4895–4909. [CrossRef]
173. Yuan, S.; Li, X.; Shi, X.; Lu, X. Preparation of multiresponsive nanogels and their controlled release properties. *Colloid Polym. Sci.* **2019**, *297*, 613–621. [CrossRef]
174. Baloch, S.G.; Shaikh, H.; Shah, S.; Memon, S.; Memon, A.A. Synthesis of an insulin intercalated graphene oxide nanogel composite: Evaluation of its release profile and stability for oral delivery of insulin. *Nanoscale Adv.* **2022**, *4*, 2303–2312. [CrossRef]
175. Wang, X.; Cheng, D.; Liu, L.; Li, X. Development of poly(hydroxyethyl methacrylate) nanogel for effective oral insulin delivery. *Pharm. Dev. Technol.* **2018**, *23*, 351–357. [CrossRef]
176. Chou, H.S.; Larsson, M.; Hsiao, M.H.; Chen, Y.C.; Röding, M.; Nydén, M.; Liu, D.M. Injectable insulin-lysozyme-loaded nanogels with enzymatically-controlled degradation and release for basal insulin treatment: In vitro characterization and in vivo observation. *J. Control. Release* **2016**, *224*, 33–42. [CrossRef] [PubMed]
177. Li, C.; Huang, F.; Liu, Y.; Lv, J.; Wu, G.; Liu, Y.; Ma, R.; An, Y.; Shi, L. Nitrilotriacetic acid-functionalized glucose-responsive complex micelles for the efficient encapsulation and self-regulated release of insulin. *Langmuir* **2018**, *34*, 12116–12125. [CrossRef]
178. Wu, L.P.; Ficker, M.; Christensen, J.B.; Trohopoulos, P.N.; Moghimi, S.M. Dendrimers in Medicine: Therapeutic Concepts and Pharmaceutical Challenges. *Bioconjug. Chem.* **2015**, *26*, 1198–1211. [CrossRef] [PubMed]

179. Andrade, R.G.D.; Reis, B.; Costas, B.; Lima, S.A.C.; Reis, S. Modulation of Macrophages M1/M2 Polarization Using Carbohydrate-Functionalized Polymeric Nanoparticles. *Polymers* **2020**, *13*, 88. [CrossRef] [PubMed] [PubMed Central]
180. Liu, X.; Li, C.; Lv, J.; Huang, F.; An, Y.; Shi, L.; Ma, R. Glucose and H_2O_2 dual-responsive polymeric micelles for the self-regulated release of insulin. *ACS Appl. Bio Mater.* **2020**, *3*, 1598–1606. [CrossRef] [PubMed]
181. Bahman, F.; Taurin, S.; Altayeb, D.; Taha, S.; Bakhiet, M.; Greish, K. Oral Insulin Delivery Using Poly (Styrene Co-Maleic Acid) Micelles in a Diabetic Mouse Model. *Pharmaceutics* **2020**, *12*, 1026. [CrossRef]
182. Han, X.; Lu, Y.; Xie, J.; Zhang, E.; Zhu, H.; Du, H.; Wang, K.; Song, B.; Yang, C.; Shi, Y.; et al. Zwitterionic micelles efficiently deliver oral insulin without opening tight junctions. *Nat. Nanotechnol.* **2020**, *15*, 605–614. [CrossRef]
183. Italiya, K.S.; Basak, M.; Mazumdar, S.; Sahel, D.K.; Shrivastava, R.; Chitkara, D.; Mittal, A. Scalable Self-Assembling Micellar System for Enhanced Oral Bioavailability and Efficacy of Lisofylline for Treatment of Type-I Diabetes. *Mol. Pharm.* **2019**, *16*, 4954–4967. [CrossRef] [PubMed]
184. Hu, W.Y.; Wu, Z.M.; Yang, Q.Q.; Liu, Y.J.; Li, J.; Zhang, C.Y. Smart pH-responsive polymeric micelles for programmed oral delivery of insulin. *Colloids Surf. B Biointerfaces* **2019**, *183*, 110443. [CrossRef] [PubMed]
185. Yan, C.; Gu, J.; Lv, Y.; Shi, W.; Wang, Y.; Liao, Y.; Deng, Y. Caproyl-modified G2 PAMAM dendrimer (G2-AC) Nanocomplexes increases the pulmonary absorption of insulin. *AAPS PharmSciTech* **2019**, *20*, 298. [CrossRef] [PubMed]
186. Wang, J.; Li, B.; Qiu, L.; Qiao, X.; Yang, H. Dendrimer-based drug delivery systems: History, challenges, and latest developments. *J. Biol. Eng.* **2022**, *16*, 18. [CrossRef]
187. Zeng, Z.; Qi, D.; Yang, L.; Liu, J.; Tang, Y.; Chen, H.; Feng, X. Stimuli-responsive self-assembled dendrimers for oral protein delivery. *J. Control. Release* **2019**, *315*, 206–213. [CrossRef] [PubMed]
188. Kamalden, T.A.; Macgregor-Das, A.M.; Kannan, S.M.; Dunkerly-Eyring, B.; Khaliddin, N.; Xu, Z.; Fusco, A.P.; Yazib, S.A.; Chow, R.C.; Duh, E.J.; et al. Exosomal microRNA-15a transfer from the pancreas augments diabetic complications by inducing oxidative stress. *Antioxid. Redox Signal.* **2017**, *27*, 913–930. [CrossRef] [PubMed]
189. Rodríguez-Morales, B.; Antunes-Ricardo, M.; González-Valdez, J. Exosome-mediated insulin delivery for the potential treatment of diabetes mellitus. *Pharmaceutics* **2021**, *13*, 1870. [CrossRef] [PubMed]
190. Xian, S.; Xiang, Y.; Liu, D.; Fan, B.; Mitrová, K.; Ollier, R.C.; Su, B.; Alloosh, M.A.; Jiráček, J.; Sturek, M.; et al. Insulin-Dendrimer Nanocomplex for Multi-Day Glucose-Responsive Therapy in Mice and Swine. *Adv. Mater.* **2024**, *36*, e2308965. [CrossRef]
191. Castaño, C.; Mirasierra, M.; Vallejo, M.; Novials, A.; Párrizas, M. Delivery of muscle-derived exosomal miRNAs induced by HIIT improves insulin sensitivity through down-regulation of hepatic FoxO1 in mice. *Proc. Natl. Acad. Sci. USA* **2020**, *117*, 30335–30343. [CrossRef]
192. Ghiasi, B.; Sefidbakht, Y.; Mozaffari-Jovin, S.; Gharehcheloo, B.; Mehrarya, M.; Khodadadi, A.; Rezaei, M.; Ranaei Siadat, S.O.; Uskoković, V. Hydroxyapatite as a biomaterial—A gift that keeps on giving. *Drug Dev. Ind. Pharm.* **2020**, *46*, 1035–1062. [CrossRef]
193. Uskoković, V. Supplementation of Polymeric Reservoirs with Redox-Responsive Metallic Nanoparticles as a New Concept for the Smart Delivery of Insulin in Diabetes. *Materials* **2023**, *16*, 786. [CrossRef] [PubMed]
194. Wu, L.; Wang, L.; Liu, X.; Bai, Y.; Wu, R.; Li, X.; Mao, Y.; Zhang, L.; Zheng, Y.; Gong, T.; et al. Milk-derived exosomes exhibit versatile effects for improved oral drug delivery. *Acta Pharm. Sin. B* **2022**, *12*, 2029–2042. [CrossRef] [PubMed]
195. Zhang, Y.; Zhang, L.; Ban, Q.; Li, J.; Li, C.H.; Guan, Y.Q. Preparation and characterization of hydroxyapatite nanoparticles carrying insulin and gallic acid for insulin oral delivery. *Nanomedicine* **2018**, *14*, 353–364. [CrossRef] [PubMed]
196. Scudeller, L.A.; Mavropoulos, E.; Tanaka, M.N.; Costa, A.M.; Braga, C.A.C.; López, E.O.; Mello, A.; Rossi, A.M. Effects on insulin adsorption due to zinc and strontium substitution in hydroxyapatite. *Mater. Sci. Eng. C Mater. Biol. Appl.* **2017**, *79*, 802–811. [CrossRef] [PubMed]
197. Kim, B.Y.S.; Rutka, J.T.; Chan, W.C.W. Nanomedicine. *N. Engl. J. Med.* **2010**, *363*, 2434–2443. [CrossRef] [PubMed]
198. Drug Products, Including Biological Products, That Contain Nanomaterials—Guidance for Industry. Available online: https://www.fda.gov/regulatory-information/search-fda-guidance-documents/drug-products-including-biological-products-contain-nanomaterials-guidance-industry (accessed on 6 July 2024).
199. Isibor, P.O. Regulations and Policy Considerations for Nanoparticle Safety. In *Environmental Nanotoxicology*; Isibor, P.O., Devi, G., Enuneku, A.A., Eds.; Springer: Cham, Germany, 2024. [CrossRef]
200. Park, J.; Lee, Y.-K.; Park, I.-K.; Hwang, S.R. Current Limitations and Recent Progress in Nanomedicine for Clinically Available Photodynamic Therapy. *Biomedicines* **2021**, *9*, 85. [CrossRef]
201. Yu, J.; Zhang, Y.; Bomba, H.; Gu, Z. Stimuli-responsive delivery of therapeutics for diabetes treatment. *Bioeng. Transl. Med.* **2016**, *1*, 323–337. [CrossRef]
202. Farokhzad, O.C.; Karnik, R.N.; Tucker, J. Development of FcRn-Targeted Nanoparticles for Efficient Oral Delivery of Insulin. RePORTER NIH. Available online: https://reporter.nih.gov/search/Nmr1_-lATEi0lvw6NUVQJg/project-details/8459384 (accessed on 6 July 2024).
203. Li, Y.; Mao, H.; Arreaza-Rubin, G. Small Intestine Targeted Fast Acting Oral Insulin Formulation. RePORTER NIH. Available online: https://reporter.nih.gov/search/Nmr1_-lATEi0lvw6NUVQJg/project-details/10385154 (accessed on 6 July 2024).
204. Majeti, R.N.; Ganugula, R.; Li, Y. Oral Delivery of Insulin Using Ligand-Directed Nanoparticles That Do Not Compete with Physiological Ligands. RePORTER NIH. Available online: https://reporter.nih.gov/search/Nmr1_-lATEi0lvw6NUVQJg/project-details/10580808 (accessed on 6 July 2024).

205. Mansoor, S.; Kondiah, P.P.D.; Choonara, Y.E.; Pillay, V. Polymer-Based Nanoparticle Strategies for Insulin Delivery. *Polymers* **2019**, *11*, 1380. [CrossRef] [PubMed]
206. Limbert, C.; Kowalski, A.J.; Danne, T.P.A. Automated Insulin Delivery: A Milestone on the Road to Insulin Independence in Type 1 Diabetes. *Diabetes Care* **2024**, *47*, 918–920. [CrossRef]
207. Xu, C.; Lei, C.; Huang, L.; Zhang, J.; Zhang, H.; Song, H.; Yu, M.; Wu, Y.; Chen, C.; Yu, C. Glucose-Responsive Nanosystem Mimicking the Physiological Insulin Secretion via an Enzyme–Polymer Layer-by-Layer Coating Strategy. *Chem. Mater.* **2017**, *29*, 7725–7732. [CrossRef]
208. Ma, Q.; Zhao, X.; Shi, A.; Wu, J. Bioresponsive Functional Phenylboronic Acid-Based Delivery System as an Emerging Platform for Diabetic Therapy. *Int. J. Nanomed.* **2021**, *16*, 297–314. [CrossRef] [PubMed]
209. Yin, R.; Bai, M.; He, J.; Nie, J.; Zhang, W. Concanavalin A-sugar affinity based system: Binding interactions, principle of glucose-responsiveness, and modulated insulin release for diabetes care. *Int. J. Biol. Macromol.* **2019**, *124*, 724–732. [CrossRef] [PubMed]
210. Yu, J.; Zhang, Y.; Wang, J.; Wen, D.; Kahkoska, A.; Buse, B.J.; Gu, Z. Glucose-responsive oral insulin delivery for postprandial glycemic regulation. *Nano Res.* **2019**, *12*, 1539–1545. [CrossRef]
211. Liu, Y.; Wang, S.; Wang, Z.; Yu, J.; Wang, J.; Buse, J.B.; Gu, Z. Recent Progress in Glucose-Responsive Insulin. *Diabetes* **2024**, *10*, db240175. [CrossRef]
212. Fuchs, S.; Caserto, J.S.; Liu, Q.; Wang, K.; Shariati, K.; Hartquist, C.M.; Zhao, X.; Ma, M. A Glucose-Responsive Cannula for Automated and Electronics-Free Insulin Delivery. *Adv. Mater.* **2024**, *36*, e2403594. [CrossRef]
213. Gu, Z.; Arreaza-Rubin, G. Towards Glucose Transporter-Mediated Glucose-Responsive Insulin Delivery with Fast Response. RePORTER NIH. Available online: https://reporter.nih.gov/search/Nmr1_-lATEi0lvw6NUVQJg/project-details/9832126 (accessed on 6 July 2024).
214. Li, S.; Arreaza-Rubin, G. Towards Glucose Transporter-Mediated Glucose-Responsive Insulin Delivery with Fast Response. RePORTER NIH. Available online: https://reporter.nih.gov/search/Nmr1_-lATEi0lvw6NUVQJg/project-details/10425401 (accessed on 6 July 2024).

Disclaimer/Publisher's Note: The statements, opinions and data contained in all publications are solely those of the individual author(s) and contributor(s) and not of MDPI and/or the editor(s). MDPI and/or the editor(s) disclaim responsibility for any injury to people or property resulting from any ideas, methods, instructions or products referred to in the content.

Review

From the INVICTUS Trial to Current Considerations: It's Not Time to Retire Vitamin K Inhibitors Yet!

Akshyaya Pradhan [1,*], Somya Mahalawat [1] and Marco Alfonso Perrone [2,*]

1. Department of Cardiology, King George's Medical University, Lucknow 226003, India; somyamahalawat@ymail.com
2. Division of Cardiology and CardioLab, Department of Clinical Sciences and Translational Medicine, University of Rome Tor Vergata, 00133 Rome, Italy
* Correspondence: akshyaya33@gmail.com (A.P.); marco.perrone@uniroma2.it (M.A.P.)

Abstract: Atrial fibrillation (AF) is a common arrhythmia in clinical practice, and oral anticoagulation is the cornerstone of stroke prevention in AF. Direct oral anticoagulants (DOAC) significantly reduce the incidence of intracerebral hemorrhage with preserved efficacy for preventing stroke compared to vitamin K antagonists (VKA). However, the pivotal randomized controlled trials (RCTs) of DOAC excluded patients with valvular heart disease, especially mitral stenosis, which remains an exclusion criterion for DOAC use. The INVICTUS study was a large multicenter global RCT aimed at evaluating the role of DOAC compared to VKA in stroke prevention among patients with rheumatic valvular AF. In this study, rivaroxaban failed to prove superiority over VKA in preventing the composite primary efficacy endpoints of stroke, systemic embolism, myocardial infarction, and death. Unfortunately, the bleeding rates were not lower with rivaroxaban either. The death and drug discontinuation rates were higher in the DOAC arm. Close to the heels of the dismal results of INVICTUS, an apixaban trial in prosthetic heart valves, PROACT-Xa, was also prematurely terminated due to futility. Hence, for AF complicating moderate-to-severe mitral stenosis or prosthetic valve VKA remains the standard of care. However, DOAC can be used in patients with surgical bioprosthetic valve implantation, TAVR, and other native valve diseases with AF, except for moderate-to-severe mitral stenosis. Factor XI inhibitors represent a breakthrough in anticoagulation as they aim to dissociate thrombosis from hemostasis, thereby indicating a potential to cut down bleeding further. Multiple agents (monoclonal antibodies—e.g., osocimab, anti-sense oligonucleotides—e.g., fesomersen, and small molecule inhibitors—e.g., milvexian) have garnered positive data from phase II studies, and many have entered the phase III studies in AF/Venous thromboembolism. Future studies on conventional DOAC and new-generation DOAC will shed further light on whether DOAC can dethrone VKA in valvular heart disease.

Keywords: oral anticoagulants; valvular heart disease; mitral stenosis; DOAC; VKA

1. Introduction

Atrial fibrillation (AF) is a common arrhythmia in clinical practice and carries a high risk of stroke/systemic embolism [1]. Oral anticoagulation (OAC) has been the cornerstone of stroke prevention in patients with AF and vitamin k antagonists have been the sheet anchor for OAC for the past 5 decades. However, their use is plagued with multiple shortcomings, such as drug interactions, food interactions, frequent INR monitoring, a narrow therapeutic range, and the risk of intracerebral hemorrhage [1]. Direct oral anticoagulants (DOAC) represent a significant advancement in OAC for AF without the need for frequent INR monitoring and with minimal food/drug interaction. More significantly, they reduce the incidence of intracerebral hemorrhage with preserved efficacy for preventing stroke compared to vitamin K antagonists (VKA), as evidenced in four large pivotal RCTs involving AF patients [2]. However, most of these trials excluded patients with valvular heart disease, especially mitral stenosis, which remains an exclusion criterion

for DOAC use. INVICTUS was a large multicenter global RCT aimed at evaluating the role of DOAC compared to VKA in stroke prevention among patients with rheumatic valvular AF. In this review, we first discuss the INVICTUS study and then provide the current perspectives in the background of the results of other studies of DOAC in valvular heart disease.

2. The INVICTUS Trial

INVICTUS was an international multicenter, double-blind trial in patients with rheumatic mitral valve disease [3]. In this trial, once-daily rivaroxaban (at a dose of 20 or 15 mg, according to renal function) was compared with a dose-adjusted vitamin K antagonist in patients with documented rheumatic mitral valve disease and atrial fibrillation (AF). The primary efficacy outcome was a composite of total stroke and systemic embolism. The secondary outcomes were acute myocardial infarction and death from vascular (cardiac or non-cardiac) causes. The primary safety outcome was major bleeding as defined by the International Society of Thrombosis and Hemostasis (ISTH) [3].

The trial was open-label, double-blinded, and used the intention-to-treat principle for the efficacy analysis. The included patients were aged ≥18 years and had echocardiography-proven rheumatic heart disease with documented atrial fibrillation or atrial flutter at any time. Additionally, they were required to have at least one of the following criteria: a CHA_2DS_2VASc score of at least 2, mitral stenosis with a mitral valve area <2 cm^2, or echocardiographic evidence of either left atrial spontaneous echo contrast or left atrial thrombus.

The trial excluded patients with a mechanical heart valve or with the likelihood of receiving one within the next 6 months, the use of dual antiplatelet therapy, treatment with dual strong inhibitors of CYP3A4 and P-glycoprotein, and the presence of severe renal disease (estimated GFR filtration rate < 15 mL/min). Women of childbearing age were excluded if they were pregnant or not using contraception.

Between August 2016 and September 2019, 4565 patients were enrolled from 138 trial sites across 24 countries. On 4 February 2022, the data and safety monitoring board recommended that the trial be terminated because the primary question addressed by the trial had been answered satisfactorily. The primary endpoint (composite of stroke, systemic embolism, myocardial infarction, or death from vascular or unknown causes) transpired in 560 of 2275 patients in the rivaroxaban group and in 446 of 2256 patients in the VKA group (proportional-hazards ratio, 1.25; 95% confidence interval [CI], 1.10 to 1.41). The restricted mean survival time was 1599 days in the rivaroxaban group and 1675 days in the VKA group (difference −76 days; 95% CI, −121 to −31 days; $p < 0.001$ for superiority). More patients in the rivaroxaban group than in the VKA group had a stroke (90 vs. 65 patients), a finding that was almost entirely driven by the higher rate of ischemic stroke in the rivaroxaban group (Figure 1). A total of 552 patients in the rivaroxaban group and 442 patients in the VKA group died (difference in restricted mean survival time, −72 days; 95% CI, −117 to −28). The difference in mortality was almost entirely due to the lower rates of sudden cardiac death and death due to mechanical or pump failure in the VKA group than in the rivaroxaban group. Between-group differences in the rates of stroke and death were similar in the on-treatment and intention-to-treat analyses. The rates of major bleeding did not differ significantly between the treatment groups.

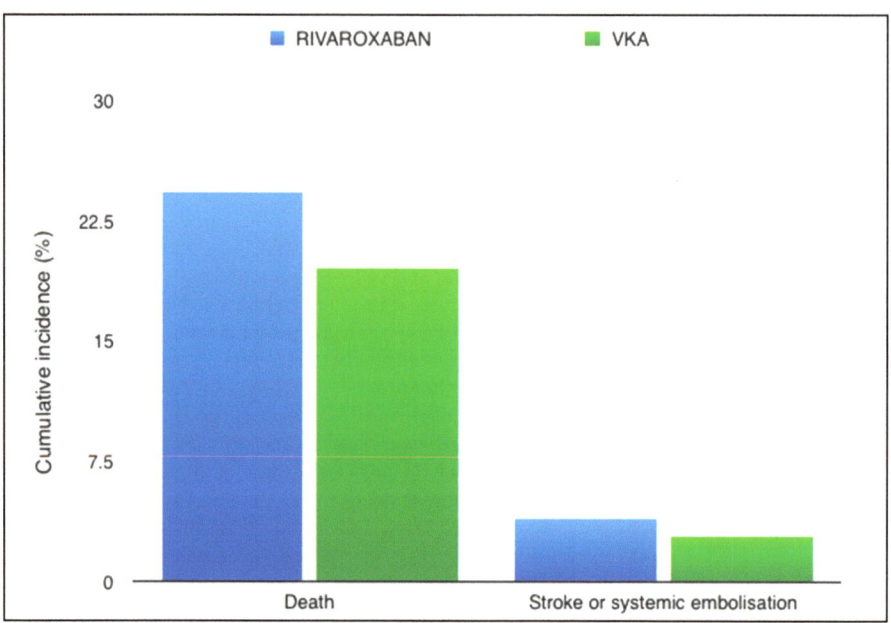

Figure 1. Rates of events in the INVICTUS trial in the Rivaroxaban and VKA arms. [VKA—Vitamin K antagonist].

3. Implications for Practice

AF may occur because of diverse pathophysiological conditions that lead to remodeling of the left atrium. Patients with atrial fibrillation are at an increased risk for embolic stroke owing to the formation of a thrombus in the left atrium, which can embolize, leading to stroke or systemic embolism [4]. Atrial fibrillation affects >30 million people worldwide. It increases the risk of stroke 5-fold and is thought to cause 15% of all strokes. Most patients with AF receive lifelong oral anticoagulation therapy to prevent strokes or systemic embolisms. During the past decade, DOAC has proven to be superior or non-inferior to warfarin for stroke prevention in non-valvular AF (NVAF), both in large randomized controlled trials and real-world observational studies, and is now the first-choice recommendation for oral anticoagulation in contemporary guidelines in preference to VKA [1,5]. Nonetheless, VKA remains the drug of choice in some conditions, such as mechanical valve prostheses, including in pregnant patients, and rheumatic mitral stenosis.

The advantage of DOAC over VKA is primarily due to their high efficacy in averting stroke in NVAF, with a lower incidence of major bleeding, especially intracranial hemorrhage. Additional advantages include ease of use, minor or no drug and food interactions, predictable pharmacokinetics and pharmacodynamics, rapid onset and offset of action, shorter half-life, and obviation of the need for strict laboratory monitoring. However, DOAC has some disadvantages. DOAC remains like higher cost, specific antidote unavailability, and limited experience with these drugs outside the purview of NVAF (Table 1). In addition, DOAC is contraindicated in patients with severe renal and hepatic disease (absence of validated monitoring tests), patients with mechanical heart valves, individuals younger than 18 years of age, and antiphospholipid antibody syndrome; however, there is a high demand for DOAC because of the convenience of administration for caregivers and patients alike [6]. On the contrary, despite regular compliance, more than half of the patients on VKA therapy fail to maintain their international normalized ratio (INR) in the therapeutic range, leading to a risk potential for thromboembolic events.

Table 1. The "Pros" and "Cons" of DOAC use for oral anticoagulation for AF. [DOAC, direct oral anticoagulant; VKA—vitamin K antagonist; ICH, intracerebral hemorrhage].

Advantage	Disadvantage
Predictable pharmacodynamics and pharmacokinetics.	Standardized tests for monitoring of DOAC activity are not easily available, when it is necessary in liver and kidney injury.
Low drug-drug and food interactions.	Currently lack a universal antidote
No dietary restriction.	High cost (in some countries)
Rapid onset and offset, short half life.	Experience limited to a decade
No need of regular laboratory monitoring unlike VKA	Needed dose modification in renal dysfunction
Wide therapeutic window.	Superiority against VKA for efficacy is yet to be achieved
DOACs can be started without LMWH due to their rapid onset.	Not indicated in pregnancy
Low rates of ICH (50% reduction)	

Rheumatic mitral valve disease continues to be a burden in low-income and middle-income countries [7]. It predominantly affects poor socio-economic strata who have poor access to healthcare facilities for myriad reasons. Because the seminal trials of DOAC excluded moderate to severe mitral stenosis, VKA remains the standard of care for rheumatic mitral valve disease with AF. Recent data also suggest overestimation of AF risk in rhematic mitral stenosis due to flawed interpretation of data [8]. Moreover, such patients are younger by a decade or two with a paucity of conventional cardiovascular risk factors compared to the typical AF population enrolled in seminal DOAC trials. Hence, the INVICTUS trial was poised to answer the clinical question of whether we can safely replace VKA with DOAC in this setting of rheumatic mitral valve disease with AF. The primer to the INVICTUS study was a meta-analysis of four pivotal RCTs of DOAC in which patients with AF with valvular heart disease other than moderate to severe mitral stenosis were included [9]. The meta-analysis found DOAC to be an alternative to VKA with respect to efficacy and safety. Subsequently, the RIVER study found that rivaroxaban was not inferior to VKA for cardiovascular events, death, and bleeding in patients with mitral valve disease and AF undergoing bioprosthetic mitral valve implantation [10]. The study enrolled close to 1000 people across 50 sites in Brazil. Death from cardiovascular causes and thromboembolic events were lower with rivaroxaban. More supporting data came from a multicenter ENVISAGE AF TAVI trial, where >1400 patients undergoing transcatheter aortic valve replacement (TAVR) with AF were randomized to VKA and DOAC [11]. DOAC (edoxaban) was not inferior to VKA for the composite endpoints of death, myocardial infarction, stroke, valve thrombosis, systemic thromboembolism, and bleeding. The small RISE-MS study ($n = 40$) in patients with moderate-to-severe mitral stenosis with AF demonstrated that rivaroxaban was safe and efficacious compared to VKA over a 12-month follow-up [12]. Magnetic resonance imaging (MRI) detected cerebral infarcts were lower with rivaroxaban, while left atrial appendage thrombogenicity was similar in both arms. The DECISIVE study enrolled 120 valvular heart disease patients with AF and randomized them to dabigatran and warfarin [13]. The primary endpoint of stroke or silent brain ischemia (on MRI) was 13% lower with DOAC, but the results were not statistically significant. Building upon the success of these studies, INVICTUS was expected to deliver favorable outcomes for DOAC.

The INVICTUS trial results suggest that, compared with rivaroxaban, vitamin K antagonist therapy results in a lower rate of ischemic stroke among patients with rheumatic heart disease-associated atrial fibrillation and lower mortality due to vascular causes without significantly increasing the rate of major bleeding. Since a large number of patients were recruited from lower-income countries, including South Asia, the results are pertinent to our practice. The results of the INVICTUS trial support the current guidelines that recommend vitamin K antagonist therapy for the prevention of stroke in patients with rheumatic heart disease in whom atrial fibrillation develops [1,5].

Whether the INVICTUS trial was a failure of DOAC or a failure of the strategy needs to be analyzed further. Despite the young age of the population, the mortality rate was quite high (22%). The warfarin group had lower rates of sudden cardiac death and death due to pump failure (combined together—2/3rd of all deaths), which are unlikely to be related to anticoagulation in any manner. Hence, the lower mortality in the VKA group cannot be ascribed to the failure or inferiority of rivaroxaban. The positive efficacy results were largely driven by lower deaths with VKA rather than attenuated strokes, which is the primary presumed benefit and indication of oral anticoagulation. Hence, we cannot conclude with certainty that rivaroxaban did not perform as expected.

The significantly higher stroke rates with DOAC were associated with four times higher drug discontinuation rates. Of the patients who discontinued DOAC, many received VKA. This is contrary to what we expected in real-world scenarios. With single daily dosing, less food and drug interaction, and freedom from INR monitoring, rivaroxaban should have had better compliance than VKA. Moreover, the incidence of gastrointestinal side effects was higher with rivaroxaban than with apixaban or edoxaban. In the large ARISTOPHANES database, apixaban was clearly superior to rivaroxaban in terms of both efficacy and bleeding [14]. Even in the DOAC meta-analysis of valvular heart disease with AF excluding mitral stenosis, rivaroxaban had a higher incidence of major bleeding than other DOAC. Hence, whether the use of another DOAC with a more favorable safety profile could improve drug compliance and outcomes remains speculative.

More negative data on DOAC in the setting of valvular heart disease have emerged from the PROACT-Xa trial, which was a study on apixaban in the setting of implantation of a metallic valve (the ON-X valve) but was stopped prematurely due to futility [15]. Previously, REDEEM failed to show the benefit of a DOAC (dabigatran) in the setting of a mechanical heart valve. However, since dabigatran is a direct thrombin inhibitor with 1:1 molar inhibition of thrombin, the use of upstream factor Xa inhibitors (rivaroxaban or apixaban) was proposed to be more beneficial in this setting, and the results of the PROACT-Xa study were eagerly awaited [16]. The failure of the INVICTUS trial along with the PROACT-Xa trial is a clear indication for VKA using oral anticoagulation in the setting of rheumatic mitral stenosis and metallic heart valves. With AF in the setting of valvular heart disease other than significant mitral stenosis and bioprosthetic valves (surgical or transcatheter), a DOAC can be used based on the patient/physician preference. Hence, appropriate clinical judgement based on contemporary evidence is needed to choose the right OAC for valvular heart disease with AF (Figure 2).

Figure 2. Current evidence-based approach to choice of oral anticoagulation in patients with AF with underlying VHD. The left panel depicts the subset of VHD, while the middle panel suggests the appropriate oral anticoagulant—DOAC/VKA. The right panel displays the evidence base for recommendation of oral anticoagulant in each scenario. [VHD, valvular heart disease; TAVR—transcatheter aortic valve replacement; MS—mitral stenosis; DOAC, direct oral anticoagulant; VKA—Vitamin K antagonist].

In this context, the results of the FRAIL-AF study are noteworthy [17]. The study aimed to answer an important question—whether elderly and frail patients with AF who are stable on VKA should be switched to DOAC given their potential for reduction in intracranial hemorrhage (ICH) and the high risk of ICH in this subset. Elderly and frail patients with AF on VKA were randomized to switch from VKA to DOAC or continue VKA therapy. Major bleeding or clinically relevant non-major bleeding at 12 months was hypothesized to be lower with the intervention (switch to DOAC). However, at the end of the study, bleeding was higher in the intervention arm, indicating that a routine switch from VKA to DOAC was not warranted. However, rivaroxaban was the DOAC used here again, as seen in the INVICTUS study. It remains to be seen whether this was a failure of the strategy or a failure of individual DOAC.

4. Future Directions

Few studies ongoing studies are comparing DOAC and VKA in valvular heart disease. The ERTEMIS (NCT05540587) study compared edoxaban 60 mg daily with warfarin for rheumatic mitral stenosis with AF [18]. The primary endpoint of the study was stroke/systemic embolism at 15 days follow up. Another ongoing trial (DAVID-MS; NCT04045093) is utilizing dabigatran in mitral stenosis with AF and comparing rates of stroke/systemic embolism at 1 year with warfarin [19]. The DIAMOND study (NCT05687448) enrolled patients with mechanical heart valves for oral anticoagulation with VKA or apixaban (5 mg BD) [20]. The NOTION 4 study (NCT06449469) evaluated the impact of DOAC on subclinical leaflet thickening on cardiac CT in patients undergoing TAVR [21]. DOACs are a heterogeneous group with respect to bleeding, as described above, but the evidence is indirect, as head-to-head studies of DOAC in AF are lacking. The COBRRA-AF study (NCT04642430) enrolled patients with AF and randomized them to receive rivaroxaban and apixaban in standard doses. The main endpoint was clinically relevant bleeding at 12 months [22].

Factor XI inhibitors represent a breakthrough in anticoagulation therapy, as they appear to dissociate thrombosis from bleeding [23–25]. Hence, these drugs offer the prospect of effective anticoagulation with minimal bleeding, which translates into practice from knowledge of hereditary factor XI deficiency leading to milder bleeding while offering protection from stroke/thromboembolism [24]. Multiple agents have successfully entered phase 3 studies in patients with AF and venous thromboembolism.

Early studies on novel anticoagulants targeting factor XI showed benefits in preventing venous thromboembolism in knee arthroplasty patients. Phase II studies are ongoing for cancer patients and hemodialysis patients with chronic kidney disease. The majority of clinical trials employing FXI inhibitors in humans are currently in phase 2, with each drug briefly outlined.

IONIS-FXI Rx/FXI-ASO is an antisense oligonucleotide (ASO) against Factor XI. The IONIS-FXI Rx/FXI-ASO molecule was the first molecule evaluated in phase II trials, showing superior efficacy in preventing venous thromboembolism in patients with total knee arthroplasty (FXI-ASO TKA) [26]. The 200 mg dose regimen was non-inferior, while the 300 mg dose was superior. The incidence of bleeding was comparable in both doses but lower than enoxaparin. Another phase II trial (NCT02553889) in ESRD patients showed a decline in FXI activity, with major bleeding events occurring in one patient [27]. A ligand-conjugated variant of IONIS-FXI Rx (Fesomersen) has greater efficacy and allows once-monthly treatment (RE-THINC ESRD) in ESRD patients undergoing hemodialysis [28].

Osocimab is a long-acting humanized monoclonal antibody against factor XIa. The FOXTROT trial compared the safety and efficacy of four different osocimab dosages to enoxaparin 40 mg once daily and apixaban 2.5 mg twice daily for preventing VTE in knee replacement patients [29]. Post-operative osocimab administration was non-inferior to enoxaparin, but 1.8 mg/kg dose was superior for preventing VTE at 10–13 days. The Osocimab arm had lower bleeding rates than the enoxaparin arm. The CONVERT trial enrolled patients with CKD and demonstrated that osocimab is generally well-tolerated in this population and corresponds to a reduced risk of bleeding [30].

Abelacimab is a humanized IgG1 monoclonal antibody that inhibits factor XI in its zymogen form and prevents factor XIIa and factor IIa from their activation. A trial (ANT-005 TKA) involving 412 patients undergoing total knee arthroplasty randomized them to either abelacimab (30 mg, 75 mg, 150 mg) or enoxaparin 40 mg subcutaneously once daily [31]. VTE rates were lower in the abelacimab group (13%, 5%, and 4%), while the 75 mg and 150 mg regimens were superior to enoxaparin. Non-major bleeding occurred in 2% of individuals receiving 30 mg or 75 mg abelacimab. The AZALEA-TIMI 71 trial evaluated the safety and efficacy of abelacimab in patients with atrial fibrillation (AF) at a moderate-to-high risk of stroke compared to rivaroxaban [32]. The study involved 1287 patients and found that both abelacimab dosages were superior to rivaroxaban 20 mg daily in decreasing bleeding episodes in individuals with AF and a high CHA2DS2-VASc score. Phase III trials are underway to evaluate the efficacy of abelacimab in preventing cancer-related thrombosis (NCT05171049; NCT05171) [33,34].

Xisomab is a recombinant antibody that selectively reduces factor XI activation through FXIIa feedback, with a differential increase in half-life with an increased dose. A randomized, placebo-controlled trial (NCT03612856) tested two doses of xisomab (0.25, 0.5 mg/kg) in 24 ESRD patients on hemodialysis [35]. The study found no significant bleeding, no significant time to hemostasis, and a single major bleeding event. However, there was a reduction in circuit occlusion rates requiring circuit exchange and decreased levels of thrombin-antithrombin complexes and C-reactive protein. A phase 2 prospective study (NCT04465760) evaluated the safety and efficacy of gruticibart, an anti-FXI monoclonal antibody, in 11 ambulatory cancer patients undergoing central line insertion [36]. The study found no significant catheter-associated thrombosis in the gruticibart group and no adverse events.

Asundexian is a direct inhibitor factor Xia. In the PACIFIC-AF study, Asundexian (20 mg or 50 mg) was compared with apixaban (5 mg) in non-valvular AF patients aged > 45 years [37]. The study enrolled 862 patients across 14 countries, and the primary end point was major or relevant non-major bleeding defined by ISTH criteria. There was a 67% reduction in bleeding with Asundexian (both doses combined) compared with apixaban at the end of the study, while side effects were similar to apixaban. The PACIFIC-AMI study compared three different dosages of Asundexian in patients with acute myocardial infarction [38]. The study involved 1600 patients randomized to three dosage groups (10 mg, 20 mg, and 50 mg), with Asundexian 50 mg showing over 90% inhibition of factor Xia levels. However, the efficacy outcome (composite of cardiovascular mortality, MI, stroke, or stent thrombosis) was identical in the maximum tested dose of Asundexian and placebo, suggesting that Asundexian 50 mg daily did not provide significant ischemic benefits compared to the placebo. The PACIFIC-STROKE study found that Asundexian, at varying doses, did not significantly reduce ischemic stroke and covert brain infarction compared to placebo, but it did reduce recurrent symptomatic strokes and TAIs in atherosclerosis patients [39]. The OCEANIC-AF trial tested asundexian against apixaban in atrial fibrillation patients at risk for stroke but was terminated early due to poor efficacy compared to the control arm [40].

Factor XI inhibitors outperformed enoxaparin in reducing VTE incidence and bleeding risk in phase II studies. However, higher doses did not reduce thrombotic events post MI, stroke, or TIA. Larger studies may establish their clinical importance in the future. Similarly, factor XII and XIII inhibitors are also in the pipeline. It will be interesting to examine the effects of these new-generation oral anticoagulants on rheumatic AF.

DOACs have emerged as a viable alternative to fill the unmet need with oral anticoagulation using VKA in AF/VTE patients. Although their ease of use and reduced bleeding events have made them popular, rheumatic mitral stenosis patients were majorly excluded from the seminal DOAC studies. The INVICTUS study was designed to evaluate the role of DOAC in rheumatic AF, but rivaroxaban failed to demonstrate superiority over VKA in this setting. Close to the heels of the INVICTUS trial, DOACs failed to prove superiority over VKA in AF, complicating the mechanical prosthetic valve in the PROACT-Xa study. Ongoing studies with contemporary DOACs and new generation OACs such as Factor XI inhibitors are expected to shed further light on the matter. The failure of DOACs in INVICTUS and PROACT-Xa studies has put an end to their unstoppable juggernaut. For patients of AF with metallic prosthetic valves, pregnancy, severe mitral stenosis, and antiphospholipid antibody syndrome, VKA's are still the standard of care, and there is no need to retire them as of yet.

5. Conclusions

DOAC are preferred over VKA for stroke prevention in AF by major international guidelines, as they have been shown to reduce intracerebral hemorrhage in pivotal trials with equivalent efficacy. In addition, DOAC use has other practical advantages, such as amelioration of the need for INR monitoring, fewer drug interactions, and minimal food restrictions, resulting in wider acceptance across the globe. However, as patients with valvular AF were excluded from these studies, VKA are the oral anticoagulants of choice in this scenario of valvular AF, especially mitral stenosis, and the INVICTUS study failed to prove the efficacy of rivaroxaban in rheumatic mitral stenosis with AF. DOAC use was not associated with lower bleeding or stroke rates than VKA therapy. In contrast, death rates increased with the use of rivaroxaban. The higher drug discontinuation rate in the DOAC arm is perplexing, given the multiple advantages of DOAC. Whether substitution with alternate DOAC could have altered the results remains unanswered. On the other hand, the failure of apixaban in the PROACT-Xa study is another limitation to the use of DOAC in patients with mechanical heart valves. However, DOAC can still be utilized in patients with bioprosthetic valve implantation, transcatheter aortic valve replacement, and other native valve diseases with AF, except for moderate to severe mitral stenosis. Pending further studies, there is no current justification for DOAC prescription for AF with mitral stenosis and prosthetic valves (Figure 3). Factor XI inhibitors are currently on the anvil and hold great promise for the future. Ongoing studies of DOAC in the valvular heart are expected to shed further light on their role.

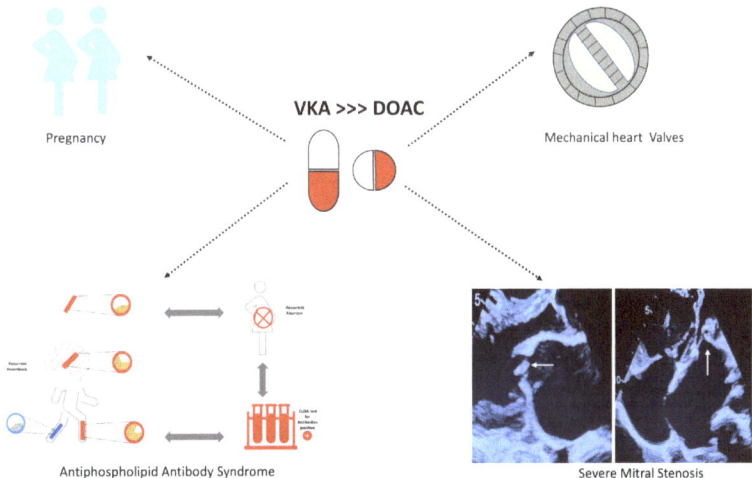

Figure 3. Current clinical scenario's in AF patients where VKA's would be still the first line agent for OAC. The upper right panel depicts a St. Jude's bi-leaflet prosthetic valve. The lower right panel depicts the 2D transthoracic echocardiogram images in patients with severe rheumatic mitral stenosis. The lower left picture represents the triad of antiphospholipid antibody syndrome—recurrent abortions, arterial (red)/venous (blue) thrombosis, and a positive ELISA test for anti-cardiolipin/anti-B_2-microglobulin antibody. [VKA—Vitamin K antagonist; DOAC—direct-acting anticoagulant; ELISA—enzyme-linked immunosorbent assay].

Funding: This research received no external funding.

Conflicts of Interest: The authors declare no conflicts of interest.

References

1. Van Gelder, I.C.; Rienstra, M.; Bunting, K.V.; Casado-Arroyo, R.; Caso, V.; Crijns, H.J.G.M.; De Potter, T.J.R.; Dwight, J.; Guasti, L.; Hanke, T.; et al. 2024 ESC Guidelines for the management of atrial fibrillation developed in collaboration with the European Association for Cardio-Thoracic Surgery (EACTS). *Eur. Heart J.* **2024**, *45*, 3314–3414. [CrossRef] [PubMed]
2. Ruff, C.T.; Giugliano, R.P.; Braunwald, E.; Hoffman, E.B.; Deenadayalu, N.; Ezekowitz, M.D.; Camm, A.J.; Weitz, J.I.; Lewis, B.S.; Parkhomenko, A.; et al. Comparison of the efficacy and safety of new oral anticoagulants with warfarin in patients with atrial fibrillation: A meta-analysis of randomized trials. *Lancet* **2014**, *383*, 955–962. [CrossRef] [PubMed]
3. Connolly, S.J.; Karthikeyan, G.; Ntsekhe, M.; Haileamlak, A.; El Sayed, A.; El Ghamrawy, A.; Damasceno, A.; Avezum, A.; Dans, A.M.L.; Gitura, B.; et al. Rivaroxaban for rheumatic heart disease-associated atrial fibrillation. *N. Engl. J. Med.* **2022**, *387*, 978–988. [CrossRef] [PubMed]
4. Zimetbaum, P. Atrial fibrillation. *Ann. Intern. Med.* **2017**, *166*, ITC33–ITC48. [CrossRef]
5. Joglar, J.A.; Chung, M.K.; Armbruster, A.L.; Benjamin, E.J.; Chyou, J.Y.; Cronin, E.M.; Deswal, A.; Eckhardt, L.L.; Goldberger, Z.D.; Gopinathannair, R.; et al. 2023 ACC/AHA/ACCP/HRS guidelines for the diagnosis and management of atrial fibrillation: A report of the American College of Cardiology/American Heart Association Joint Committee on Clinical Practice Guidelines. *Circulation* **2024**, *149*, e1–e156. [CrossRef]
6. Mekaj, Y.H.; Mekaj, A.Y.; Duci, S.B.; Miftari, E.I. New oral anticoagulants: Their advantages and disadvantages compared with vitamin K antagonists in the prevention and treatment of patients with thromboembolic events. *Ther. Clin. Risk Manag.* **2015**, *11*, 967–977. [CrossRef]
7. Negi, P.C.; Sondhi, S.; Asotra, S.; Mahajan, K.; Mehta, A. Current status of rheumatic heart disease in India. *Indian. Heart J.* **2019**, *71*, 85–90. [CrossRef]
8. Karthikeyan, G.; Connolly, S.J.; Yusuf, S. Overestimation of Stroke Risk in Rheumatic Mitral Stenosis and the Implications for Oral Anticoagulation. *Circulation* **2020**, *142*, 1697–1699. [CrossRef]
9. Pan, K.L.; Singer, D.E.; Ovbiagele, B.; Wu, Y.L.; Ahmed, M.A.; Lee, M. Effects of Non-Vitamin K Antagonist Oral Anticoagulants Versus Warfarin in Patients with Atrial Fibrillation and Valvular Heart Disease: A Systematic Review and Meta-Analysis. *J. Am. Heart Assoc.* **2017**, *6*, e005835. [CrossRef]
10. Guimarães, H.P.; Lopes, R.D.; de Barros ESilva, P.G.M.; Liporace, I.L.; Sampaio, R.O.; Tarasoutchi, F.; Hoffmann-Filho, C.R.; de Lemos Soares Patriota, R.; Leiria, T.L.L.; Lamprea, D.; et al. Rivaroxaban in Patients with Atrial Fibrillation and a Bioprosthetic Mitral Valve. *N. Engl. J. Med.* **2020**, *383*, 2117–2126. [CrossRef]
11. Van Mieghem, N.M.; Unverdorben, M.; Hengstenberg, C.; Möllmann, H.; Mehran, R.; López-Otero, D.; Nombela-Franco, L.; Moreno, R.; Nordbeck, P.; Thiele, H.; et al. Edoxaban versus vitamin K antagonists for atrial fibrillation after TAVR. *N. Engl. J. Med.* **2021**, *385*, 2150–2160. [CrossRef] [PubMed]
12. Sadeghipour, P.; Pouraliakbar, H.; Parsaee, M.; Shojaeifard, M.; Farrashi, M.; Jamal Khani, S.; Tashakori Beheshti, A.; Rostambeigi, S.; Ebrahimi Meimand, S.; Firouzi, A.; et al. RIvaroxaban in mitral stenosis (RISE MS): A pilot randomized clinical trial. *Int. J. Cardiol.* **2022**, *356*, 83–86. [CrossRef] [PubMed]
13. Cho, M.S.; Kim, M.; Lee, S.A.; Lee, S.; Kim, D.H.; Kim, J.; Song, J.M.; Nam, G.B.; Kim, S.J.; Kang, D.H.; et al. Comparison of Dabigatran Versus Warfarin Treatment for Prevention of New Cerebral Lesions in Valvular Atrial Fibrillation. *Am. J. Cardiol.* **2022**, *175*, 58–64. [CrossRef] [PubMed]
14. Lip, G.Y.H.; Keshishian, A.; Li, X.; Hamilton, M.; Masseria, C.; Gupta, K.; Luo, X.; Mardekian, J.; Friend, K.; Nadkarni, A.; et al. Effectiveness and Safety of Oral Anticoagulants Among Nonvalvular Atrial Fibrillation Patients. *Stroke* **2018**, *49*, 2933–2944. [CrossRef]
15. Wang, T.Y.; Svensson, L.G.; Wen, J.; Vekstein, A.; Gerdisch, M.; Rao, V.U.; Moront, M.; Johnston, D.; Lopes, R.D.; Chavez, A.; et al. Apixaban or Warfarin in Patients with an On-X Mechanical Aortic Valve. *NEJM Evid.* **2023**, *2*, EVIDoa2300067. [CrossRef]
16. Narain, V.S.; Pradhan, A.; Bhandari, M. Mechanical Prosthetic valve on Oral Anticoagulants—How to improve safety. In *Cardiology Update Book 2021—Cardiological Society of India*; Banerjee, P.S., Das, M.K., Roy, D., Eds.; Evangel Publishers: Gurgaon, India, 2021; pp. 889–898.
17. Joosten, L.P.; van Doorn, S.; van de Ven, P.M.; Köhlen, B.T.; Nierman, M.C.; Koek, H.L.; Hemels, M.E.; Huisman, M.V.; Kruip, M.; Faber, L.M.; et al. Safety of Switching from Vitamin K Antagonist to Non-Vitamin K Antagonist Oral Anticoagulant in Frail Older Patients with Atrial Fibrillation: Results of the Frail-AF Randomized Control Trial. *Circulation* **2023**, *149*, 279–289. [CrossRef]
18. Efficacy and Safety of Edoxaban in Patients with Atrial Fibrillation and Mitral Stenosis (ERTEMIS). Available online: https://clinicaltrials.gov/study/NCT05540587 (accessed on 16 October 2024).
19. Zhou, M.; Chan, E.W.; Hai, J.J.; Wong, C.K.; Lau, Y.M.; Huang, C.C.; Lam, C.C.; Tam, C.C.F.; Wong, Y.T.A.; Yung, S.Y.A.; et al. Protocol, rationale and design of DAbigatran for Stroke PreVention in Atrial Fibrillation in MoDerate or Severe Mitral Stenosis (DAVID-MS): A randomised, open-label study. *BMJ Open* **2020**, *10*, e038194. [CrossRef] [PubMed] [PubMed Central]
20. DIrect Oral Anticoagulation and mechaNical Aortic Valve (DIAMOND). Available online: https://clinicaltrials.gov/study/NCT05687448 (accessed on 16 October 2024).

21. The Nordic Aortic Valve Intervention Trial 4 (NOTION-4). Available online: https://clinicaltrials.gov/study/NCT06449469 (accessed on 16 October 2024).
22. COmparison of Bleeding Risk Between Rivaroxaban and Apixaban in Patients with Atrial Fibrillation (COBRRA-AF). Available online: https://clinicaltrials.gov/study/NCT04642430 (accessed on 16 October 2024).
23. Harrington, J.; Piccini, J.P.; Alexander, J.H.; Granger, C.B.; Patel, M.R. Clinical Evaluation of Factor XIa Inhibitor Drugs: JACC Review Topic of the Week. *J. Am. Coll. Cardiol.* **2023**, *81*, 771–779. [CrossRef]
24. Preis, M.; Hirsch, J.; Kotler, A.; Zoabi, A.; Stein, N.; Rennert, G.; Saliba, W. Factor XI deficiency is associated with lower risk for cardiovascular and venous thromboembolism events. *Blood* **2017**, *129*, 1210–1215. [CrossRef]
25. De Caterina, R.; Prisco, D.; Eikelboom, J.W. Factor XI inhibitors: Cardiovascular perspectives. *Eur. Heart J.* **2023**, *44*, 280–292. [CrossRef]
26. Büller, H.R.; Bethune, C.; Bhanot, S.; Gailani, D.; Monia, B.P.; Raskob, G.E.; Segers, A.; Verhamme, P.; Weitz, J.I. Factor XI Antisense Oligonucleotide for Prevention of Venous Thrombosis. *N. Engl. J. Med.* **2015**, *372*, 232–240. [CrossRef] [PubMed]
27. Walsh, M.; Bethune, C.; Smyth, A.; Tyrwhitt, J.; Jung, S.W.; Yu, R.Z.; Wang, Y.; Geary, R.S.; Weitz, J.; Bhanot, S. Phase 2 Study of the Factor XI Antisense Inhibitor IONIS-FXIRx in Patients with ESRD. *Kidney Int. Rep.* **2021**, *7*, 200–209. [CrossRef] [PubMed]
28. Winkelmayer, W.C.; Lensing, A.W.A.; Thadhani, R.I.; Mahaffey, K.W.; Walsh, M.; Pap, Á.F.; Willmann, S.; Thelen, K.; Hodge, S.; Solms, A.; et al. A Phase II randomized controlled trial evaluated antithrombotic treatment with fesomersen in patients with kidney failure on hemodialysis. *Kidney Int.* **2024**, *106*, 145–153. [CrossRef] [PubMed]
29. Weitz, J.I.; Bauersachs, R.; Becker, B.; Berkowitz, S.D.; Freitas, M.C.; Lassen, M.R.; Metzig, C.; Raskob, G.E. Effect of Osocimab in Preventing Venous Thromboembolism Among Patients Undergoing Knee Arthroplasty: The FOXTROT Randomized Clinical Trial. *JAMA* **2020**, *323*, 130–139. [CrossRef] [PubMed]
30. Weitz, J.I.; Tankó, L.B.; Floege, J.; Fox, K.A.A.; Bhatt, D.L.; Thadhani, R.; Hung, J.; Pap, Á.F.; Kubitza, D.; Winkelmayer, W.C.; et al. Anticoagulation with osocimab in patients with kidney failure undergoing hemodialysis: A randomized phase 2 trial. *Nat. Med.* **2024**, *30*, 435–442. [CrossRef]
31. Verhamme, P.; Yi, B.A.; Segers, A.; Salter, J.; Bloomfield, D.; Büller, H.R.; Raskob, G.E.; Weitz, J.I.; ANT-005 TKA Investigators. Abelacimab for Prevention of Venous Thromboembolism. *N. Engl. J. Med.* **2020**, *385*, 609–617. [CrossRef]
32. Kumbhani, D.J. A Multicenter, RandomiZed, Active-ControLled Study to Evaluate the Safety and Tolerability of Two Blinded Doses of Abelacimab Compared with Open-Label Rivaroxaban in Patients with Atrial Fibrillation—AZALEA-TIMI 78. In: American College of Cardiology. [Internet]. Available online: https://www.acc.org/Latest-in-Cardiology/Clinical-Trials/2023/11/10/22/46/azalea-timi-71 (accessed on 8 October 2024).
33. A Study Comparing Abelacimab to Apixaban in the Treatment of Cancer-Associated VTE (ASTER). Available online: https://clinicaltrials.gov/study/NCT05171049 (accessed on 16 October 2024).
34. A Study Comparing Abelacimab to Dalteparin in the Treatment of Gastrointestinal/Genitourinary Cancer and Associated VTE (MAGNOLIA). Available online: https://clinicaltrials.gov/study/NCT05171075 (accessed on 16 October 2024).
35. Lorentz, C.U.; Tucker, E.I.; Verbout, N.G.; Shatzel, J.J.; Olson, S.R.; Markway, B.D.; Wallisch, M.; Ralle, M.; Hinds, M.T.; McCarty, O.J.T.; et al. The contact activation inhibitor AB023 in heparin-free hemodialysis: Results of a randomized phase 2 clinical trial. *Blood* **2021**, *138*, 2173–2184. [CrossRef]
36. Pfeffer, M.A.; Kohs, T.C.L.; Vu, H.H.L.; Jordan, K.R.; Wang, J.S.H.; Lorentz, C.U.; Tucker, E.I.; Puy, C.; Olson, S.R.; DeLoughery, T.G.; et al. Factor XI Inhibition for the Prevention of Catheter-Associated Thrombosis in Patients with Cancer Undergoing Central Line Placement: A Phase 2 Clinical Trial. *Arterioscler. Thromb. Vasc. Biol.* **2024**, *44*, 290–299. [CrossRef]
37. Piccini, J.P.; Caso, V.; Connolly, S.J.; Fox, K.A.A.; Oldgren, J.; Jones, W.S.; A Gorog, D.; Viethen, T.; Neumann, C.; Mundl, H.; et al. Safety of the oral factor XIa inhibitor asundexian compared with apixaban in patients with atrial fibrillation (PACIFIC-AF): A multicentre, randomised, double-blind, double-dummy, dosefinding phase 2 study. *Lancet* **2022**, *399*, 1383–1390. [CrossRef]
38. Rao, S.V.; Kirsch, B.; Bhatt, D.L.; Budaj, A.; Coppolecchia, R.; Eikelboom, J.; James, S.K.; Jones, W.S.; Merkely, B.; Keller, L.; et al. A Multicenter, Phase 2, Randomized, Placebo-Controlled, Double-Blind, Parallel-Group, Dose-Finding Trial of the Oral Factor XIa Inhibitor Asundexian to Prevent Adverse Cardiovascular Outcomes After Acute Myocardial Infarction. *Circulation* **2022**, *146*, 1196–1206. [CrossRef]
39. Shoamanesh, A.; Mundl, H.; Smith, E.E.; Masjuan, J.; Milanov, I.; Hirano, T.; Agafina, A.; Campbell, B.; Caso, V.; Mas, J.-L.; et al. Factor XIa inhibition with asundexian after acute non-cardioembolic ischaemic stroke (PACIFIC-Stroke): An international, randomised, double-blind, placebo-controlled, phase 2b trial. *Lancet* **2022**, *400*, 997–1007. [CrossRef]
40. Piccini, J.P.; Patel, M.R.; Steffel, J.; Ferdinand, K.; Van Gelder, I.C.; Russo, A.M.; Ma, C.-S.; Goodman, S.G.; Oldgren, J.; Hammett, C.; et al. Asundexian Versus Apixaban in Patients with Atrial Fibrillation. *N. Engl. J. Med. Epub ahead of print.* **2024**. [CrossRef]

Disclaimer/Publisher's Note: The statements, opinions and data contained in all publications are solely those of the individual author(s) and contributor(s) and not of MDPI and/or the editor(s). MDPI and/or the editor(s) disclaim responsibility for any injury to people or property resulting from any ideas, methods, instructions or products referred to in the content.

Review

Sodium-Glucose Cotransporter-2 Inhibitors in Diabetic Patients with Heart Failure: An Update

Nicia I. Profili [1], Roberto Castelli [1], Antonio Gidaro [2], Roberto Manetti [1], Margherita Maioli [3] and Alessandro P. Delitala [1,*]

[1] Department of Medicine, Surgery and Pharmacy, University of Sassari, 07100 Sassari, Italy; rmanetti@uniss.it (R.M.)
[2] Department of Biomedical and Clinical Sciences Luigi Sacco, Luigi Sacco Hospital, University of Milan, 20157 Milan, Italy; gidaro.antonio@asst-fbf-sacco.it
[3] Department of Biochemical Science, University of Sassari, 07100 Sassari, Italy; mmaioli@uniss.it
* Correspondence: aledelitala@uniss.it

Abstract: Diabetes mellitus and heart failure are two diseases that are commonly found together, in particular in older patients. High blood glucose has a detrimental effect on the cardiovascular system, and worse glycemic control contributes to the onset and the recrudesce of heart failure. Therefore, any specific treatment aimed to reduce glycated hemoglobin may, in turn, have a beneficial effect on heart failure. Sodium-glucose cotransporter-2 inhibitors have been initially developed for the treatment of type 2 diabetes mellitus, and their significant action is to increase glycosuria, which in turn causes a reduction in glucose blood level and contributes to the reduction of cardiovascular risk. However, recent clinical trials have progressively demonstrated that the glycosuric effect of the sodium-glucose cotransporter-2 inhibitors also have a diuretic effect, which is a crucial target in the management of patients with heart failure. Additional studies also documented that sodium-glucose cotransporter-2 inhibitors improve the therapeutical management of heart failure, independently by the glycemic control and, therefore, by the presence of diabetes mellitus. In this review, we analyzed studies and trials demonstrating the efficacy of sodium-glucose cotransporter-2 inhibitors in treating chronic and acute heart failure.

Keywords: SGLT2i; Na$^+$-glucose cotransporter-2 inhibitors; chronic heart failure; acute heart failure; diabetes; cardiovascular disease

1. Introduction

Diabetes is a heterogeneous disease with an increasing prevalence worldwide. Some regions of the world are experiencing rapid growth in frequency, particularly in eastern countries [1]. The causes of this epidemiological trend can be mainly found in unhealthy obesogenic diets and reduced physical activity. Patients with diabetes have an increased risk of developing cardiovascular disease (CVD): heart failure (HF), coronary artery disease, atrial fibrillation, and stroke. Diabetic patients are also at increased risk of developing chronic kidney disease (CKD) and have a higher risk of all-cause mortality, mainly due to the combination of cardiovascular and kidney disease. The treatment of type 2 diabetes has been profoundly changed in the last years due to the possibility of prescribing the inhibitors of Na+-glucose cotransporter-2 (SGLT2i), which has a good effect on glycated hemoglobin reduction. Further studies also demonstrated a broader impact, not limited to diabetes. Indeed, due to its glycosuric action, it has been shown that SGLT2i could also be used for the treatment of HF as well as CKD. In this narrative review, we focused on the cardiac effect of this drug, analyzing the central studies that demonstrated the positive impact that SGLT2i has on the management of HF.

2. Mechanism of Action of Sodium-Glucose Cotransporter-2 Inhibitors

The kidney plays a critical role in glucose homeostasis. Indeed, it contributes to gluconeogenesis (15–55 g/die) and reabsorbs the glucose filtered into the glomerular filtrate. Glycosuria occurs when blood glucose exceeds 180 mg/dL, thus usually reabsorbing all the daily glucose filtered in healthy individuals [2]. Na^+-glucose cotransporter-2 (SGLT2) is the principal cotransporter responsible for the reabsorption of filtered glucose and is expressed in the luminal membrane in the early portion of the kidney's proximal tubule, where 80–90% of the filtered glucose is physiologically absorbed [3]. The remaining 10–20% is adsorbed by Na^+-glucose cotransporter-1 (SGLT1) in the distal segment of the renal proximal tubule. SGTL2 has a high capacity and low affinity for glucose transport, which occurs against a concentration gradient. Further, its transport is coupled with the downhill of Na^+, which is then actively extruded in the basolateral surface of the cell. Glucose transporter 2 (GLUT2) carries glucose in the blood by facilitated diffusion [3]. The proximal tubule can increase the glucose reabsorption along with the rising of plasma glucose level until the transport maximum for glucose is reached, which is usually set at 260–350 mg/min. Once past this threshold, which is roughly equal to 180–200 mg/dL of blood glucose, the SGTL capacity is saturated, and glucose begins to be excreted via the urine. The blood glucose threshold has been demonstrated to be higher in diabetic patients (e.g., 220 mg/dL).

Inhibition of SGLT2 can thus increase glucose excretion in the urine by lowering the renal threshold for glucose excretion. Indeed, SGLT2i reduces the reabsorption of 30–50% of the glucose filtered by the kidney. This action is independent from insulin [2].

3. Heart Failure

Cardiovascular disease is the leading cause of morbidity and mortality, and different causes may contribute to its development [4,5].

HF is a clinical syndrome characterized by specific symptoms (breathlessness, ankle swelling, and fatigue) and signs (pulmonary crackles, peripheral edema, and elevated jugular venous pressure). Several conditions may lead to HF, and its diagnosis—and specific treatment—is mandatory. Incidence of HF is dramatically increasing worldwide: in Europe, it is about 5/1000 person-years in adults [6], with an apparent increase with age (>10% in subjects aged over 70). But the real-world prevalence is likely higher than those reported in the studies that included diagnosed HF [7].

Traditionally, HF is divided into three different phenotypes related to the value of left ventricular ejection fraction (LVEF) [6]. Reduced LVEF is defined as $\leq 40\%$ and designated as HFrEF, while patients with LVEF between 41% and 49% had mildly reduced left ventricular function (NFmrEF). SGLT2i with preserved ejection fraction (HFpEF) is defined as the presence of clinical diagnosis of HF without evidence of structural and/or functional cardiac abnormalities and/or raised natriuretic peptides and with LVEF $\geq 50\%$.

4. SGLT2i in Heart Failure: Mechanisms of Action

SGLT2i exhibits pleiotropic effects on different physiological systems, some of which are independent of the anti-hyperglycemic effect (Figure 1). The most plausible hypothesis is the augmented diuretic effect secondary to glucosuria and natriuresis. The effect starts within 24 h, which leads to a 300 mL/day increase in urinary output [6] and decreases after 12 weeks of treatment [8]. Studies also reported that the diuretic effect is more efficient when SGTL2 inhibitors are associated with loop diuretics [9], although other reports argue against the possible diuretic effect. Indeed, the EMPA-RESPONSE-AHF study reported that Empagliflozin did not reduce dyspnea scores [10]. Further, another study showed that Dapagliflozin was not associated with an acute drop in NT-proBNP in patients with SGLT2i with reduced ejection fraction [11]. Studies that supported the possible diuretic effect also reported to low blood pressure and reduced renal deterioration. Indeed, studies have reported a reduction of intraglomerular hypertension, modulation of the sympathetic nervous system, and reduction of oxidative stress and inflammation [12]. However, it should be noted that the effect of SGTL2i on the cardiorenal axis was higher in hospitalized patients with HFrEF than in those with HFpEF [13]. SGTL2i also had a beneficial effect on the cardiomyocytes of mouse models that developed a reduction of cardiac fibrosis through different pathways: reduced expression of fibronectin 1, collage type I and III, and transforming growth factor-β [14]. Another critical mechanism of SGLT2i is the reduced oxidative stress and inflammation, as showed by some studies which reported a reduction of circulating pro-inflammatory factors (C-reactive protein, interleukin-6, and tumor necrosis factor-α) [15] and is mostly multifactorial. In addition, SGLT2i inhibits the NLRP3 inflammasome, which has a clear role in chronic inflammation in some CVD. This effect leads to decreased macrophage infiltration and favours the release of specific cytokine [16]. Lastly, some studies reported an increased autophagic flux, which is a measure of autophagic degradation activity. Patients with HF had an impaired autophagy, which physiologically contributes to the degradation of dysfunctional mitochondria, thus reducing the oxidative stress [17]. Recent studies showed that SGLT2i can induce autophagy, mainly through an increase of AMPK, sirtuins, and HIF, and a reduction of mTOR [18].

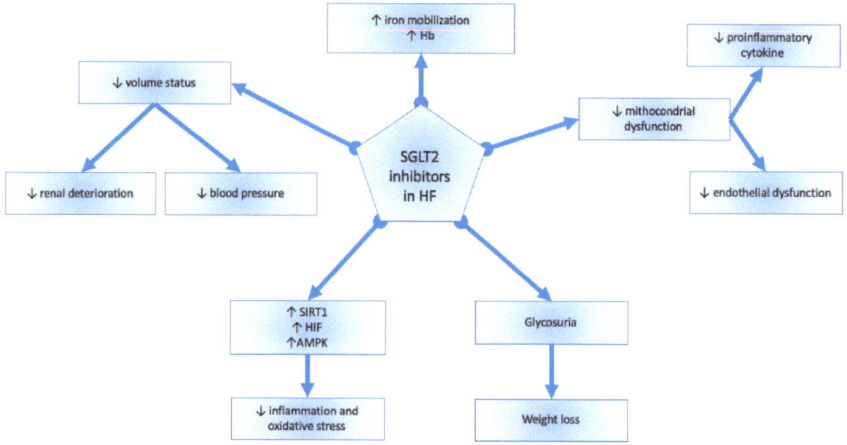

Figure 1. Mechanisms of SGLT2 inhibitors in heart failure.

5. SGLT2i and Chronic Heart Failure with Reduced Left Ventricular Ejection Fraction

The Dapagliflozin and Prevention of Adverse Outcomes in SGLT2i (DAPA-HF) trial was the first study that tested the efficacy of Dapagliflozin in patients with chronic HFrEF [19], as reported in Table 1. This study randomly assigned 4744 patients with HFrEF to receive either 10 mg/day Dapagliflozin or placebo in addition to recommended therapy, aiming to evaluate, as a primary outcome, a composite of worsening SGLT2i or cardiovascular death. At the end of the follow-up period (median 18.2 months), the authors found a lower frequency of primary outcome in patients treated with SGTL2 inhibitors compared to placebo (16.3% vs. 21.2%, HR 0.74; 95%CI 0.65–0.85). Similarly, the frequency of death from cardiovascular causes was lower in patients treated with Dapagliflozin (9.6% vs. 11.5%, HR 0.82; 95%CI 0.69–0.98). Interestingly, the main findings were comparable between patients with diabetes and those without diabetes. In addition, the frequency of adverse effects was similar between the groups. Post hoc analyses from the same trial also revealed that SGLT2i improved specific clinical outcomes regardless of frailty [20], race [21], and atrial fibrillation [22].

Similarly, the EMPEROR-Reduced trial is another randomized, double-blind, parallel-group, placebo-controlled trial that focused on patients with chronic HF (functional class II, III, or IV) with a left ventricular ejection fraction of 40% or less [23]. The sample consisted of 3730 patients who received either Empagliflozin 10 mg/daily or placebo and were followed for a median of 16 months. The risk of composite outcome (cardiovascular death or hospitalization for worsening SGLT2i) was reduced in patients who received SGLT2i compared to placebo (HR 0.75, 95%CI 0.58–0.85, $p < 0.001$) regardless of the presence of diabetes. Again, post hoc analyses of this trial revealed additional positive effects: reduced worsening of SGLT2i even in outpatients, with benefits seen early after initiation [24].

Another multicenter, randomized, double-blind clinical trial enrolled 90 patients with HFrEF, randomly assigned to receive either Dapagliflozin or placebo [25]. Follow-up was set at 1 and 3 months to evaluate changes in maximal functional capacity. The authors found a significant improvement in peak VO2 at 1 and 3 months, thus resulting in an early improvement in maximal exercise capacity. Finally, the DEFINE-HF trial demonstrated that patients treated with Dapagliflozin experienced an improvement in lung fluid volumes [26].

Table 1. Studies that tested SGLT2i in chronic heart failure with reduced left ventricular ejection fraction.

Study	n	Diabetes	SGTL2i	Follow Up	Outcomes	Events P.O. SGLT2i	Events P.O. Placebo	Result
DAPA-HF [19]	4744	45.0%	D	18.2 months	Composite outcome of worsening SGLT2i or death from CV causes	386	502	Reduced risk
					Composite of hospitalization for HF or CV death	382	495	Reduced risk
					Composite of number of hospitalizations for HF and CV death	567	742	Reduced risk
					Composite of worsening renal function (decline in the eGFR or renal death)	28	39	No effect
					Death from any cause	276	329	No effect
				8 months	Change from baseline of KCCQ	N/A	N/A	Improved patient-reported symptoms
DAPA-HF [20]	4742			18.2 months	Worsening of HF or CV death accordingly to the frailty index	N/A	N/A	Reduced risk regardless of frailty status. Absolute reductions were larger in more frail patients.
Palau et al. [25]	90	54.5%	D	1 and 3 months	Change from baseline in mean peakVO2	N/A	N/A	Improvement in peakVO2 at 1 and 3 months
EMPIRE HF [27]	190	20.0%	E	90 days	Change of N-terminal pro-brain natriuretic peptide (NT-proBNP)	N/A	N/A	No change
EMPIRE HF [28]	190	20.0%	E	12 weeks	Changes in erythropoiesis and iron metabolism	N/A	N/A	Increased erythropoiesis and augmented early iron utilization
EMPEROR-Reduced 32865377	3730	49.8%	E	18 months	Composite of cardiovascular death or hospitalization for worsening SGLT2i	361	462	Reduced risk
EMPEROR-Reduced [29]	3730	49.8%	E	12, 32, and 52 weeks	Changes in body weight	N/A	N/A	Benefits of SGTL2i were present across all BMI categories. Weight loss was associated with higher risk of all-cause mortality, regardless of treatment group.
DEFINE_HF [26]	85	75.6%	D	12 weeks	Changes in lung fluid volumes	N/A	N/A	Reduced lung congestion

Abbreviations: D, Dapagliflozin; E, Empagliflozin; heart failure, HF; CV, cardiovascular; N/A, not applicable; KCCQ-CS, Kansas City Cardiomyopathy Questionnaire; eGFR, estimated glomerular filtrate rate; SGLT2i, Na$^+$-glucose cotransporter-2 inhibitorsSGLT2i and chronic heart failure with mildly reduced left ventricular ejection fraction.

The DELIVER trial was a double-blind, randomized, controlled study that tested the efficacy of Dapagliflozin in patients with NFmrEF or HFrEF, as reported in Table 2. The study randomly assigned 6263 patients to Dapagliflozin 10 mg/day or placebo in addition to usual therapy to test the primary outcome (composite of worsening of HF-defined as unplanned hospitalization or urgent visit for HF or cardiovascular death) [30]. Results of the analyses revealed that worsening of SGLT2i occurred less frequently in patients treated with SGTL2 inhibitors (11.8% vs. 14.5%, HR 0.79, 95%CI, 0.69–0.91) as well as cardiovascular death (7.4% vs. 8.3%, HR 0.88, 95%CI 0.74–1.05). The presence of NFmrEF or HFpEF did not affect the results, as well as the presence of diabetes. The incidence of adverse effects was comparable between the two groups. The same trial also found patients treated with Dapagliflozin had a mild decline in estimated glomerular filtration rate, which was not associated with subsequent risk of cardiovascular event of acute kidney injury [31]. Further, chronic obstructive pulmonary disease, which is common in patients with HF, did not affect the beneficial effect of Dapagliflozin [32].

Table 2. Studies that tested SGLT2i in chronic heart failure with mildly reduced left ventricular ejection fraction and preserved left ventricular ejection fraction.

Study	n	Type of HF	Diabetes	SGTL2i	Follow Up	Outcomes	Events P.O. SGLT2i	Events P.O. Placebo	Result
DELIVER [31]	5788	NFmrEF HFpEF	42–50%	D	1 month	Kidney composite outcome (first occurrence of ≥50% decline in eGFR within 1 month, development of end-stage kidney disease, or death to kidney cause)	N/A	N/A	Initial eGFR decline after Dapagliflozin, which was not associated with subsequent risk of cardiovascular or kidney events.
DELIVER [30]	2216	NFmrEF	44.7% *		2.3 years	Composite of worsening HF or CV death	207	229	Reduced risk
	2064	HFpEF	44.7% *		2.3 years	Composite of worsening HF or CV death	305	381	Reduced risk
	3131	NFmrEF HFpEF	44.7% *	D	2.3 years	Total number of worsening SGLT2i events and cardiovascular death	815	1057	Reduced number
			44.7% *		8 months	Change from baseline of KCCQ	N/A	N/A	Improved patient-reported symptoms
			44.7% *		2.3 years	Cardiovascular death and from any cause	497	526	No difference
DELIVER [32]	NO COPD 5567	NFmrEF HFpEF	44.6%	D	8 months	Composite of worsening heart or cardiovascular death	N/A	N/A	Mild to moderate COPD is associated with worse outcomes but did not affect the beneficial effects of Dapagliflozin
	COPD 694	NFmrEF HFpEF	46.5%	D	8 months		N/A	N/A	
EMPERIAL-Preserved Trial [33]	315	HFpEF	51.1%	E	12 weeks	Change from baseline in 6MWT	N/A	N/A	Neutral effect on exercise ability
						Change from baseline in KCCQ-TSS and CHQ-SAS dyspnoea score	N/A	N/A	No effect on specific dyspnoea score

Table 2. Cont.

Study	n	Type of HF	Diabetes	SGTL2i	Follow Up	Outcomes	Events P.O. SGTL2i	Events P.O. Placebo	Result
EMPEROR-Preserved trial [34]	5988	HFpEF	44.8%	D	26.2 months	Combined risk of CV death or hospitalization for HF	415	511	Reduced risk, regardless of diabetes
						Occurrence of all adjudicated hospitalizations for HF	407	541	Reduction of hospitalization for HF
						Rate of decline in the eGFR during treatment	N/A	N/A	Reduction of rate of decline
PRESERVED-HF [35]	324	HFpEF	56.6%	D	12 weeks	Change in KCCQ-CS at 12 weeks	N/A	N/A	Improved patient-reported symptoms
						Meaningful (five points or greater) change in KCCQ-CS and -OS	N/A	N/A	Magnitude of benefit higher in patients treated with SGTL2i
						Change in 6MWT distance	N/A	N/A	Improved exercise function

* Frequency of diabetes in whole sample (heart failure with mildly reduced left ventricular ejection fraction and heart failure with preserved left ventricular ejection fraction). Abbreviations: D, Dapagliflozin; E, Empagliflozin; HFrEF, heart failure with reduced left ventricular ejection fraction; NFmrEF, heart failure with mildly reduced left ventricular ejection fraction; HFpEF, heart failure with preserved left ventricular ejection fraction; heart failure, HF; CV, cardiovascular; COPD, chronic obstructive lung disease; N/A, not applicable; KCCQ-CS, Kansas City Cardiomyopathy Questionnaire Clinical Summary Score; KCCQ-OS, Kansas City Cardiomyopathy Questionnaire Overall Summary Score; 6MWT, 6-min walk test; CHQ-SAS, Chronic Heart Failure Questionnaire Self-Administered Standardized format; eGFR, estimated glomerular filtrate rat; SGLT2i, Na^+-glucose cotransporter-2 inhibitors.

6. SGLT2i and Chronic Heart Failure with Preserved Left Ventricular Ejection Fraction

HFpEF accounts for at least half of the patients with SGLT2i. Clinical trials for HFpEF gave different results. Indeed, while some previous studies reported no positive effect on mortality and limited impact on HF hospitalizations, recent guidelines suggested the prescription of SGTL2 inhibitors for the management of HF. The EMPERIAL-Preserved Trial tested the effect of Empagliflozin on exercise ability and HF symptoms [33]. The authors enrolled HFpEF and HFrEF, with or without type 2 diabetes mellitus, and treated with Empagliflozin 10 mg/day or placebo for 12 weeks to assess a change in the 6-min walk test distance. The authors also evaluated symptoms of SGLT2i through the Kansas City Cardiomyopathy Questionnaire Total Symptom Score (KCCQ-TSS) and Chronic SGLT2i Questionnaire Self-Administered Standardized format (CHQ-SAS) dyspnea score. Analyses showed no effect on exercise ability or specific dyspnea score in both types of HF. This study was somewhat limited by the small sample size and by the short follow-up, which did not allow the evaluation of specific outcomes (mortality and hospitalization). The EMPEROR-Preserved trial was a multicenter, double-blinded, placebo-controlled randomized trial that assessed Empagliflozin's effects on a composite of cardiovascular death or hospitalization for SGLT2i [34]. The study included 5988 patients with HFpEF (defined as ejection fraction > 40%), who were treated with Empagliflozin 10 mg/daily or placebo in addition to the usual therapy for a median of 26.2 months. Primary outcomes showed a 21% reduction in patients treated with SGTL2 inhibitors, mainly driven by the reduction of hospitalization for SGLT2i (HR 0.79%; 95%CI 0.69–0.90, $p < 0.001$). Indeed, the incidence of cardiovascular death was lower but not significant (HR 0.91, 95%CI 0.76–1.0). This trial also showed a benefit of the use of Empagliflozin in the secondary outcomes: reduction of eGFR decline and total SGLT2i hospitalization, and a modest improvement in quality of life, regardless of the presence of diabetes. Another trial, PRESERVED-HF, focused on the effects of SGTL2 inhibitors on symptoms and exercise function in HFpEF patients [35]. The study, which was a multicenter, double-blinded, placebo-controlled study, included 324 patients treated either with Empagliflozin 10 mg/daily or placebo. At 12 weeks, using SGTL2 inhibitor improved the Kansas City Cardiomyopathy Questionnaire Clinical Summary Score (KCCQ-CS) and weight, natriuretic peptides, glycated hemoglobin, and systolic blood pressure. The benefit of Empagliflozin on HFpEF patients was independent of diabetes.

The results of the DELIVER trial were in line with previous studies. This multicenter, double-blind, placebo-controlled, randomized study assessed whether Dapagliflozin 10 mg/daily would improve worsening SGLT2i or cardiovascular death (primary composite outcome) in symptomatic stable HFpEF patients, with or without diabetes mellitus [30]. The study enrolled 6263 patients randomized to either Dapagliflozin 10 mg/daily or placebo in addition to the usual therapy. Over a median of 2.3 years, the primary outcome occurred less frequently in the Dapagliflozin group (16.4% vs. 19.5%, HR 0.73, 95%CI 0.73–0.92, $p < 0.001$), mainly driven by the reduction of worsening of SGLT2i (11.8% vs. 14.5%, HR 0.79, 95%CI 0.69–0.91) as compared to cardiovascular death (7.4% vs. 8.3%, HR 0.88, 95%CI 0.74–1.05).

7. SGLT2i and Acute Heart Failure

Studies tested the benefit of SGTL2 inhibitors for treating acute decompensated SGLT2i in hospitalized patients (Table 3). The EMPULSE trial analyzed 530 patients hospitalized for decompensated SGLT2i irrespective of left ventricular ejection fraction [36]. Compared to placebo, subjects treated with Empagliflozin 10 mg daily within 5 days of admission had a significant clinical benefit, defined as a hierarchical composite of death from any cause, number of SGLT2i events, and Kansas City Cardiomyopathy Questionnaire total symptom score. The EMPAG-HF focused on cumulative urine output over 5 days [37]. This single-center prospective, double-blind, placebo-controlled study randomized 59 patients within 12 h of hospitalization for acute decompensated HF. Patients, in addition to the standard decongestive treatments, were randomly assigned to Empagliflozin 25 mg daily or placebo. The authors reported a 25% increase in cumulative urine output without a decline of glomerular filtration rate. Similar results were obtained in the EMPA-RESPONSE-AHF, which demonstrated the beneficial effect of Empagliflozin on patients treated within 24 h of the presentation to the hospital [10]. Albeit the authors did not find changes in Visual Analogue Scale dyspnea, diuretic response, and length of hospital stay, they reported an increased urinary output and reduced combined endpoint (worsening SGLT2i, hospitalization for SGLT2i, or death at 60 days). It should be noted that all these studies tested SGTL2 inhibitors during low doses of intravenous furosemide. The DAPA-RESIST study tested the effect of SGTL2 inhibitors in patients with diuretic resistance, defined as insufficient decongestion despite treatment, with a high dose of intravenous furosemide ($\geq$160 mg/day) [38]. Patients were randomized to Dapagliflozin 10 mg/day or Metolazone 5–10 mg/day for a 3-day treatment period. The authors found that patients treated with Dapagliflozin received a larger cumulative dose of furosemide but without a more efficient relief of pulmonary congestion. A significant weight reduction at up to 96 h of Dapagliflozin was also documented.

Recent studies also focused on early prescriptions of SGTL2 inhibitors. The SOLOIST-WHF, a multicenter, double-blind trial, randomized 608 patients to Sotagliflozin and 614 to placebo and administered before discharge (48.8%) and a median of 2 days after discharge (51.2%). The primary endpoint was death from cardiovascular causes and hospitalization or urgent visits for SGLT2i [39]. Patients were followed for a median of 9 months, and 600 primary endpoints occurred. The rate of cardiovascular death was lower in the Sotagliflozin group (HR 0.84; 95%CI 0.58–1.22), while the frequency of acute kidney injury was similar to those who had a placebo. A post hoc analysis also demonstrated that starting Sotagliflozin before the discharge significantly decreased cardiovascular death and SGLT2i events 30 and 90 days after the discharge [40]. Another retrospective analysis by Burgos et al. pointed out that in-hospital initiation of SGLT-2 inhibitors was associated with significantly higher prescription rates and lower prevalence of hospitalization or urgent visits for acute SGLT2i or all-cause mortality at 90 days [41].

Table 3. Studies that tested SGLT2i in acute decompensated heart failure.

Study	n	% Reduced LVEF	Treatment Type	Treatment Start	Treatment Duration	Follow-Up	Results
EMPULSE [36]	530	67% [1]	Empagliflozin 10 mg/day vs. placebo	1–5 days after hospital admission	90 days	90 days	Clinical benefit (hierarchical composite of death from any cause, number of SGLT2i events, and KCCQ-SC)
EMPAG-HF [37]	59	20.7% [2]	Empagliflozin 10 mg/day vs. placebo	Within 12 h	5 days	30 days	25% increase in cumulative urine output without affection of renal function
EMPA-REPONSE-AHF [10]	79	100% [3]	Empagliflozin 10 mg/day vs. placebo	Within 24 h	30 days	60 days	Increased urinary output and reduced combined endpoint (worsening HF, rehospitalization for HF or death at 60 days). No effect on VAS dyspnea, diuretic response, NT-pro-BNP, or length of hospital stay.
DAPA-RESIST [38]	54	44% [4]	Dapagliflozin 10 mg/day vs. Metalozone 5–10 mg/day	Within 24 h	5 days	90 days	Weight reduction at up to 96 h

[1] LVEF < 40%; [2] LVEF < 30%; [3] LVEF < 50%; [4] LVEF < 40%. Abbreviation: LVEF, Left ventricular ejection fraction, HF heart failure; VAS, Visual Analogue Scale; KCCQ-SC, Kansas City Cardiomyopathy Questionnaire total symptom score; HF, heart failure; SGLT2i, Na$^+$-glucose cotransporter-2 inhibitors.

Current data suggest a role for SGTL2 inhibitors in the treatment of acute decompensated SGLT2i. However, they must be considered an additional therapy that cannot replace the loop diuretic, which is the landmark for the treatment of acute SGLT2i. Studies also suggest starting SGLT2 early, which is substantially equivalent when administered within 12 h of the onset of acute SGLT2i or the days before discharge from the hospital.

8. Additional Effects of SGTL2 Inhibitors

8.1. SGLT2i and Left Ventricular Mass

The EMPA-HEART CardioLink-6 study evaluated the effect of Empagliflozin on left ventricular mass in patients with coronary artery disease and type 2 diabetes mellitus. The authors recruited 97 subjects randomized to Empagliflozin 10 mg/day or placebo. The primary outcome was the 6-month change in left ventricular mass indexed to the body surface assessed by cardiac resonance imaging. Authors reported a regression of mean left ventricular mass in patients treated with SGTL2i compared to placebo (-2.6 g/m^2 vs. 0.01 g/m^2, $p = 0.01$). Further, the Empagliflozin group had a reduction of systolic and diastolic blood pressure (respectively, -6.8 mmHg vs. -2.3 mmHg $p = 0.003$ and -3.2 mmHg vs. -0.6 mmHg $p = 0.02$) and an elevation of hematocrit ($p = 0.0003$) [42]. To rule out the anti-hyperglycemic effect, the EMPA-TROPISM study evaluated the left ventricular mass on 84 nondiabetic patients with SGLT2i with reduced ejection fraction, which were randomized to Empagliflozin 10 mg/daily or placebo. After 6 months, these patients showed a significant decrease in left ventricular mass, other than an improvement in ejection fraction and 6-min walk [43].

8.2. SGLT2i and Acid Uric Metabolism

The association between hyperuricemia and HF is well acknowledged. Serum uric acid is an oxidative stress index that contributes to endothelial dysfunction by impairing nitric oxide production, considered a prognostic index in patients with preexisting HF [44]. The exact pathophysiological link between the two diseases is not clear, but it has been demonstrated that serum acid uric concentration is related to greater activity of superoxide dismutase and endothelium-dependent vasodilatation [40], and other studies showed a possible further link with inflammation. Indeed, hyperuricemia is associated with interleukin-6, neutrophil count, and C-reactive protein, all specific markers of proinflammatory state associated with an increased risk of HF [44–46]. In addition, hyperuricemia has been associated with an increased risk of developing hypertension and coronary heart disease [47,48], which can further explain the link between increased uric acid and HF.

SGTL2i act on different pathways of uric acid metabolism. Indeed, SGTL2i decrease purine synthesis, downregulate different enzymes of the pentose phosphate pathways, and reduce intracellular levels of hypoxanthine, reducing NADPH oxidase activity [49,50]. The effect of SGTL2i is not limited to lowering uric acid synthesis but also promotes its excretion. Indeed, these drugs show a uricosuric effect, which is strictly connected to the glycosuric effect. Indeed, SGTL2i may upregulate ABCG2 and can downregulate URAT1, which is a major protein involved in uric acid reabsorption [51,52].

8.3. SGLT2i and Iron Metabolism

Absolute or relative iron deficiency is commonly found in patients with HF. Reduced dietary intake, chronic blood loss, and impaired absorption, which can be secondary to gut edema, the use of specific drugs, and/or chronic inflammation, are all factors that can contribute to iron deficiency. Anemia is less frequent but recognizes the same causes [53]. It is well acknowledged that iron deficiency has a worse impact in patients with HF, and several randomized trials have documented an improvement after treatment with ferric Carboxymaltose [54]. An increased hematocrit and hemoglobin are frequently found in HF patients treated with SGLT2i, but it is not clear whether this effect is secondary to hemoconcentration or increased erythropoiesis. The DAPA-HF examined iron deficiency's prevalence and consequences in patients with HF treated with SGLT2i [55]. The study found that 43.7% of the sample had iron deficiency, with an increased rate of worsening SGLT2i (hospitalization or urgent visit requiring intravenous treatment) compared to those with normal blood iron. Analyses showed a trend for the beneficial effect of Dapagliflozin on worsening SGLT2i was greater in patients with iron deficiency compared to those iron-repleted (HR 0.74 95%CI 0.58–0.92 vs. HR 0.81, 95%CI 0.63–1.03, p-interaction = 0.59). In addition, the authors also found that patients treated with Dapagliflozin also had a reduction of transferrin saturation, ferritin, and hepcidin. In contrast, iron-binding capacity and soluble transferrin receptors increased compared to the placebo. Therefore, the authors pointed out that Dapagliflozin increased iron use and, at the same time, improved clinical outcomes regardless of baseline iron status. Another trial aimed to investigate the early effect of Empagliflozin on iron metabolism and erythropoiesis in patients with HFrEF [28]. Patients were randomly assigned to either Empagliflozin or placebo for 12 weeks, and the analyses suggested that the use of SGLT2i increased erythropoiesis and augmented early iron utilization, contributing to a cardioprotective effect. This result is somewhat consistent with findings reported by Docherty et al. [56]. The study was a post hoc exploratory analysis of the IRONMAN trial, which randomized patients with SGLT2i with iron deficiency to intravenous ferric Derisomaltose or usual care. Analyses reported a trend of a more significant increase in hemoglobin in patients treated with ferric Derisomaltose, which had SGLT2i at baseline.

Overall, it is still not clear whether anemia simply reflects a marker of poor prognosis in patients with HFrEF or is a therapeutic goal that needs to be treated. Indeed, while some studies reported Dapagliflozin corrected anemia more frequently than placebo, thus improving outcomes, other studies failed to demonstrate improvement in cardiovascular outcomes in anemic patients with SGLT2i treated with Darbepoetin alfa [57]. Therefore, it is still to be elucidated whether SGLT2i has a synergistic effect with iron therapy in some patients with HFrEF.

9. Conclusions

SGLT2i have dramatically improved the management of diabetes, and their use for the treatment of HF is now strengthened by the growing evidence that suggested beneficial effects in almost all types of HF. The mechanisms underlying the cardioprotective effect of SGLT2i are multiple: lowering blood pressure, enhancing diuresis, improving glycemic control, and preventing inflammation. Possible adverse effects may limit their prescription in selected patients, in particular in those who experienced genital mycotic infection, pyelonephritis, and acute kidney injury. Evidence suggests that all these mechanisms collectively contribute to the benefits documented by the different clinical trials in patients with HF.

Author Contributions: N.I.P. and A.P.D. wrote the manuscript; R.C., A.G., R.M. and M.M. reviewed and edited the manuscript. All authors have read and agreed to the published version of the manuscript.

Funding: This research received no external funding.

Informed Consent Statement: Not applicable.

Data Availability Statement: Not applicable.

Conflicts of Interest: The authors declare no conflicts of interest.

References

1. Sinclair, A.; Saeedi, P.; Kaundal, A.; Karuranga, S.; Malanda, B.; Williams, R. Diabetes and global ageing among 65–99-year-old adults: Findings from the International Diabetes Federation Diabetes Atlas, 9th edition. *Diabetes Res. Clin. Pract.* **2020**, *162*, 108078. [CrossRef] [PubMed]
2. Mather, A.; Pollock, C. Glucose handling by the kidney. *Kidney Int. Suppl.* **2011**, *79*, S1–S6. [CrossRef] [PubMed]
3. Wright, E.M.; Loo, D.D.; Hirayama, B.A. Biology of human sodium glucose transporters. *Physiol. Rev.* **2011**, *91*, 733–794. [CrossRef] [PubMed]
4. Profili, N.I.; Castelli, R.; Gidaro, A.; Manetti, R.; Maioli, M.; Petrillo, M.; Capobianco, G.; Delitala, A.P. Possible Effect of Polycystic Ovary Syndrome (PCOS) on Cardiovascular Disease (CVD): An Update. *J. Clin. Med.* **2024**, *13*, 698. [CrossRef]
5. Castelli, R.; Gidaro, A.; Casu, G.; Merella, P.; Profili, N.I.; Donadoni, M.; Maioli, M.; Delitala, A.P. Aging of the Arterial System. *Int. J. Mol. Sci.* **2023**, *24*, 6910. [CrossRef]
6. McDonagh, T.A.; Metra, M.; Adamo, M.; Gardner, R.S.; Baumbach, A.; Bohm, M.; Burri, H.; Butler, J.; Celutkiene, J.; Chioncel, O.; et al. Corrigendum to: 2021 ESC Guidelines for the diagnosis and treatment of acute and chronic heart failure: Developed by the Task Force for the diagnosis and treatment of acute and chronic heart failure of the European Society of Cardiology (ESC) with the special contribution of the Heart Failure Association (HFA) of the ESC. *Eur. Heart J.* **2021**, *42*, 4901. [CrossRef]
7. van Riet, E.E.; Hoes, A.W.; Limburg, A.; Landman, M.A.; van der Hoeven, H.; Rutten, F.H. Prevalence of unrecognized heart failure in older persons with shortness of breath on exertion. *Eur. J. Heart Fail.* **2014**, *16*, 772–777. [CrossRef]
8. Sha, S.; Polidori, D.; Heise, T.; Natarajan, J.; Farrell, K.; Wang, S.S.; Sica, D.; Rothenberg, P.; Plum-Morschel, L. Effect of the sodium glucose co-transporter 2 inhibitor canagliflozin on plasma volume in patients with type 2 diabetes mellitus. *Diabetes Obes. Metab.* **2014**, *16*, 1087–1095. [CrossRef]
9. Griffin, M.; Rao, V.S.; Ivey-Miranda, J.; Fleming, J.; Mahoney, D.; Maulion, C.; Suda, N.; Siwakoti, K.; Ahmad, T.; Jacoby, D.; et al. Empagliflozin in Heart Failure: Diuretic and Cardiorenal Effects. *Circulation* **2020**, *142*, 1028–1039. [CrossRef]
10. Damman, K.; Beusekamp, J.C.; Boorsma, E.M.; Swart, H.P.; Smilde, T.D.J.; Elvan, A.; van Eck, J.W.M.; Heerspink, H.J.L.; Voors, A.A. Randomized, double-blind, placebo-controlled, multicentre pilot study on the effects of empagliflozin on clinical outcomes in patients with acute decompensated heart failure (EMPA-RESPONSE-AHF). *Eur. J. Heart Fail.* **2020**, *22*, 713–722. [CrossRef]
11. Nassif, M.E.; Windsor, S.L.; Tang, F.; Khariton, Y.; Husain, M.; Inzucchi, S.E.; McGuire, D.K.; Pitt, B.; Scirica, B.M.; Austin, B.; et al. Dapagliflozin Effects on Biomarkers, Symptoms, and Functional Status in Patients with Heart Failure with Reduced Ejection Fraction: The DEFINE-HF Trial. *Circulation* **2019**, *140*, 1463–1476. [CrossRef] [PubMed]
12. Margonato, D.; Galati, G.; Mazzetti, S.; Cannistraci, R.; Perseghin, G.; Margonato, A.; Mortara, A. Renal protection: A leading mechanism for cardiovascular benefit in patients treated with SGLT2 inhibitors. *Heart Fail. Rev.* **2021**, *26*, 337–345. [CrossRef] [PubMed]
13. Kato, E.T.; Silverman, M.G.; Mosenzon, O.; Zelniker, T.A.; Cahn, A.; Furtado, R.H.M.; Kuder, J.; Murphy, S.A.; Bhatt, D.L.; Leiter, L.A.; et al. Effect of Dapagliflozin on Heart Failure and Mortality in Type 2 Diabetes Mellitus. *Circulation* **2019**, *139*, 2528–2536. [CrossRef] [PubMed]

14. Heerspink, H.J.L.; Perco, P.; Mulder, S.; Leierer, J.; Hansen, M.K.; Heinzel, A.; Mayer, G. Canagliflozin reduces inflammation and fibrosis biomarkers: A potential mechanism of action for beneficial effects of SGLT2 inhibitors in diabetic kidney disease. *Diabetologia* 2019, *62*, 1154–1166. [CrossRef] [PubMed]
15. Theofilis, P.; Sagris, M.; Oikonomou, E.; Antonopoulos, A.S.; Siasos, G.; Tsioufis, K.; Tousoulis, D. The impact of SGLT2 inhibitors on inflammation: A systematic review and meta-analysis of studies in rodents. *Int. Immunopharmacol.* 2022, *111*, 109080. [CrossRef]
16. Byrne, N.J.; Matsumura, N.; Maayah, Z.H.; Ferdaoussi, M.; Takahara, S.; Darwesh, A.M.; Levasseur, J.L.; Jahng, J.W.S.; Vos, D.; Parajuli, N.; et al. Empagliflozin Blunts Worsening Cardiac Dysfunction Associated with Reduced NLRP3 (Nucleotide-Binding Domain-Like Receptor Protein 3) Inflammasome Activation in Heart Failure. *Circ. Heart Fail.* 2020, *13*, e006277. [CrossRef]
17. Gao, Y.M.; Feng, S.T.; Wen, Y.; Tang, T.T.; Wang, B.; Liu, B.C. Cardiorenal protection of SGLT2 inhibitors-Perspectives from metabolic reprogramming. *EBioMedicine* 2022, *83*, 104215. [CrossRef]
18. Pandey, A.K.; Bhatt, D.L.; Pandey, A.; Marx, N.; Cosentino, F.; Pandey, A.; Verma, S. Mechanisms of benefits of sodium-glucose cotransporter 2 inhibitors in heart failure with preserved ejection fraction. *Eur. Heart J.* 2023, *44*, 3640–3651. [CrossRef]
19. McMurray, J.J.V.; Solomon, S.D.; Inzucchi, S.E.; Kober, L.; Kosiborod, M.N.; Martinez, F.A.; Ponikowski, P.; Sabatine, M.S.; Anand, I.S.; Belohlavek, J.; et al. Dapagliflozin in Patients with Heart Failure and Reduced Ejection Fraction. *N. Engl. J. Med.* 2019, *381*, 1995–2008. [CrossRef]
20. Butt, J.H.; Dewan, P.; Merkely, B.; Belohlavek, J.; Drozdz, J.; Kitakaze, M.; Inzucchi, S.E.; Kosiborod, M.N.; Martinez, F.A.; Tereshchenko, S.; et al. Efficacy and Safety of Dapagliflozin according to Frailty in Heart Failure with Reduced Ejection Fraction: A Post Hoc Analysis of the DAPA-HF Trial. *Ann. Intern. Med.* 2022, *175*, 820–830. [CrossRef]
21. Docherty, K.F.; Ogunniyi, M.O.; Anand, I.S.; Desai, A.S.; Diez, M.; Howlett, J.G.; Nicolau, J.C.; O'Meara, E.; Verma, S.; Inzucchi, S.E.; et al. Efficacy of Dapagliflozin in Black Versus White Patients with Heart Failure and Reduced Ejection Fraction. *JACC Heart Fail.* 2022, *10*, 52–64. [CrossRef] [PubMed]
22. Butt, J.H.; Docherty, K.F.; Jhund, P.S.; de Boer, R.A.; Bohm, M.; Desai, A.S.; Howlett, J.G.; Inzucchi, S.E.; Kosiborod, M.N.; Martinez, F.A.; et al. Dapagliflozin and atrial fibrillation in heart failure with reduced ejection fraction: Insights from DAPA-HF. *Eur. J. Heart Fail.* 2022, *24*, 513–525. [CrossRef] [PubMed]
23. Packer, M.; Anker, S.D.; Butler, J.; Filippatos, G.; Pocock, S.J.; Carson, P.; Januzzi, J.; Verma, S.; Tsutsui, H.; Brueckmann, M.; et al. Cardiovascular and Renal Outcomes with Empagliflozin in Heart Failure. *N. Engl. J. Med.* 2020, *383*, 1413–1424. [CrossRef] [PubMed]
24. Packer, M.; Anker, S.D.; Butler, J.; Filippatos, G.; Ferreira, J.P.; Pocock, S.J.; Carson, P.; Anand, I.; Doehner, W.; Haass, M.; et al. Effect of Empagliflozin on the Clinical Stability of Patients with Heart Failure and a Reduced Ejection Fraction: The EMPEROR-Reduced Trial. *Circulation* 2021, *143*, 326–336. [CrossRef]
25. Palau, P.; Amiguet, M.; Dominguez, E.; Sastre, C.; Mollar, A.; Seller, J.; Garcia Pinilla, J.M.; Larumbe, A.; Valle, A.; Gomez Doblas, J.J.; et al. Short-term effects of dapagliflozin on maximal functional capacity in heart failure with reduced ejection fraction (DAPA-VO(2)): A randomized clinical trial. *Eur. J. Heart Fail.* 2022, *24*, 1816–1826. [CrossRef]
26. Nassif, M.E.; Windsor, S.L.; Tang, F.; Husain, M.; Inzucchi, S.E.; McGuire, D.K.; Pitt, B.; Scirica, B.M.; Austin, B.; Fong, M.W.; et al. Dapagliflozin effects on lung fluid volumes in patients with heart failure and reduced ejection fraction: Results from the DEFINE-HF trial. *Diabetes Obes. Metab.* 2021, *23*, 1426–1430. [CrossRef]
27. Jensen, J.; Omar, M.; Kistorp, C.; Poulsen, M.K.; Tuxen, C.; Gustafsson, I.; Kober, L.; Gustafsson, F.; Faber, J.; Fosbol, E.L.; et al. Twelve weeks of treatment with empagliflozin in patients with heart failure and reduced ejection fraction: A double-blinded, randomized, and placebo-controlled trial. *Am. Heart J.* 2020, *228*, 47–56. [CrossRef]
28. Fuchs Andersen, C.; Omar, M.; Glenthoj, A.; El Fassi, D.; Moller, H.J.; Lindholm Kurtzhals, J.A.; Styrishave, B.; Kistorp, C.; Tuxen, C.; Poulsen, M.K.; et al. Effects of empagliflozin on erythropoiesis in heart failure: Data from the Empire HF trial. *Eur. J. Heart Fail.* 2023, *25*, 226–234. [CrossRef]
29. Anker, S.D.; Khan, M.S.; Butler, J.; Ofstad, A.P.; Peil, B.; Pfarr, E.; Doehner, W.; Sattar, N.; Coats, A.J.S.; Filippatos, G.; et al. Weight change and clinical outcomes in heart failure with reduced ejection fraction: Insights from EMPEROR-Reduced. *Eur. J. Heart Fail.* 2023, *25*, 117–127. [CrossRef]
30. Solomon, S.D.; McMurray, J.J.V.; Claggett, B.; de Boer, R.A.; DeMets, D.; Hernandez, A.F.; Inzucchi, S.E.; Kosiborod, M.N.; Lam, C.S.P.; Martinez, F.; et al. Dapagliflozin in Heart Failure with Mildly Reduced or Preserved Ejection Fraction. *N. Engl. J. Med.* 2022, *387*, 1089–1098. [CrossRef]
31. Mc Causland, F.R.; Claggett, B.L.; Vaduganathan, M.; Desai, A.; Jhund, P.; Vardeny, O.; Fang, J.C.; de Boer, R.A.; Docherty, K.F.; Hernandez, A.F.; et al. Decline in Estimated Glomerular Filtration Rate after Dapagliflozin in Heart Failure with Mildly Reduced or Preserved Ejection Fraction: A Prespecified Secondary Analysis of the DELIVER Randomized Clinical Trial. *JAMA Cardiol.* 2024, *9*, 144–152. [CrossRef] [PubMed]
32. Butt, J.H.; Lu, H.; Kondo, T.; Bachus, E.; de Boer, R.A.; Inzucchi, S.E.; Jhund, P.S.; Kosiborod, M.N.; Lam, C.S.P.; Martinez, F.A.; et al. Heart failure, chronic obstructive pulmonary disease and efficacy and safety of dapagliflozin in heart failure with mildly reduced or preserved ejection fraction: Insights from DELIVER. *Eur. J. Heart Fail.* 2023, *25*, 2078–2090. [CrossRef] [PubMed]
33. Abraham, W.T.; Lindenfeld, J.; Ponikowski, P.; Agostoni, P.; Butler, J.; Desai, A.S.; Filippatos, G.; Gniot, J.; Fu, M.; Gullestad, L.; et al. Effect of empagliflozin on exercise ability and symptoms in heart failure patients with reduced and preserved ejection fraction, with and without type 2 diabetes. *Eur. Heart J.* 2021, *42*, 700–710. [CrossRef] [PubMed]

34. Anker, S.D.; Butler, J.; Filippatos, G.; Ferreira, J.P.; Bocchi, E.; Bohm, M.; Brunner-La Rocca, H.P.; Choi, D.J.; Chopra, V.; Chuquiure-Valenzuela, E.; et al. Empagliflozin in Heart Failure with a Preserved Ejection Fraction. *N. Engl. J. Med.* **2021**, *385*, 1451–1461. [CrossRef] [PubMed]
35. Nassif, M.E.; Windsor, S.L.; Borlaug, B.A.; Kitzman, D.W.; Shah, S.J.; Tang, F.; Khariton, Y.; Malik, A.O.; Khumri, T.; Umpierrez, G.; et al. The SGLT2 inhibitor dapagliflozin in heart failure with preserved ejection fraction: A multicenter randomized trial. *Nat. Med.* **2021**, *27*, 1954–1960. [CrossRef]
36. Tromp, J.; Kosiborod, M.N.; Angermann, C.E.; Collins, S.P.; Teerlink, J.R.; Ponikowski, P.; Biegus, J.; Ferreira, J.P.; Nassif, M.E.; Psotka, M.A.; et al. Treatment effects of empagliflozin in hospitalized heart failure patients across the range of left ventricular ejection fraction—Results from the EMPULSE trial. *Eur. J. Heart Fail.* **2024**, *26*, 963–970. [CrossRef]
37. Schulze, P.C.; Bogoviku, J.; Westphal, J.; Aftanski, P.; Haertel, F.; Grund, S.; von Haehling, S.; Schumacher, U.; Mobius-Winkler, S.; Busch, M. Effects of Early Empagliflozin Initiation on Diuresis and Kidney Function in Patients with Acute Decompensated Heart Failure (EMPAG-HF). *Circulation* **2022**, *146*, 289–298. [CrossRef]
38. Yeoh, S.E.; Osmanska, J.; Petrie, M.C.; Brooksbank, K.J.M.; Clark, A.L.; Docherty, K.F.; Foley, P.W.X.; Guha, K.; Halliday, C.A.; Jhund, P.S.; et al. Dapagliflozin vs. metolazone in heart failure resistant to loop diuretics. *Eur. Heart J.* **2023**, *44*, 2966–2977. [CrossRef]
39. Bhatt, D.L.; Szarek, M.; Steg, P.G.; Cannon, C.P.; Leiter, L.A.; McGuire, D.K.; Lewis, J.B.; Riddle, M.C.; Voors, A.A.; Metra, M.; et al. Sotagliflozin in Patients with Diabetes and Recent Worsening Heart Failure. *N. Engl. J. Med.* **2021**, *384*, 117–128. [CrossRef]
40. Pitt, B.; Bhatt, D.L.; Szarek, M.; Cannon, C.P.; Leiter, L.A.; McGuire, D.K.; Lewis, J.B.; Riddle, M.C.; Voors, A.A.; Metra, M.; et al. Effect of Sotagliflozin on Early Mortality and Heart Failure-Related Events: A Post Hoc Analysis of SOLOIST-WHF. *JACC Heart Fail.* **2023**, *11*, 879–889. [CrossRef]
41. Burgos, L.M.; Ballari, F.N.; Spaccavento, A.; Ricciardi, B.; Suarez, L.L.; Baro Vila, R.C.; De Bortoli, M.A.; Conde, D.; Diez, M. In-hospital initiation of sodium-glucose cotransporter-2 inhibitors in patients with heart failure and reduced ejection fraction: 90-day prescription patterns and clinical implications. *Curr. Probl. Cardiol.* **2024**, *49*, 102779. [CrossRef] [PubMed]
42. Verma, S.; Mazer, C.D.; Yan, A.T.; Mason, T.; Garg, V.; Teoh, H.; Zuo, F.; Quan, A.; Farkouh, M.E.; Fitchett, D.H.; et al. Effect of Empagliflozin on Left Ventricular Mass in Patients with Type 2 Diabetes Mellitus and Coronary Artery Disease: The EMPA-HEART CardioLink-6 Randomized Clinical Trial. *Circulation* **2019**, *140*, 1693–1702. [CrossRef] [PubMed]
43. Santos-Gallego, C.G.; Vargas-Delgado, A.P.; Requena-Ibanez, J.A.; Garcia-Ropero, A.; Mancini, D.; Pinney, S.; Macaluso, F.; Sartori, S.; Roque, M.; Sabatel-Perez, F.; et al. Randomized Trial of Empagliflozin in Nondiabetic Patients with Heart Failure and Reduced Ejection Fraction. *J. Am. Coll. Cardiol.* **2021**, *77*, 243–255. [CrossRef]
44. Kanellis, J.; Kang, D.H. Uric acid as a mediator of endothelial dysfunction, inflammation, and vascular disease. *Semin. Nephrol.* **2005**, *25*, 39–42. [CrossRef] [PubMed]
45. Coutinho Tde, A.; Turner, S.T.; Peyser, P.A.; Bielak, L.F.; Sheedy, P.F., 2nd; Kullo, I.J. Associations of serum uric acid with markers of inflammation, metabolic syndrome, and subclinical coronary atherosclerosis. *Am. J. Hypertens* **2007**, *20*, 83–89. [CrossRef]
46. Ruggiero, C.; Cherubini, A.; Ble, A.; Bos, A.J.; Maggio, M.; Dixit, V.D.; Lauretani, F.; Bandinelli, S.; Senin, U.; Ferrucci, L. Uric acid and inflammatory markers. *Eur. Heart J.* **2006**, *27*, 1174–1181. [CrossRef]
47. Baker, J.F.; Krishnan, E.; Chen, L.; Schumacher, H.R. Serum uric acid and cardiovascular disease: Recent developments, and where do they leave us? *Am. J. Med.* **2005**, *118*, 816–826. [CrossRef]
48. Sundstrom, J.; Sullivan, L.; D'Agostino, R.B.; Levy, D.; Kannel, W.B.; Vasan, R.S. Relations of serum uric acid to longitudinal blood pressure tracking and hypertension incidence. *Hypertension* **2005**, *45*, 28–33. [CrossRef]
49. Xing, Y.J.; Liu, B.H.; Wan, S.J.; Cheng, Y.; Zhou, S.M.; Sun, Y.; Yao, X.M.; Hua, Q.; Meng, X.J.; Cheng, J.H.; et al. A SGLT2 Inhibitor Dapagliflozin Alleviates Diabetic Cardiomyopathy by Suppressing High Glucose-Induced Oxidative Stress in vivo and in vitro. *Front. Pharmacol.* **2021**, *12*, 708177. [CrossRef]
50. Lu, Y.P.; Zhang, Z.Y.; Wu, H.W.; Fang, L.J.; Hu, B.; Tang, C.; Zhang, Y.Q.; Yin, L.; Tang, D.E.; Zheng, Z.H.; et al. SGLT2 inhibitors improve kidney function and morphology by regulating renal metabolic reprogramming in mice with diabetic kidney disease. *J. Transl. Med.* **2022**, *20*, 420. [CrossRef]
51. Suijk, D.L.S.; van Baar, M.J.B.; van Bommel, E.J.M.; Iqbal, Z.; Krebber, M.M.; Vallon, V.; Touw, D.; Hoorn, E.J.; Nieuwdorp, M.; Kramer, M.M.H.; et al. SGLT2 Inhibition and Uric Acid Excretion in Patients with Type 2 Diabetes and Normal Kidney Function. *Clin. J. Am. Soc. Nephrol.* **2022**, *17*, 663–671. [CrossRef] [PubMed]
52. Lu, Y.H.; Chang, Y.P.; Li, T.; Han, F.; Li, C.J.; Li, X.Y.; Xue, M.; Cheng, Y.; Meng, Z.Y.; Han, Z.; et al. Empagliflozin Attenuates Hyperuricemia by Upregulation of ABCG2 via AMPK/AKT/CREB Signaling Pathway in Type 2 Diabetic Mice. *Int. J. Biol. Sci.* **2020**, *16*, 529–542. [CrossRef] [PubMed]
53. Alnuwaysir, R.I.S.; Hoes, M.F.; van Veldhuisen, D.J.; van der Meer, P.; Grote Beverborg, N. Iron Deficiency in Heart Failure: Mechanisms and Pathophysiology. *J. Clin. Med.* **2021**, *11*, 125. [CrossRef] [PubMed]
54. Ponikowski, P.; Kirwan, B.A.; Anker, S.D.; McDonagh, T.; Dorobantu, M.; Drozdz, J.; Fabien, V.; Filippatos, G.; Gohring, U.M.; Keren, A.; et al. Ferric carboxymaltose for iron deficiency at discharge after acute heart failure: A multicentre, double-blind, randomised, controlled trial. *Lancet* **2020**, *396*, 1895–1904. [CrossRef] [PubMed]
55. Docherty, K.F.; Welsh, P.; Verma, S.; De Boer, R.A.; O'Meara, E.; Bengtsson, O.; Kober, L.; Kosiborod, M.N.; Hammarstedt, A.; Langkilde, A.M.; et al. Iron Deficiency in Heart Failure and Effect of Dapagliflozin: Findings From DAPA-HF. *Circulation* **2022**, *146*, 980–994. [CrossRef]

56. Docherty, K.F.; McMurray, J.J.V.; Kalra, P.R.; Cleland, J.G.F.; Lang, N.N.; Petrie, M.C.; Robertson, M.; Ford, I. Intravenous iron and SGLT2 inhibitors in iron-deficient patients with heart failure and reduced ejection fraction. *ESC Heart Fail.* **2024**, *11*, 1875–1879. [CrossRef]
57. Swedberg, K.; Young, J.B.; Anand, I.S.; Cheng, S.; Desai, A.S.; Diaz, R.; Maggioni, A.P.; McMurray, J.J.; O'Connor, C.; Pfeffer, M.A.; et al. Treatment of anemia with darbepoetin alfa in systolic heart failure. *N. Engl. J. Med.* **2013**, *368*, 1210–1219. [CrossRef]

Disclaimer/Publisher's Note: The statements, opinions and data contained in all publications are solely those of the individual author(s) and contributor(s) and not of MDPI and/or the editor(s). MDPI and/or the editor(s) disclaim responsibility for any injury to people or property resulting from any ideas, methods, instructions or products referred to in the content.

 *pharmaceuticals*

Systematic Review

Evaluating the Impact of Novel Incretin Therapies on Cardiovascular Outcomes in Type 2 Diabetes: An Early Systematic Review

Teodor Salmen [1], Claudia-Gabriela Potcovaru [1], Ioana-Cristina Bica [1], Rosaria Vincenza Giglio [2,3], Angelo Maria Patti [4], Roxana-Adriana Stoica [1], Marcello Ciaccio [2,3], Mohamed El-Tanani [5], Andrej Janež [6], Manfredi Rizzo [5,7], Florentina Gherghiceanu [8,*] and Anca Pantea Stoian [9]

1. Doctoral School, "Carol Davila" University of Medicine and Pharmacy, 050474 Bucharest, Romania; claudia-gabriela.potcovaru@drd.umfcd.ro (C.-G.P.); ioana-cristina.bica@drd.umfcd.ro (I.-C.B.); roxana-adriana.stoica@drd.umfcd.ro (R.-A.S.)
2. Department of Biomedicine, Neuroscience and Advanced Diagnostics, University of Palermo, 90133 Palermo, Italy; rosariavincenza.giglio@unipa.it (R.V.G.); marcello.ciaccio@unipa.it (M.C.)
3. Department of Laboratory Medicine, University Hospital, 90133 Palermo, Italy
4. Internal Medicine Unit, "Vittorio Emanuele II" Hospital, 91022 Castelvetrano, Italy; pattiangelomaria@gmail.com
5. College of Pharmacy, Ras Al Khaimah Medical and Health Sciences University, Ras Al Khaimah 11172, United Arab Emirates; eltanani@rakmhsu.ac.ae (M.E.-T.); manfredi.rizzo@unipa.it (M.R.)
6. Department of Endocrinology, Diabetes and Metabolic Diseases, University Medical Center Ljubljana, 1000 Ljubljana, Slovenia; andrej.janez@kclj.si
7. School of Medicine, Promise Department of Health Promotion Sciences Maternal and Infantile Care, Internal Medicine and Medical Specialties, University of Palermo, 90133 Palermo, Italy
8. Department of Marketing and Medical Technology, "Carol Davila" University of Medicine and Pharmacy, 050474 Bucharest, Romania
9. Department of Diabetes, Nutrition and Metabolic Disease, "Carol Davila" University of Medicine and Pharmacy, 050474 Bucharest, Romania; anca.stoian@umfcd.ro
* Correspondence: f.gherghiceanu@umfcd.ro

Abstract: *Background* This systematic review is registered with CRD42024507397 protocol number and aims to compare the known data about retatrutide on long-term cardiovascular (CV) protection with tirzepatide, an incretin with recent proven CV benefits. *Material and Methods* The inclusion criteria were (i) original full-text articles that are randomized control or clinical trials; (ii) published within the last ten years; (iii) published in English; and (iv) conducted on adult human populations. The exclusion criteria were articles deruled on cell cultures or mammals. Studies were selected if they (1) included patients with type 2 diabetes mellitus (DM) and CV risk; (2) patients that received either tirzepatide or retatrutide; and (3) provided sufficient information such as the corresponding 95% confidence intervals or at least a sufficient *p*-value. Studies were excluded if they were a letter to the editor, expert opinions, case reports, meeting abstracts, or reviews; redundant publications; or needed more precise or complete data. *Results* The seven included studies were assessed for bias with the Newcastle Ottawa scale, heterogenous, and emphasized the potential CV beneficial effect of type 2 DM (T2DM) therapies (glycemia, glycated A1c hemoglobin, body weight, lipid profile, blood pressure and renal parameter). *Discussions* Further, longer follow-up studies are necessary to verify the long-term CV protection, standardize the specific aspects of CV risk, and compare with subjects without T2DM for a more integrative interpretation of the CV effects independent of the improvement of metabolic activity.

Keywords: retatrutide; tirzepatide; type 2 diabetes mellitus; cardiovascular disease

1. Introduction

Diabetes mellitus (DM), whose hallmark is elevated blood sugar, is a health challenge, with a growing number of new cases of type 2 DM (T2DM) worldwide contributing to the pandemic of non-communicable chronic diseases. As a major threat to vascular health, T2DM increases the risk of cardiovascular disease (CVD) and also affects various organs, including the eyes, kidneys, and nerves. Damage to these organs results in deficits associated with diabetes, including vascular, neurological, cardiac, and renal impairments. Consequently, individuals with T2DM often have reduced participation in both basic and instrumental activities of daily living, leading to an increase rate of disability- adjusted life years among those affected. Considering this, T2DM imposes a substantial burden not only on healthcare systems but also on society [1,2].

The formation of advanced glycation end products (AGEs), promoted by high glucose levels, accumulates in tissues and leads to endothelial dysfunction, inflammation, oxidative stress, and fibrosis, significantly contributing to DM complications, particularly CV ones. Effective medication management is crucial for addressing DM complications and mitigate their impact. For comprehensive management, medication must be combined with physical activity and diet, all cantered around the patients' needs, with the ultimate goal of improving their quality of life [2–4].

Patients with poorly controlled T2DM often present with microvascular complications that affect small blood vessels, such as chronic kidney disease (CKD), diabetic neuropathy, and retinopathy, all of which lead to increased mortality and morbidity rates in this category of patients [5]. CKD is the first cause of kidney failure or kidney transplant among patients with DM, retinopathy is the leading cause of vision-threatening, while diabetic neuropathy can lead to foot ulcers and amputations [6–8].

CVD exacerbates macrovascular complications and maintains a bidirectional relationship with T2DM [9]. In addition, several complications can arise from T2DM that make it difficult for most patients to achieve their treatment goals. These complications include conditions like peripheral artery disease, coronary heart disease, cardiomyopathy, arrhythmias, and cerebrovascular disease. They are often caused by risk factors such as high blood pressure (BP), dyslipidemia, and obesity, which not only worsen the condition but also lead to disabilities [10].

Polypharmacy in patients with T2DM is due to the disease itself, along with complications and comorbidities that should be addressed. Moreover, the association between CVD and T2DM highlights the importance of comprehensive care and management strategies targeting both DM and CV risk factors [11]. All these medications add up with the consequences of long-term hyperglycemic status and can lead to non-adherence, keeping the patients from attaining their therapeutic goals, so there is a severe need to integrate new medications into the existing schemes [12]. In addition to polypharmacy, incorporating non-pharmacological strategies such as nutritional therapy, psychological interventions, physical therapies, social interventions, self-blood glucose monitoring in non-insulin-treated T2DM, health coaching, and usual care can improve medication and associated treatment adherence. These interventions can potentially lower CVD risk and serve as adjunctive measures in managing T2DM, enhancing functioning and reducing disability [13]. Knowing that obesity is a significant CV risk factor, the scientific world is investigating incretin role in weight management, starting with glucagon-like peptide-1 receptor agonists (GLP-1 Ra) [14–20], to dual incretin tirzepatide in SURMOUNT program [21], to the GLP-1 and glucagon receptor agonist cotatutide [22], or even the triple incretin retatrutide [23], and the first results are promising in how clinicians could take action in the obesity pandemic.

Over the years, various classes of medications have been developed to better manage the course of T2DM. Some of these medications were designed to reduce body weight (BW) by suppressing the central sensation of hunger. However, they were later withdrawn from the market due to insufficient data on their long-term CV safety and the lack of evidence of CV benefits [24,25]. The first classes of incretins were dipeptidyl peptidase four inhibitors (DPP-4i) and GLP-1 Ra, and the latter has demonstrated CV benefits through different mechanisms, including BW loss, BP reduction, lipid levels amelioration, and endothelial function improvement [26]. Their benefits were observed in patients with T2DM, with or without obesity, highlighting the newly demonstrated CV benefits along with significant BW reduction (BWR) [27]. Newly described agents, such as the dual glucose-dependent insulinotropic polypeptide (GIP) and GLP-1 Ra tirzepatide, demonstrated significant BWR and better metabolic control compared to other treatment options, including semaglutide, insulin degludec, and insulin glargine [28–30]. The future is still under construction with the development of the triple hormone receptor agonist retatrutide (LY3437943), which is a GIP agonist, GLP-1 Ra, and glucagon receptor agonist, and shows promise as an upgrade to the available treatment options, while in a phase 2 trial, it demonstrated remarkable BWR compared to placebo in a dose-dependent manner [23].

This systematic review intends to identify the effect of new incretin therapies on CV risk in individuals with DM. For this purpose, we have compared tirzepatide, an incretin that has shown CV benefits in recent studies, with retatrutide, a molecule that is currently being researched in phase 2 studies and awaiting approval, as seen in Figure 1.

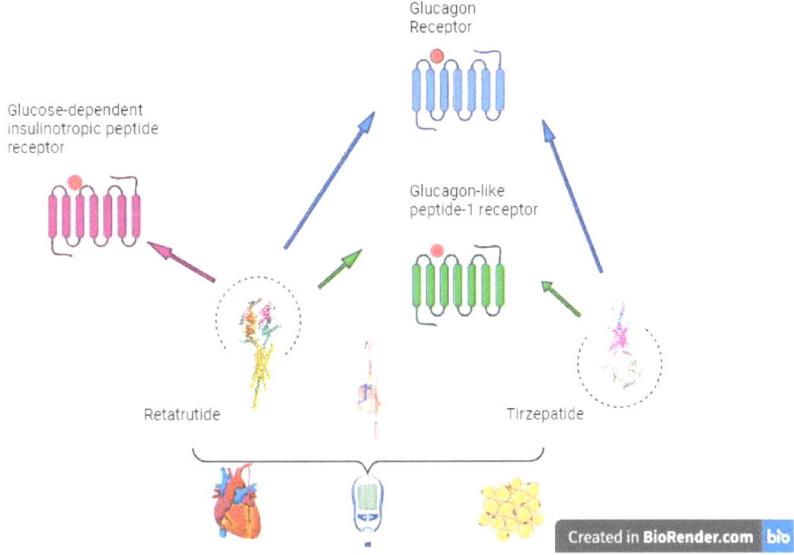

Figure 1. Retatrutide and its potential benefits. Created in BioRender.com.

2. Results

This early systematic review encompasses seven studies published within the last 3 years. Table 1 summarizes the information extracted from the selected studies, as described below.

The study heterogeneity is emphasized by the comparison between retatrutide and older GLP-1 Ras, such as dulaglutide, the placebo control group and tirzepatide with placebo, and with glargine insulin, following the various duration of follow-up ranging from 10.2 to 72 weeks. The study samples include relatively young patients, with a mean age of 59.39 ± 2.05 years.

Glycated A1c hemoglobin (HbA1c) and BW were assessed in all studies. Other parameters evaluated only in some studies were fasting plasma glucose (FPG), systolic BP (SBP), diastolic BP (DBP), heart rate (HR), and BW, while parameters such as fasting triglycerides (TG), high-density lipoprotein cholesterol (HDL-C), non-high-density lipoprotein cholesterol (non-HDL-C), 10-year predicted atherosclerotic CVD risk score, estimated glomerular filtration rate (eGFR) decline per year, and urine albumin–creatinine ratio (uACR) change from baseline were evaluated in only one study each.

Table 1. Included studied characteristics and parameters of interest.

First Author, Year of Publication	Trial Phase	Duration of the Study	Sample Size	Mean Age	Treatment Arm Comparator	Clinical Outcome	Statistical Power
Doggrell et al. [31], 2023	2	24 weeks	281	56.2	Retatrutide 12 mg	HbA1c = −2.02%	
					Dulaglutide 1.5 mg	HbA1c = −1.41%	
					Placebo group	HbA1c = −0.01%	
					Retatrutide 12 mg	BWR = NR, but differences were registered	
					Dulaglutide 1.5 mg	No effect of BWR	
					Placebo group	BWR NR	
					Retatrutide 12 mg	SBP (−8.3, −12.1)/DBP (−4.6, −8.1) mmHg	
					Dulaglutide 1.5 mg	SBP/DBP reduction NR	
					Placebo group	SBP/DBP reduction NR	
					Retatrutide 12 mg	HR = −6.7 beat/min	
					Dulaglutide 1.5 mg	HR reduction NR	
					Placebo group	HR reduction NR	
					Retatrutide 12 mg	↓LDL-C ↓VLDL-C ↓TG ↑HDL-C	
					Dulaglutide 1.5 mg	LDL-C, VLDL-C, TG, HDL-C differences NR	
					Placebo group	LDL-C, VLDL-C, TG, HDL-C differences NR	
Doggrell et al. [32], 2023	1b	78 days	34	NR	Retatrutide 12 mg	HbA1c = −1.9%	
					Dulaglutide 1.5 mg	HbA1c = −1%	
					Retatrutide 12 mg	BWR = −9kg	
					Dulaglutide 1.5 mg	BWR NR	
					Retatrutide 12 mg	↓LDL-C ↓VLDL-C ↓TG ↓HDL-C	
					Dulaglutide 1.5 mg	LDL-C, VLDL-C, TG, HDL-C differences NR	
					Retatrutide 12 mg	↓SBP ↓DBP ↑HR	
					Dulaglutide 1.5 mg	SBP, DBP, HR differences NR	

Table 1. *Cont.*

First Author, Year of Publication	Trial Phase	Duration of the Study	Sample Size	Mean Age	Treatment Arm Comparator	Clinical Outcome	Statistical Power
Rosenstock et al. [33], 2023	2	36 weeks	281	56.2	Retatrutide 12 mg Dulaglutide 1.5 mg Placebo group Retatrutide 12 mg Dulaglutide 1.5 mg Placebo group Retatrutide 12 mg Dulaglutide 1.5 mg Placebo group	HbA1c = −2.16% HbA1c = −1.36% HbA1c = −0.3% FPG = −67.87 mg/dL FPG = −27.53 mg/dL FPG = −17.26 mg/dL BWR = −17.18 kg BWR = −1.97 kg BWR = −3.28 kg	$p < 0.0001$ $p < 0.0001$ $p = 0.2091$ $p < 0.0001$ $p = 0.0024$ $p = 0.1126$ $p < 0.0001$ $p = 0.0242$ $p = 0.0004$
Hankosky et al. [34], 2023		72 weeks	2539	44.5 ± 12.3	Tirzepatide 15 mg Placebo	10-year predicted ASCVD risk score variability = −22.4% 10-year predicted ASCVD risk score variability = 12.7%	OR 2.4, 95% CI (1.7, 3.5), $p < 0.001$
Garvey et al. [35], 2023	3	72 weeks	938	54.2	Tirzepatide 15 mg Placebo Tirzepatide 15 mg Placebo Tirzepatide 15 mg Placebo Tirzepatide 15 mg Placebo Tirzepatide 15 mg Placebo Tirzepatide 15 mg Placebo Tirzepatide 15 mg Placebo	BWR = −14.8 kg BWR = −3.2 kg HbA1c = −2.1% ± 0.07 HbA1c = −0.5% ± 0.07 SBP = −6.3 mmHg SBP = −1.2 mmHg DBP = −2.5 mmHg DBP = −0.3 mmHg Fasting TG = −27.2% Fasting TG = −3.3% HDL-C = −9% HDL-C = −0.2% Non-HDL-C = −5.9% Non-HDL-C = 3.7%	−11.6%, 95% CI (−13, −10.1, $p < 0.0001$) $p < 0.0001$ $p < 0.0001$ $p = 0.0012$ $p < 0.0001$ $p < 0.0001$ $p < 0.0001$

Table 1. Cont.

First Author, Year of Publication	Trial Phase	Duration of the Study	Sample Size	Mean Age	Treatment Arm Comparator	Clinical Outcome	Statistical Power
Heerspink et al. [36], 2022	3	52 weeks	1995	63.6	Tirzepatide 15 mg	eGFR decline per year = −1.4 mL/min/ 1.73 m² ± 0.2	2.2 mL/min/ 1.73 m², 95% CI (1.6, 2.8, p = 0.55)
					Glargin insulin	eGFR decline per year = −3.6 mL/min/ 1.73 m² ± 0.2	
					Tirzepatide 15 mg	uACR change from baseline = −6.8%, 95% CI (−14.1, 1.1)	−31.9%, 95% CI (−37.7, −25.7, p = 0.99)
					Glargin insulin	uACR change from baseline = 36.9%, 95% CI (26, 48.7)	
Del Prato et al. [30], 2021	3	52 weeks	1995	63.6	Tirzepatide 15 mg	HbA1c = −2.58 ± 0.05%	
					Glargin insulin	HbA1c = −1.44 ± 0.03%	
					Tirzepatide 15 mg	BWR = −11.7 ± 0.33	
					Glargin insulin	BWR = −1.9 ± 0.19	
					Tirzepatide 15 mg	FPG = −3.29 ± 0.115	
					Glargin insulin	FPG = −2.84 ± 0.066	
					Tirzepatide 15 mg	SBP = −4.8 ± 0.74	
					Glargin insulin	SBP = −1.3 ± 0.44	
					Tirzepatide 15 mg	DBP = −1 ± 0.44	
					Glargin insulin	DBP = −0.7 ± 0.26	
					Tirzepatide 15 mg	HR = −4.1 ± 0.48	
					Glargin insulin	HR = −1.2 ± 0.29	

HbA1c—glycated hemoglobin A1C, BWR—body weight reduction, SBP—systolic blood pressure, DBP—diastolic blood pressure, HR—heart rate, LDL-C—low-density lipoprotein cholesterol, VLDL-C—very low-density lipoprotein cholesterol, HDL-C—high-density lipoprotein cholesterol, TG—triglycerides, non-HDL-C—non-high-density lipoprotein cholesterol, FPG—fasting plasma glucose, ↓—a decrease was reported; ↑—an increase was reported; ASCVD—atherosclerotic cardiovascular disease, eGFR—estimated glomerular filtration rate, uACR—urine albumin-creatinine ratio.

2.1. Comparative Dosage Strategies in Retatrutide and Tirzepatide Studies

The studies incorporating retatrutide employed specific dosage strategies, categorizing participants into groups based on their retatrutide dosage. In the study by Doggrell et al. [31,32], participants were allocated diverse doses of retatrutide (1, 2, or 4 mg) that were progressively escalated to a maximum of 8 or 12 mg or a placebo. This diversified dosage strategy aimed to assess the impact of different retatrutide doses on the study participants.

In contrast, Rosenstock et al. [33], divided the patients into various treatment groups, ensuring an even distribution. These groups included placebo, 1.5 mg dulaglutide, and various retatrutide dosages: 0.5 mg, 4 mg escalation, 4 mg, 8 mg slow escalation, 8 mg fast escalation, and 12 mg escalation.

The dosage strategies employed in studies utilizing tirzepatide were structured as follows: in the study led by Hankosky et al. [34], participants underwent random allocation to receive subcutaneous tirzepatide, with dosages ranging between 5, 10, 15 mg, or a placebo. The protocol involved gradually increasing the dosage at 72 weeks, with the medication being administered once weekly. Similarly, Garvey et al. [35], initiated participants with tirzepatide or a placebo, beginning with a weekly dosage of 2.5 mg, that progressively increased by 2.5 mg every four weeks until reaching the targeted dose of either 10 mg or 15 mg at 12 and 20 weeks, respectively. Heerspink et al. [36], randomized participants in a 1:1:1:3 ratio involving tirzepatide dosages of 5 mg, 10 mg, or 15 mg, along with insulin glargine. The study format accounted for differences in dosing schedules between once-per-week tirzepatide and once-per-day insulin glargine. Del Prato et al. [30], utilized random assignment (1:1:1:3) to evaluate tirzepatide at 5 mg, 10 mg, and 15 mg alongside insulin glargine.

The consistent approach to tirzepatide dosing across these studies facilitates potential comparative analyses and enhances the generalizability of findings.

2.2. Side Effects in Incretin Therapy

The most notable side effects in patients treated with incretin therapies were gastrointestinal (GI) ones [17,19,22,28,33,35,37]. STEP trials noted the most frequent adverse events (AE) were nausea, diarrhea, vomiting, and constipation, with the majority being transient and of mild or moderate severity [37]. SCALE trial also reported GI AEs in patients with T2DM treated with injectable liraglutide versus placebo as an add-on to insulin basal therapy (71% vs. 49%), the most frequent being nausea, vomiting, constipation, diarrhea, or abdominal discomfort [19]. SURPASS-2, a phase 3 trial, demonstrated the superiority of tirzepatide, compared to semaglutide in patients with T2DM, with an elevated rate of mild to moderate GI AE in the tirzepatide vs. semaglutide group (nausea, 17 to 22% and 18%; diarrhea, 13 to 16% and 12%; and vomiting, 6 to 10% and 8%, respectively), with a greater AE risk in the 15 mg group compared to the 10 mg, 5 mg, or semaglutide group [28]. Cotadutide, a dual GLP-1 RA and glucagon receptor agonist was also associated with GI AEs when compared to liraglutide 1.8 mg or placebo, and the risk was proportional to the dose escalation [22]. Also, the triple incretin therapy, retatrutide, has been reported to have mild to moderate GI side effects, and the proportion of nausea, vomiting and constipation was elevated in the escalated dosage group (13% vs. 50% in 0.5 mg group compared to the 8 mg one) [33].

3. Discussion

GLP-1 Ra agents demonstrated CV benefits in several trials, including semaglutide, both in injectable [38] and oral administration [39], liraglutide [40], and dulaglutide [41]. These trials pioneered the CV protection point of view, a required attribute for every new antidiabetic medication.

Currently, injectable semaglutide is the only GLP-1Ra that has an additional approved indication for the reduction in major adverse CV events in adults, with established CVD and either overweight or obesity, approved by United States Food and Drug Administration. The approval decision was supported by findings from the placebo-controlled SELECT trial, which showed that the injectable semaglutide given on top of standard therapy significantly reduced the risk of CV death, myocardial infarction, or stroke (6.5% vs. 8.0%; hazard ratio 0.80; 95% confidence interval 0.72–0.90) in patients with established CVD, no prior history of DM, and a body mass index of at least 27 kg/m^2. The exact mechanism of the CV risk reduction observed in the trial is unclear [42].

Because incretins significantly reduced CV events (CVEs), researchers have searched for the physiopathology behind these benefits, considering their metabolic implications [43]. Tuttolomondo et al. demonstrated that dulaglutide significantly decreased DBP, BW, total cholesterol and low-density lipoprotein cholesterol (LDL-C), FPG, HbA1c, microalbuminuria, and pulse wave velocity, while significantly increasing the reactive hyperemia index after nine-month follow-up in patients with T2DM compared to traditional antidiabetic treatment alone [44]. All these parameters represent valuable markers of vascular health, with effects on endothelial and arterial stiffness indexes. A Study of Tirzepatide (LY3298176) Compared with Dulaglutide on Major Cardiovascular Events in Participants with Type 2 Diabetes (SURPASS-CVOT) is an ongoing trial trying to evaluate the noninferiority or even superiority of the dual incretin, tirzepatide, in comparison to the GLP-1 Ra dulaglutide in terms of major adverse CVEs in patients with T2DM [45]. These research trends emphasize the importance of better understanding CV benefits in novel antidiabetic therapies.

Another phase 2b placebo-controlled trial, which investigated an oral GLP-1 Ra, danuglipron, demonstrated improved HbA1c, basal glycemia, and BWR in patients with T2DM in the active group [46]. Further studies are needed to determine whether this new oral GLP-1 Ra also has CV protection or not in future trials with well-established CV outcomes.

The studies evaluating retatrutide and tirzepatide are either in phase 1 or phase 2, emphasizing their developmental stages where safety and efficacy are assessed. However, tirzepatide is in a phase 3 trial, being compared to placebo or glargine insulin, reflecting its comprehensive evaluation, pivotal for regulatory considerations, and potential approval. This distinction underscores the evolving nature and maturity of the investigated therapies.

The variety of study durations observed ranges from 10.2 to 72 weeks, enabling researchers and clinicians to assess the subtleties of these interventions, potentially revealing trends, sustained benefits, and emerging long-term outcomes. This consideration is especially relevant in chronic conditions such as DM, where long-term management and sustained efficacy are crucial.

Moreover, the mean age of 59.39 $\pm$ 2.05 years reflects young participants, an age demographic group that is facing increased challenges in DM management, making the outcomes particularly relevant for understanding the effectiveness of retatrutide and tirzepatide in addressing the specific needs and complexities associated with an older age group.

The consistent use of tirzepatide doses across Hankosky et al. [34], Heerspink et al. [36] and Del Prato et al. [30] study reflects a standardized approach in their respective evaluations, facilitating potential comparative analyses and enhances the generalizability of the findings.

The efficacy results on HbA1c for retatrutide [31–33], compared to other antidiabetic classes or placebo, and for tirzepatide compared to placebo [28] or insulin [30], were also observed for the most recent once-weekly GLP-1 and glucagon receptor dual agonist, mazdutide. These results are also present for BWR. So, the efficacy is not limited to race, as mazdutide is being evaluated on Chinese patients, with T2DM leading the way to future research on CV protection in patients treated with dual incretin therapies [47]. Other adjuvants of the antidiabetic treatment with benefits in CV protection are supplements that are reported to lower FPG levels and protein glycation (AGEs and HbA1c), ameliorates LDL-C level, and improve HDL-C functionality [48,49]. If we talk about the HbA1c value, in our study, for retatrutide, it ranges between 1.9 and 2.16%, which is inferior to the value for tirzepatide of 2.1–2.58% and superior to GLP-1 Ra −0.78–1.9%. For BW, retatrutide with 9–17.87 kg is like tirzepatide with 11.7–14.8 kg, but higher than GLP-1

Ra with 0.64–5.8 kg [50]. Similar data are reported for the FPG, which, in our study, is decreased by retatrutide and also for tirzepatide, with a greater effect for the first class.

Moreover, for BP, in our study, the reduction in SBP ranges between 8.3 and 12.1 mmHg for retatrutide, which is superior to the reduction of 4.8–6.3 mmHg for tirzepatide. Similarly, the effect on DBP of 4.6–8.1 mmHg was superior to the reduction of 0.7–2.5 mmHg for tirzepatide; retatrutide demonstrated a reduction in HR of 6.7 beat per minutes, which is higher than 4.1 beat per minutes for tirzepatide.

Regarding the lipidic profile, there are reported data about a decrement of LDL-C, VLDL-C, TG, and HDL-C only for retatrutide. Data that are reported only for tirzepatide are about its efficacy in reducing the 10-year predicted ASCVD risk score; the level of fasting TG, HDL-C, and non-HDL-C; eGFR decline per year; and uACR change from baseline.

Our study limitations are linked to the scarcity of data on long-term CV protection for these new dual and triple incretins. These agents need longer follow-up trials to establish their benefit in reducing the risk for major adverse CVEs. Also, an indirect limitation is that the included studies design does not provide a distribution of the drug's effects regarding patient's parameters, such as BMI. These new data could be a promising research area for future investigations. Also, both tirzepatide and retatrutide can be administered through injections, which could represent a barrier to the patient's adherence in the first place despite their promising CV benefits.

Furthermore, the Semaglutide Effects on Cardiovascular Outcomes in People with Overweight or Obesity (SELECT) trial, a randomized controlled trial that evaluates the CV benefit of a once-weekly injection of semaglutide in patients with overweight or obesity but without T2DM, will lead to a more comprehensive interpretation of incretins' CV effects, unattached to the metabolic improvement in patients with T2DM, which can represent a major confounding factor [41].

On the other hand, the trial extension with once-weekly Semaglutide in overweight or obese adults (STEP-1) found that participants regained two-thirds of their BWR, with similar results in other cardiometabolic parameters, such as lipid profile, C-reactive protein levels, or BP values, after one year of withdrawal from semaglutide [51,52]. These findings emphasize the complexity and chronic character of obesity and its involvement in CV risk modulation [53,54].

These CV benefits, the reason why the GLP-1 Ra class is recommended, have been researched in several real-life studies. Liraglutide and semaglutide (the injectable administration form) demonstrated significant effects in reducing carotid intima-media thickness and lipid profile amelioration, in addition to their effects on glucose metabolism [29,34]. These results emphasize the need for this class in the early stages of CVD to achieve the best results in preventing CVEs.

Despite the well-known CV protective effects of this therapy class, some patients still do not achieve their therapeutic targets, partly because of therapeutic inertia, and partly because of their unavailability in different areas, including developing countries [3,55,56]. These latest data should raise awareness of the importance of CV risk stratification and proper therapeutic approaches, in accordance with guidelines. New molecules included in the incretins class could represent a chance for patients with T2DM to achieve their metabolic goals and reduce their CV risk.

4. Materials and Methods

The present systematic review followed the Preferred Reporting Items for Systematic Reviews and Meta-Analysis (PRISMA) checklist and guidelines. The review protocol has been registered with the identifier CRD42024507397.

4.1. Research Question and Search Strategy

An electronic search for relevant publications was performed using PubMed and Web of Science library databases and was conducted from January 2013 to February 2024. The following search strategy was used: Retatrutide or Tirzepatide and cardiovascular disease and (diabetes mellitus or type 2 diabetes mellitus OR T2DM). After this search, 93 articles were found (63 from PubMed and 30 from Web of Science). After applying filters for language (English), publication type (original articles), and date range (2013 to the date of the search), 15 articles remained. These articles underwent initial title screening, followed by an abstract review by two independent reviewers, and 13 articles remained for a full assessment.

Because the titles and abstracts lacked data about the retatrutide molecule, we performed another search of PubMed and Web of Science library databases with "retatrutide" and obtained 48 articles (26 on PubMed and 22 on Web of Science, respectively). After applying filters for language (English) and publication type (original articles), 9 articles remained. After removing the duplicates, 7 articles remained for full assessment.

As seen in Figure 2, 7 studies were included from the 20 articles that were fully assessed.

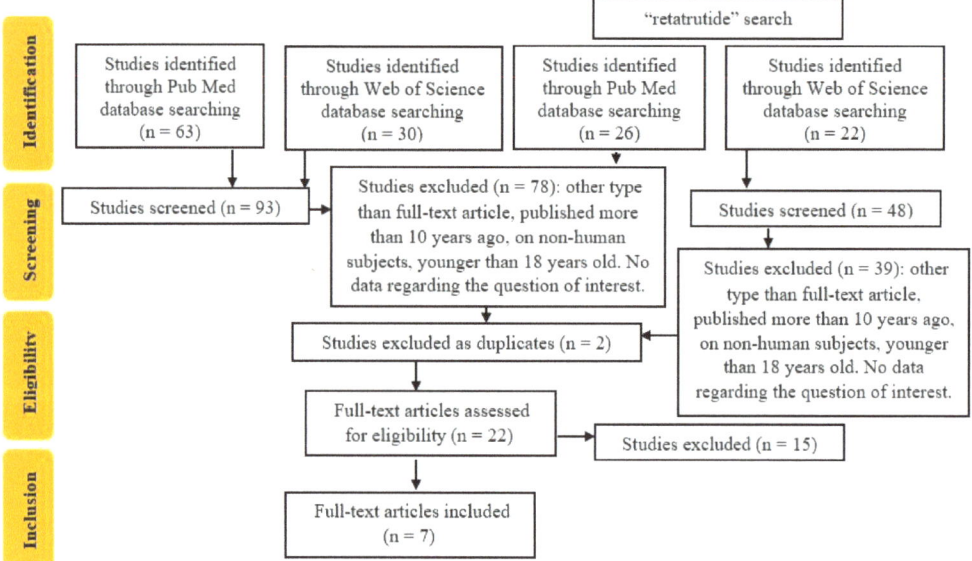

Figure 2. Flowchart of the study selection.

The research question was framed using the Population, Intervention, Comparison, and Outcome (PICO) method. The population was represented by patients with T2DM and CV risk, and intervention was represented by the new incretin therapy (tirzepatide and retatrutide). The comparison was against patients who received treatment other than the agent of interest. The outcome was defined by BW, HbA1c, FPG level, lipid profile, BP, and HR or other CV risk benefits in patients with T2DM and new incretin therapy, and the effect was measured by percentage, confidence intervals, odds ratios, and relative risk.

4.2. Inclusion Criteria

The studies had to meet specific publication criteria to be included in this review. These criteria are as follows: (i) the studies should be original full-text articles that are randomized control or clinical trials; (ii) the articles should be published within the last ten years; (iii) the articles should be published in English; and (iv) the studies should be conducted on adult human populations.

4.3. Exclusion Criteria

Studies were excluded from the analysis if they were derulated on (i) cell cultures or (ii) mammals.

4.4. Selection of Studies

Studies that met the following eligibility criteria were selected: (1) included patients with T2DM and CV risk; (2) patients received either tirzepatide or retatrutide; (3) provided sufficient information such as the corresponding 95% CIs or at least *p*-value. Studies were excluded if they (1) were a letter to the editor, expert opinions, case reports, meeting abstracts, or reviews; (2) were redundant publications; or (3) needed more precise or complete data.

4.5. Data Extraction

Two authors (T.S. and C.-G.P.) used a self-made data extraction table to individually evaluate and extract the following data for each included article: the first author and year of publication, study period, sample size, the average age of participants, treatment details, outcomes, confounding factors adjusted, reported outcomes, and risk estimates with their corresponding 95% CIs. Whenever there were differences in opinion, we resolved them by having a discussion or by consulting with a third author (I.-C.B.).

4.6. Risk of Bias Assessment

Two reviewers (T.S. and C.-G.P.) independently assessed the quality of the studies according to Newcastle–Ottawa Scale criteria and provided their classification in Table 2.

Table 2. Newcastle–Ottawa Scale analysis of the included articles.

Author (Reference)	Selection				Comparability		Outcome			Total Score	Quality
	Representativeness of the Exposed Cohort	Selection of the Non-Exposed Cohort	Ascertainment of Exposure	Demonstration That Outcome of Interest Was Not Present at the Start of the Study	Comparability of Cohorts Based on the Design or Analysis	Assessment of Outcome	Was Follow-Up Long Enough for Outcomes to Occur	Adequacy of Follow-Up of Cohorts			
Doggrell et al. [31], 2023	*	-	*	*	-	*	*	*		6	good
Doggrell et al. [32], 2023	*	-	*	*	-	*	*	*		6	good
Rosenstock et al. [33], 2023	-	*	*	*	*	*	*	*		7	good
Hankosky et al. [34], 2023	-	*	*	*	*	*	*	*		7	good
Garvey et al. [35], 2023	-	*	*	*	*	*	*	*		7	good
Heerspink et al. [36], 2022	-	*	*	*	*	*	*	*		7	good
Del Prato et al. [30], 2021	-	*	*	*	*	*	*	*		7	good

"*" indicates that the article meets the criteria mentioned above; "-" indicates that the article does not meet the abovementioned criteria.

4.7. Strategy for Data Synthesis

A narrative synthesis of the findings in the studies, centered around each class of incretin therapy and benefit on CV risk, where a minimum of two studies were identified, was performed.

5. Conclusions

DM is a disease that can have serious CV, renal, nervous, and ophthalmic complications if it is not diagnosed and treated adequately. For this reason, it is essential to follow a therapy that aims to maintain blood glucose levels within normal values. In recent years, new innovative drugs have been introduced on the market for the treatment of T2DM, which have been shown to have beneficial effects not only on glycemic control, but also on the prevention of CV complications.

A new molecule, called tirzepatide, has recently been identified, capable of acting simultaneously on the GLP-1 and GIP receptors, thus making its hypoglycemic activity even more effective compared to drugs already in use. Furthermore, a characteristic of this class of drugs is that they also achieve a significant BWR, which is why they have also been studied as a treatment for obesity in patients both with and without T2DM.

In the study phase, we have retatrutide, characterized by a more complex mechanism of action than its potential competitors as it acts on three receptor targets: GIP, GLP-1, and glucagon. Its action as a triple receptor agonist has proven effective in inducing BWR and metabolic effects with results comparable to those of bariatric surgery.

These drugs represent a new frontier in the treatment of T2DM, as they allow us to modify the course of the disease and aim for its remission. In fact, some studies have shown that these drugs are able to preserve or improve the function of pancreatic beta cells, reduce chronic inflammation, and promote the regeneration of tissues damaged by DM. In this way, functional healing and a reduction in long-term complications could be achieved. The arrival of new innovative therapies offers new opportunities and hopes to patients with T2DM, who can thus pursue the goal of better quality and greater life expectancy. Much research still needs to be conducted to apply these therapies in a personalized manner to the various metabolic alteration phenotypes.

Author Contributions: Conceptualization, T.S. and I.-C.B.; methodology, C.-G.P.; software, A.M.P.; validation, M.C., R.-A.S. and A.J.; formal analysis, T.S.; investigation, C.-G.P. and F.G.; resources, I.-C.B.; data curation, R.V.G.; writing—original draft preparation, T.S.; writing—review and editing, R.V.G.; visualization, M.R.; supervision, A.P.S.; project administration, T.S.; funding acquisition, M.E.-T. All authors have read and agreed to the published version of the manuscript.

Funding: This research received no external funding.

Acknowledgments: Publication of this paper was supported by the University of Medicine and Pharmacy Carol Davila through the institutional program Publish not Perish.

Conflicts of Interest: The authors declare no conflicts of interest.

References

1. Ye, J.; Wu, Y.; Yang, S.; Zhu, D.; Chen, F.; Chen, J.; Ji, X.; Hou, K. The global, regional and national burden of type 2 diabetes mellitus in the past, present and future: A systematic analysis of the Global Burden of Disease Study 2019. *Front. Endocrinol.* **2023**, *14*, 1192629. [CrossRef] [PubMed]
2. Oyewole, O.O.; Ale, A.O.; Ogunlana, M.O.; Gurayah, T. Burden of Disability in Type 2 Diabetes Mellitus and the Moderating Effects of Physical Activity. *World J. Clin. Cases* **2023**, *11*, 3128. [CrossRef]
3. Reurean-Pintilei, D.; Potcovaru, C.-G.; Salmen, T.; Mititelu-Tartau, L.; Cinteză, D.; Lazăr, S.; Pantea Stoian, A.; Timar, R.; Timar, B. Assessment of Cardiovascular Risk Categories and Achievement of Therapeutic Targets in European Patients with Type 2 Diabetes. *J. Clin. Med.* **2024**, *13*, 2196. [CrossRef] [PubMed]
4. Potcovaru, C.-G.; Salmen, T.; Bîgu, D.; Săndulescu, M.I.; Filip, P.V.; Diaconu, L.S.; Pop, C.; Ciobanu, I.; Cinteză, D.; Berteanu, M. Assessing the Effectiveness of Rehabilitation Interventions through the World Health Organization Disability Assessment Schedule 2.0 on Disability—A Systematic Review. *J. Clin. Med.* **2024**, *13*, 1252. [CrossRef]
5. An, J.; Nichols, G.A.; Qian, L.; Munis, M.A.; Harrison, T.N.; Li, Z.; Wei, R.; Weiss, T.; Rajpathak, S.; Reynolds, K. Prevalence and incidence of microvascular and macrovascular complications over 15 years among patients with incident type 2 diabetes. *BMJ Open Diabetes Res. Care.* **2021**, *9*, e001847. [CrossRef]
6. de Boer, I.H.; Khunti, K.; Sadusky, T.; Tuttle, K.R.; Neumiller, J.J.; Rhee, C.M.; Rosas, S.E.; Rossing, P.; Bakris, G. Diabetes Management in Chronic Kidney Disease: A Consensus Report by the American Diabetes Association (ADA) and Kidney Disease: Improving Global Outcomes (KDIGO). *Diabetes Care* **2022**, *45*, 3075–3090. [CrossRef]
7. Hou, X.; Wang, L.; Zhu, D.; Guo, L.; Weng, J.; Zhang, M.; Zhou, Z.; Zou, D.; Ji, Q.; Guo, X.; et al. Prevalence of diabetic retinopathy and vision-threatening diabetic retinopathy in adults with diabetes in China. *Nat. Commun.* **2023**, *14*, 4296. [CrossRef] [PubMed]
8. Galiero, R.; Caturano, A.; Vetrano, E.; Beccia, D.; Brin, C.; Alfano, M.; Di Salvo, J.; Epifani, R.; Piacevole, A.; Tagliaferri, G.; et al. Peripheral Neuropathy in Diabetes Mellitus: Pathogenetic Mechanisms and Diagnostic Options. *Int. J. Mol. Sci.* **2023**, *24*, 3554. [CrossRef]

9. Blonde, L.; Umpierrez, G.E.; Reddy, S.S.; McGill, J.B.; Berga, S.L.; Bush, M.; Chandrasekaran, S.; DeFronzo, R.A.; Einhorn, D.; Galindo, R.J.; et al. American Association of Clinical Endocrinology Clinical Practice Guideline: Developing a Diabetes Mellitus Comprehensive Care Plan-2022 Update. *Endocr. Pract.* **2022**, *28*, 923–1049.
10. American Diabetes Association Professional Practice Committee. 10. Cardiovascular Disease and Risk Management: Standards of Care in Diabetes—2024. *Diabetes Care* **2024**, *47*, S179–S218. [CrossRef]
11. Salmen, T.; Serbanoiu, L.-I.; Bica, I.-C.; Serafinceanu, C.; Muzurović, E.; Janez, A.; Busnatu, S.; Banach, M.; Rizvi, A.A.; Rizzo, M.; et al. A Critical View over the Newest Antidiabetic Molecules in Light of Efficacy—A Systematic Review and Meta-Analysis. *Int. J. Mol. Sci.* **2023**, *24*, 9760. [CrossRef] [PubMed]
12. Sahoo, J.; Mohanty, S.; Kundu, A.; Epari, V. Medication Adherence Among Patients of Type II Diabetes Mellitus and Its Associated Risk Factors: A Cross-Sectional Study in a Tertiary Care Hospital of Eastern India. *Cureus* **2022**, *14*, e33074. [CrossRef]
13. Raveendran, A.V.; Chacko, E.C.; Pappachan, J.M. Non-pharmacological Treatment Options in the Management of Diabetes Mellitus. *Eur. Endocrinol.* **2018**, *14*, 31–39. [CrossRef] [PubMed]
14. Garvey, W.T.; Batterham, R.L.; Bhatta, M.; Buscemi, S.; Christensen, L.N.; Frias, J.P.; Jódar, E.; Kandler, K.; Rigas, G.; Wadden, T.A.; et al. Two-year effects of semaglutide in adults with overweight or obesity: The STEP 5 trial. *Nat. Med.* **2022**, *28*, 2083–2091. [CrossRef]
15. Wadden, T.A.; Bailey, T.S.; Billings, L.K.; Davies, M.; Frias, J.P.; Koroleva, A.; Lingvay, I.; O'Neil, P.M.; Rubino, D.M.; Skovgaard, D.; et al. Effect of Subcutaneous Semaglutide vs Placebo as an Adjunct to Intensive Behavioral Therapy on Body Weight in Adults With Overweight or Obesity: The STEP 3 Randomized Clinical Trial. *JAMA* **2021**, *325*, 1403–1413. [CrossRef] [PubMed]
16. Rubino, D.M.; Greenway, F.L.; Khalid, U.; O'Neil, P.M.; Rosenstock, J.; Sørrig, R.; Wadden, T.A.; Wizert, A.; Garvey, W.T.; STEP 8 Investigators. Effect of Weekly Subcutaneous Semaglutide vs Daily Liraglutide on Body Weight in Adults With Overweight or Obesity Without Diabetes: The STEP 8 Randomized Clinical Trial. *JAMA* **2022**, *327*, 138–150. [CrossRef] [PubMed]
17. Kushner, R.F.; Calanna, S.; Davies, M.; Dicker, D.; Garvey, W.T.; Goldman, B.; Lingvay, I.; Thomsen, M.; Wadden, T.A.; Wharton, S.; et al. Semaglutide 2.4 mg for the Treatment of Obesity: Key Elements of the STEP Trials 1 to 5. *Obesity* **2020**, *28*, 1050–1061. [CrossRef]
18. Silver, H.J.; Olson, D.; Mayfield, D.; Wright, P.; Nian, H.; Mashayekhi, M.; Koethe, J.R.; Niswender, K.D.; Luther, J.M.; Brown, N.J. Effect of the glucagon-like peptide-1 receptor agonist liraglutide, compared to caloric restriction, on appetite, dietary intake, body fat distribution and cardiometabolic biomarkers: A randomized trial in adults with obesity and prediabetes. *Diabetes Obes. Metab.* **2023**, *25*, 2340–2350. [CrossRef]
19. Wadden, T.A.; Tronieri, J.S.; Sugimoto, D.; Lund, M.T.; Auerbach, P.; Jensen, C.; Rubino, D. Liraglutide 3.0 mg and Intensive Behavioral Therapy (IBT) for Obesity in Primary Care: The SCALE IBT Randomized Controlled Trial. *Obesity* **2020**, *28*, 529–536. [CrossRef]
20. Garvey, W.T.; Birkenfeld, A.L.; Dicker, D.; Mingrone, G.; Pedersen, S.D.; Satylganova, A.; Skovgaard, D.; Sugimoto, D.; Jensen, C.; Mosenzon, O. Efficacy and Safety of Liraglutide 3.0 mg in Individuals With Overweight or Obesity and Type 2 Diabetes Treated With Basal Insulin: The SCALE Insulin Randomized Controlled Trial. *Diabetes Care* **2020**, *43*, 1085–1093. [CrossRef]
21. le Roux, C.W.; Zhang, S.; Aronne, L.J.; Kushner, R.F.; Chao, A.M.; Machineni, S.; Dunn, J.; Chigutsa, F.B.; Ahmad, N.N.; Bunck, M.C. Tirzepatide for the treatment of obesity: Rationale and design of the SURMOUNT clinical development program. *Obesity* **2023**, *31*, 96–110. [CrossRef] [PubMed]
22. Nahra, R.; Wang, T.; Gadde, K.M.; Oscarsson, J.; Stumvoll, M.; Jermutus, L.; Hirshberg, B.; Ambery, P. Effects of Cotadutide on Metabolic and Hepatic Parameters in Adults With Overweight or Obesity and Type 2 Diabetes: A 54-Week Randomized Phase 2b Study. *Diabetes Care* **2021**, *44*, 1433–1442. [CrossRef] [PubMed]
23. Jastreboff, A.M.; Kaplan, L.M.; Frías, J.P.; Wu, Q.; Du, Y.; Gurbuz, S.; Coskun, T.; Haupt, A.; Milicevic, Z.; Hartman, M.L.; et al. Triple-Hormone-Receptor Agonist Retatrutide for Obesity—A Phase 2 Trial. *NEJM* **2023**, *389*, 514–526. [CrossRef] [PubMed]
24. Lupianez-Merly, C.; Dilmaghani, S.; Vosoughi, K.; Camilleri, M. Review article: Pharmacologic management of obesity—Updates on approved medications, indications and risks. *Aliment. Pharmacol. Ther.* **2024**, *59*, 475–491. [CrossRef] [PubMed]
25. Alobaida, M.; Alrumayh, A.; Oguntade, A.S.; Al-Amodi, F.; Bwalya, M. Cardiovascular Safety and Superiority of Anti-Obesity Medications. *Diabetes Metab. Syndr. Obes. Targets Ther.* **2021**, *14*, 3199–3208. [CrossRef]
26. Ferhatbegović, L.; Mršić, D.; Macić-Džanković, A. The benefits of GLP1 receptors in cardiovascular diseases. *Front. Clin. Diabetes Healthc.* **2023**, *4*, 1293926. [CrossRef]
27. Ussher, J.R.; Drucker, D.J. Glucagon-like peptide 1 receptor agonists: Cardiovascular benefits and mechanisms of action. *Nat. Rev. Cardiol.* **2023**, *20*, 463–474. [CrossRef]
28. Frías, J.P.; Davies, M.J.; Rosenstock, J.; Pérez Manghi, F.C.; Fernández Landó, L.; Bergman, B.K.; Liu, B.; Cui, X.; Brown, K.; SURPASS-2 Investigators. Tirzepatide versus Semaglutide Once Weekly in Patients with Type 2 Diabetes. *NEJM* **2021**, *385*, 503–515. [CrossRef]
29. Ludvik, B.; Giorgino, F.; Jódar, E.; Frias, J.P.; Fernández Landó, L.; Brown, K.; Bray, R.; Rodríguez, Á. Once-weekly tirzepatide versus once-daily insulin degludec as add-on to metformin with or without SGLT2 inhibitors in patients with type 2 diabetes (SURPASS-3): A randomised, open-label, parallel-group, phase 3 trial. *Lancet* **2021**, *398*, 583–598. [CrossRef]
30. Del Prato, S.; Kahn, S.E.; Pavo, I.; Weerakkody, G.J.; Yang, Z.; Doupis, J.; Aizenberg, D.; Wynne, A.G.; Riesmeyer, J.S.; Heine, R.J.; et al. Tirzepatide versus insulin glargine in type 2 diabetes and increased cardiovascular risk (SURPASS-4): A randomised, open-label, parallel-group, multicentre, phase 3 trial. *Lancet* **2021**, *398*, 1811–1824. [CrossRef]

31. Doggrell, S.A. Retatrutide showing promise in obesity (and type 2 diabetes). *Expert Opin. Investig. Drugs* **2023**, *32*, 997–1001. [CrossRef] [PubMed]
32. Doggrell, S.A. Is retatrutide (LY3437943), a GLP-1, GIP, and glucagon receptor agonist a step forward in the treatment of diabetes and obesity? *Expert. Opin. Investig. Drugs* **2023**, *32*, 355–359. [CrossRef] [PubMed]
33. Rosenstock, J.; Frias, J.; Jastreboff, A.M.; Du, Y.; Lou, J.; Gurbuz, S.; Thomas, M.K.; Hartman, M.L.; Haupt, A.; Milicevic, Z.; et al. Retatrutide, a GIP, GLP-1 and glucagon receptor agonist, for people with type 2 diabetes: A randomised, double-blind, placebo and active-controlled, parallel-group, phase 2 trial conducted in the USA. *Lancet* **2023**, *402*, 529–544. [CrossRef] [PubMed]
34. Hankosky, E.R.; Wang, H.; Neff, L.M.; Kan, H.; Wang, F.; Ahmad, N.N.; Griffin, R.; Stefanski, A.; Garvey, W.T. Tirzepatide reduces the predicted risk of atherosclerotic cardiovascular disease and improves cardiometabolic risk factors in adults with obesity or overweight: SURMOUNT-1 post hoc analysis. *Diabetes Obes. Metab.* **2024**, *26*, 319–328. [CrossRef] [PubMed]
35. Garvey, W.T.; Frias, J.P.; Jastreboff, A.M.; le Roux, C.W.; Sattar, N.; Aizenberg, D.; Mao, H.; Zhang, S.; Ahmad, N.N.; Bunck, M.C.; et al. Tirzepatide once weekly for the treatment of obesity in people with type 2 diabetes (SURMOUNT-2): A double-blind, randomised, multicentre, placebo-controlled, phase 3 trial. *Lancet* **2023**, *402*, 613–626. [CrossRef]
36. Heerspink, H.J.L.; Sattar, N.; Pavo, I.; Haupt, A.; Duffin, K.L.; Yang, Z.; Wiese, R.J.; Tuttle, K.R.; Cherney, D.Z.I. Effects of tirzepatide versus insulin glargine on kidney outcomes in type 2 diabetes in the SURPASS-4 trial: Post-hoc analysis of an open-label, randomised, phase 3 trial. *Lancet Diabetes Endocrinol.* **2022**, *10*, 774–785. [CrossRef]
37. Bergmann, N.C.; Davies, M.J.; Lingvay, I.; Knop, F.K. Semaglutide for the treatment of overweight and obesity: A review. *Diabetes Obes. Metab.* **2023**, *25*, 18–35. [CrossRef]
38. Marso, S.P.; Bain, S.C.; Consoli, A.; Eliaschewitz, F.G.; Jódar, E.; Leiter, L.A.; Lingvay, I.; Rosenstock, J.; Seufert, J.; Warren, M.L.; et al. Semaglutide and Cardiovascular Outcomes in Patients with Type 2 Diabetes. *NEJM* **2016**, *375*, 1834–1844. [CrossRef]
39. Husain, M.; Birkenfeld, A.L.; Donsmark, M.; Dungan, K.; Eliaschewitz, F.G.; Franco, D.R.; Jeppesen, O.K.; Lingvay, I.; Mosenzon, O.; Pedersen, S.D.; et al. Oral Semaglutide and Cardiovascular Outcomes in Patients with Type 2 Diabetes. *NEJM* **2019**, *381*, 841–851. [CrossRef]
40. Marso, S.P.; Daniels, G.H.; Brown-Frandsen, K.; Kristensen, P.; Mann, J.F.; Nauck, M.A.; Nissen, S.E.; Pocock, S.; Poulter, N.R.; Ravn, L.S.; et al. Liraglutide and Cardiovascular Outcomes in Type 2 Diabetes. *NEJM* **2016**, *375*, 311–322. [CrossRef]
41. Riddle, M.C.; Gerstein, H.C.; Xavier, D.; Cushman, W.C.; Leiter, L.A.; Raubenheimer, P.J.; Atisso, C.M.; Raha, S.; Varnado, O.J.; Konig, M.; et al. Efficacy and Safety of Dulaglutide in Older Patients: A post hoc Analysis of the REWIND trial. *J. Clin. Endocrinol. Metab.* **2021**, *106*, 1345–1351. [CrossRef] [PubMed]
42. Lingvay, I.; Brown-Frandsen, K.; Colhoun, H.M.; Deanfield, J.; Emerson, S.S.; Esbjerg, S.; Hardt-Lindberg, S.; Hovingh, G.K.; Kahn, S.E.; Kushner, R.F.; et al. Semaglutide for cardiovascular event reduction in people with overweight or obesity: SELECT study baseline characteristics. *Obesity* **2023**, *31*, 111–122. [CrossRef] [PubMed]
43. Rizzo, M.; Nikolic, D.; Patti, A.M.; Mannina, C.; Montalto, G.; McAdams, B.S.; Rizvi, A.A.; Cosentino, F. GLP-1 receptor agonists and reduction of cardiometabolic risk: Potential underlying mechanisms. *Biochim. Biophys. Acta Mol. Basis Dis.* **2018**, *1864 Pt B*, 2814–2821. [CrossRef]
44. Tuttolomondo, A.; Cirrincione, A.; Casuccio, A.; Del Cuore, A.; Daidone, M.; Di Chiara, T.; Di Raimondo, D.; Corte, V.D.; Maida, C.; Simonetta, I.; et al. Efficacy of dulaglutide on vascular health indexes in subjects with type 2 diabetes: A randomized trial. *Cardiovasc. Diabetol.* **2021**, *20*, 1. [CrossRef] [PubMed]
45. Nicholls, S.J.; Bhatt, D.L.; Buse, J.B.; Prato, S.D.; Kahn, S.E.; Lincoff, A.M.; McGuire, D.K.; Nauck, M.A.; Nissen, S.E.; Sattar, N.; et al. Comparison of tirzepatide and dulaglutide on major adverse cardiovascular events in participants with type 2 diabetes and atherosclerotic cardiovascular disease: SURPASS-CVOT design and baseline characteristics. *Am. Heart J.* **2024**, *267*, 1–11. [CrossRef] [PubMed]
46. Saxena, A.R.; Frias, J.P.; Brown, L.S.; Gorman, D.N.; Vasas, S.; Tsamandouras, N.; Birnbaum, M.J. Efficacy and Safety of Oral Small Molecule Glucagon-Like Peptide 1 Receptor Agonist Danuglipron for Glycemic Control Among Patients With Type 2 Diabetes: A Randomized Clinical Trial. *JAMA Netw. Open.* **2023**, *6*, e2314493. [CrossRef]
47. Zhang, B.; Cheng, Z.; Chen, J.; Zhang, X.; Liu, D.; Jiang, H.; Ma, G.; Wang, X.; Gan, S.; Sun, J.; et al. Efficacy and Safety of Mazdutide in Chinese Patients With Type 2 Diabetes: A Randomized, Double-Blind, Placebo-Controlled Phase 2 Trial. *Diabetes Care* **2024**, *47*, 160–168. [CrossRef]
48. Oprea, E.; Berteanu, M.; Cinteză, D.; Manolescu, B.N. The effect of the ALAnerv nutritional supplement on some oxidative stress markers in postacute stroke patients undergoing rehabilitation. *Appl. Physiol. Nutr. Metab.* **2013**, *38*, 613–620. [CrossRef]
49. Yousefian, M.; Abedimanesh, S.; Yadegar, A.; Nakhjavani, M.; Bathaie, S.Z. Co-administration of "L-Lysine, Vitamin C, and Zinc" increased the antioxidant activity, decreased insulin resistance, and improved lipid profile in streptozotocin-induced diabetic rats. *Biomed. Pharmacother.* **2024**, *174*, 116525. [CrossRef]
50. Hu, E.-H.; Tsai, M.-L.; Lin, Y.; Chou, T.-S.; Chen, T.-H. A Review and Meta-Analysis of the Safety and Efficacy of Using Glucagon-like Peptide-1 Receptor Agonists. *Medicina* **2024**, *60*, 357. [CrossRef]
51. Wilding, J.P.H.; Batterham, R.L.; Davies, M.; Van Gaal, L.F.; Kandler, K.; Konakli, K.; Lingvay, I.; McGowan, B.M.; Oral, T.K.; Rosenstock, J.; et al. Weight regain and cardiometabolic effects after withdrawal of semaglutide: The STEP 1 trial extension. *Diabetes Obes. Metab.* **2022**, *24*, 1553–1564. [CrossRef] [PubMed]

52. Giorgino, F.; Bhana, S.; Czupryniak, L.; Dagdelen, S.; Galstyan, G.R.; Janež, A.; Lalić, N.; Nouri, N.; Rahelić, D.; Stoian, A.P.; et al. Management of patients with diabetes and obesity in the COVID-19 era: Experiences and learnings from South and East Europe, the Middle East, and Africa. *Diabetes Res. Clin. Pract.* **2021**, *172*, 108617. [CrossRef] [PubMed]
53. Patti, A.M.; Rizvi, A.A.; Giglio, R.V.; Stoian, A.P.; Ligi, D.; Mannello, F. Impact of Glucose-Lowering Medications on Cardiovascular and Metabolic Risk in Type 2 Diabetes. *J. Clin. Med.* **2020**, *9*, 912. [CrossRef] [PubMed]
54. Giglio, R.V.; Pantea Stoian, A.; Al-Rasadi, K.; Banach, M.; Patti, A.M.; Ciaccio, M.; Rizvi, A.A.; Rizzo, M. Novel Therapeutical Approaches to Managing Atherosclerotic Risk. *Int. J. Mol. Sci.* **2021**, *22*, 4633. [CrossRef]
55. Salmen, T.; Pietrosel, V.-A.; Reurean-Pintilei, D.; Iancu, M.A.; Cimpeanu, R.C.; Bica, I.-C.; Dumitriu-Stan, R.-I.; Potcovaru, C.-G.; Salmen, B.-M.; Diaconu, C.-C.; et al. Assessing Cardiovascular Target Attainment in Type 2 Diabetes Mellitus Patients in Tertiary Diabetes Center in Romania. *Pharmaceuticals* **2024**, *17*, 1249. [CrossRef]
56. Cardio-Metabolic Academy Europe East. Adoption of the ADA/EASD guidelines in 10 Eastern and Southern European countries: Physician survey and good clinical practice recommendations from an international expert panel. *Diabetes Res. Clin. Pract.* **2021**, *172*, 108535. [CrossRef]

Disclaimer/Publisher's Note: The statements, opinions and data contained in all publications are solely those of the individual author(s) and contributor(s) and not of MDPI and/or the editor(s). MDPI and/or the editor(s) disclaim responsibility for any injury to people or property resulting from any ideas, methods, instructions or products referred to in the content.

Systematic Review

Examining the Impact of Ertugliflozin on Cardiovascular Outcomes in Patients with Diabetes and Metabolic Syndrome: A Systematic Review of Clinical Trials

Silvius Alexandru Pescariu [1], Ahmed Elagez [2], Balaji Nallapati [3], Felix Bratosin [4], Adina Bucur [5,*], Alina Negru [1], Laura Gaita [6], Ioana Mihaela Citu [7], Zoran Laurentiu Popa [8] and Paula Irina Barata [9,10]

1. Department of Cardiology, Victor Babes University of Medicine and Pharmacy, 300041 Timisoara, Romania; pescariu.alexandru@umft.ro (S.A.P.); eivanica@yahoo.com (A.N.)
2. Department of General Medicine, Misr University for Science & Technology, Giza 3236101, Egypt; ahmeddmahmouudd@gmail.com
3. Department of General Medicine, Katuri Medical College and Hospital, Katuri City 522019, India; dr.balajinallapati@gmail.com
4. Department of Infectious Disease, Victor Babes University of Medicine and Pharmacy, 300041 Timisoara, Romania; felix.bratosin@umft.ro
5. Department III Functional Sciences, Division of Public Health and Management, University of Medicine and Pharmacy Victor Babes Timisoara, 300041 Timisoara, Romania
6. Second Department of Internal Medicine, University of Medicine and Pharmacy Victor Babes Timisoara, 300041 Timisoara, Romania; gaita.laura@umft.ro
7. First Department of Internal Medicine, University of Medicine and Pharmacy Victor Babes Timisoara, 300041 Timisoara, Romania; citu.ioana@umft.ro
8. Department of Obstetrics and Gynecology, University of Medicine and Pharmacy Victor Babes Timisoara, 300041 Timisoara, Romania; popa.zoran@umft.ro
9. Center for Research and Innovation in Precision Medicine of Respiratory Diseases, University of Medicine and Pharmacy Victor Babes Timisoara, 300041 Timisoara, Romania; barata.paula@student.uvvg.ro
10. Department of Physiology, Faculty of Medicine, "Vasile Goldis" Western University of Arad, 310025 Arad, Romania
* Correspondence: adina.bucur@umft.ro

Abstract: Cardiovascular diseases (CVDs) constitute a significant cause of morbidity and mortality globally, particularly among individuals with type 2 diabetes mellitus (T2DM). Ertugliflozin, a Sodium-Glucose Co-transporter-2 (SGLT2) inhibitor, is hypothesized to confer cardiovascular protection; however, long-term follow-up studies are necessary to support the hypothesis. This systematic review was conducted to evaluate the cardiovascular effects of ertugliflozin in diabetic versus non-diabetic cohorts, focusing on major adverse cardiovascular events (MACEs), hospitalizations for heart failure, and cardiovascular mortality. Adhering to PRISMA guidelines, the review encompassed studies indexed in PubMed, Scopus, and Web of Science up to March 2024. Eligibility was restricted to studies involving T2DM patients undergoing ertugliflozin treatment with reported outcomes relevant to cardiovascular health. Out of 767 initially identified articles, 6 met the inclusion criteria. Data concerning hazard ratios (HR) and confidence intervals (CI) were extracted to compare the effects of ertugliflozin with those of a placebo or other standard therapies. The collective sample size across these studies was 8246 participants. Ertugliflozin was associated with a significant reduction in hospitalizations for heart failure relative to a placebo (HR 0.70, 95% CI 0.54–0.90, $p < 0.05$). Furthermore, when combined with metformin, ertugliflozin potentially reduced MACEs (HR 0.92, 95% CI 0.79–1.07), although this finding did not reach statistical significance. Importantly, for patients with pre-existing heart failure, ertugliflozin significantly decreased the exacerbations of heart failure (HR 0.53, 95% CI 0.33–0.84, $p < 0.01$). Overall, ertugliflozin markedly reduces hospitalizations due to heart failure in T2DM patients and may improve additional cardiovascular outcomes. These results endorse the integration of ertugliflozin into therapeutic protocols for T2DM patients at elevated cardiovascular risk and substantiate its efficacy among SGLT2 inhibitors. Continued investigations are recommended to delineate its long-term cardiovascular benefits in diverse patient populations, including the potential impact on arrhythmias.

Citation: Pescariu, S.A.; Elagez, A.; Nallapati, B.; Bratosin, F.; Bucur, A.; Negru, A.; Gaita, L.; Citu, I.M.; Popa, Z.L.; Barata, P.I. Examining the Impact of Ertugliflozin on Cardiovascular Outcomes in Patients with Diabetes and Metabolic Syndrome: A Systematic Review of Clinical Trials. *Pharmaceuticals* **2024**, *17*, 929. https://doi.org/10.3390/ph17070929

Academic Editor: Alfredo Caturano

Received: 7 June 2024
Revised: 8 July 2024
Accepted: 10 July 2024
Published: 11 July 2024

Copyright: © 2024 by the authors. Licensee MDPI, Basel, Switzerland. This article is an open access article distributed under the terms and conditions of the Creative Commons Attribution (CC BY) license (https:// creativecommons.org/licenses/by/ 4.0/).

Keywords: cardiology; diabetes; systematic review; heart failure

1. Introduction

Cardiovascular diseases (CVDs) remain the leading cause of mortality worldwide, accounting for approximately 17.9 million deaths annually, a figure that represents a significant public health challenge [1]. The burden of CVD is particularly pronounced among individuals with metabolic disorders such as type 2 diabetes mellitus (T2DM), where the risk of developing cardiovascular complications is markedly elevated [2,3]. Diabetes itself is a global epidemic, with current estimates indicating that over 460 million adults live with the condition, a number expected to rise past 700 million by 2045 [4,5].

The pathophysiological link between T2DM and CVD is complex, involving a multifaceted interplay of hyperglycemia, insulin resistance, and a host of metabolic derangements, including dyslipidemia and hypertension [6,7]. These factors contribute to a heightened state of inflammation and increased atherosclerotic burden, propelling the progression of cardiovascular pathology [8]. Accordingly, the management of hyperglycemia in diabetic patients is aimed not only at controlling blood glucose levels but also at mitigating cardiovascular risk, which has propelled the development of antidiabetic therapies with cardioprotective properties [9,10].

In this regard, Sodium-Glucose Cotransporter 2 (SGLT2) inhibitors have emerged as a significant advancement in the treatment landscape of T2DM, with implications for CVD management [11,12]. SGLT2 inhibitors, by facilitating glucose excretion through kidney filtration, not only improve glycemic control but also confer benefits on body weight, blood pressure, and metabolic control [13]. Emerging evidence from clinical trials suggests that SGLT2s can significantly reduce hospitalization for heart failure and may lower the incidence of major adverse cardiovascular events in patients with T2DM [14,15]. Since their initial introduction in clinical practice, the hypotheses regarding the mechanisms of action of SGLT2 inhibitors have evolved significantly: initially viewed as straightforward glycosuric agents that lower glucose levels, enhance erythropoiesis, and stimulate ketogenesis, this class of drugs is now recognized as one containing intracellular sodium-lowering compounds. This property underlies their observed cardioprotective effects, which contribute to a substantial reduction in cardiovascular events, particularly in high-risk populations. From a molecular standpoint, the administration of gliflozins to patients induces conditions akin to nutrient and oxygen deprivation, thereby triggering autophagy to uphold cellular homeostasis via diverse degradative mechanisms [16]. In recent years, SGLT2 inhibitor uses have also expanded into the realm of renal protection [17].

Ertugliflozin was approved by the U.S. Food and Drug Administration (FDA) in December 2017 and subsequently received marketing authorization from the European Medicines Agency (EMA) in March 2018 [18]. Animal studies of ertugliflozin prior to these approvals demonstrated promising outcomes, particularly in the context of glucose control and potential cardiovascular benefits [19]. In rodent models, ertugliflozin effectively reduced blood glucose levels, body weight, and visceral adiposity [20]. Moreover, these studies suggested improvements in cardiac function and structure, indicating potential protective effects against heart failure—a common complication in diabetic populations.

While other SGLT2 inhibitors have enjoyed the spotlight regarding their efficacy against heart failure, the hypothesis of this study is that ertugliflozin also provides a significant cardiovascular benefit, an effect which it would share with the other drugs in the SGLT2 inhibitor class. The primary objective of this systematic review is to evaluate the extent of cardiovascular outcomes associated with the use of ertugliflozin in the diabetic and non-diabetic population. This investigation aims to inform clinical practice and guide future research in the management of cardiovascular risk. While ertugliflozin has shown promise in initial studies for its cardiovascular benefits, comprehensive analysis is required to consolidate these findings and assess their robustness across diverse patient populations.

This study can provide a thorough evaluation of existing evidence, addressing gaps in knowledge about the long-term efficacy and safety of ertugliflozin's action on heart failure

2. Materials and Methods

2.1. Eligibility Criteria

This systematic review selected studies for inclusion based on the following criteria and the existing literature [16–18]: (1) patients diagnosed with type 2 diabetes mellitus (T2DM) and treated with the SGLT2 inhibitor ertugliflozin either as monotherapy or in combination with other antidiabetic agents; (2) research specifically investigating the cardiovascular outcomes associated with the use of ertugliflozin, with a particular focus on major adverse cardiovascular events, hospitalization for heart failure, and overall cardiovascular mortality; (3) studies that were defined as clinical trials; (4) studies employing validated methods or clearly defined parameters to evaluate cardiovascular outcomes, efficacy, safety profiles, and patient adherence; and (5) only peer-reviewed articles published in English.

The exclusion criteria were the following: (1) studies not involving human participants, such as in vitro or animal studies, except where preclinical data were used to support clinical findings; (2) research not specifically examining patients treated with ertugliflozin or studies that did not differentiate the effects of ertugliflozin from other SGLT2 inhibitors; (3) studies that failed to provide clear, quantifiable outcomes related to cardiovascular health or lacked sufficient detail for a comprehensive assessment; (4) grey literature, including non-peer-reviewed articles, preprints, conference abstracts, general reviews, commentaries, and editorials, to avoid potential biases and inconsistencies; (5) studies deemed low quality based on predetermined, quantifiable metrics; and (6) studies with different designs other than clinical trials.

2.2. Information Sources

The primary information sources for this systematic review were the electronic databases PubMed, Scopus, and the Web of Science. The literature search specifically targeted publications up to the initial search date, 17 March 2024. The search was designed to capture a wide array of studies, including those that evaluated clinical outcomes, patient demographics, treatment modalities, and the long-term cardiovascular effects associated with ertugliflozin.

The PICOS statement of the following study was considered as follows:

P (Population): Patients diagnosed with type 2 diabetes mellitus (T2DM), including both diabetic and non-diabetic individuals being treated with ertugliflozin either as monotherapy or in combination with other antidiabetic agents.

I (Intervention): Use of the Sodium-Glucose Cotransporter 2 (SGLT2) inhibitor, ertugliflozin, focusing on its impact on cardiovascular outcomes.

C (Comparison): Comparison of ertugliflozin with placebo or other standard diabetes treatments, assessing differential impacts on cardiovascular outcomes.

O (Outcomes): Primary outcomes include major adverse cardiovascular events (MACEs), such as myocardial infarction, stroke, cardiovascular death, coronary revascularization, hospitalization for unstable angina, sudden cardiac arrest, significant arrhythmias, and changes in renal function measured by eGFR levels. Secondary outcomes involve patient demographic data, duration of diabetes, baseline HbA1c levels, lipid profiles, and overall patient safety and efficacy of the treatment.

S (Study Design): Randomized controlled trials (RCTs), observational studies, cohort studies, case-control studies, and cross-sectional studies published in peer-reviewed journals and written in English.

2.3. Search Strategy

The search strategy for this systematic review was crafted using keywords and phrases directly relevant to the study's objectives, focusing on the interplay between ertugliflozin, cardiovascular disease, and diabetes management. The keywords included "type 2 diabetes mellitus", "T2DM", "cardiovascular disease", "CVD", "heart failure", "major adverse cardiovascular events", "blood pressure", "SGLT2 inhibitors", "ertugliflozin", "clinical outcomes", "cardiovascular mortality", "hospitalization", "glycemic control", "metabolic effects", "patient safety", and "efficacy".

To maximize the precision and coverage of the literature search, these terms were combined using Boolean operators (AND, OR, NOT) along with relevant Medical Subject Headings (MeSH) to refine the search further. The formulated search string included combinations such as: (("diabetes" OR "type 2 diabetes mellitus" OR "T2DM") AND ("ertugliflozin" OR "SGLT2 inhibitors") AND ("cardiovascular disease" OR "heart failure" OR "heart disease") AND ("clinical outcomes" OR "hospitalization" OR "cardiovascular mortality") AND ("safety" OR "efficacy")).

2.4. Selection Process

In line with the Preferred Reporting Items for Systematic Reviews and Meta-Analyses (PRISMA) guidelines [21], the selection process of this systematic review was structured to ensure transparency and reproducibility. Initially, all the retrieved records underwent an independent screening by two reviewers (S.A.P. and F.B.) to ascertain their eligibility according to the predefined inclusion and exclusion criteria. Any discrepancies between the reviewers at this stage were addressed through a consensus meeting or, if required, by consulting a third reviewer. The protocol for this review, including detailed methodologies and criteria, has been registered and is publicly accessible on the Open Science Framework (OSF) with the registration code osf.io/ve7q3.

2.5. Data Collection Process

The data collection process commenced with the removal of duplicate entries from the initial dataset. Subsequently, two independent reviewers (S.A.P. and F.B.) performed a detailed screening of the abstracts to evaluate the relevance of each study against the established inclusion and exclusion criteria. In cases of discrepancies between the reviewers, discussions were held to reach a consensus, and if necessary, a third reviewer was consulted to make a final determination.

2.6. Data Items

The primary outcomes assessed included major adverse cardiovascular events (MACEs), such as myocardial infarction (heart attack), stroke, cardiovascular death, coronary revascularization, hospitalization for unstable angina, sudden cardiac arrest, significant arrhythmias, and hospitalization due to heart failure, as well as changes in renal function, quantified through estimated Glomerular Filtration Rate (eGFR) levels. Specific hazard ratios (HR) and confidence intervals (CI) were recorded to measure the effect size of ertugliflozin compared to various controls, including a placebo and other standard treatments. Sensitivity analyses were performed in the case of missing data, in order to estimate the impact of the missing data.

Secondary outcomes involved patient demographic data, such as age, body mass index (BMI), and race, details on the duration of diabetes, and baseline HbA1c levels; lipid profiles were also collected to further contextualize the impact of ertugliflozin on long-term glucose control and lipid metabolism. We also collected data on study design, quality, and geographic location from the referenced studies. Patient characteristics, including sample size, age distribution, and comparison groups, were documented to understand the population diversity and the comparative efficacy of the treatment under different demographic settings.

We utilized standardized data extraction forms designed to systematically capture all relevant data elements from each RCT, including study design, participant demographics, intervention details, outcomes, and any reported biases. To ensure reliability and validity in our data extraction process, these forms were pilot tested on a subset of studies before full implementation. The pilot testing helped refine the forms based on the initial findings, allowing for adjustments that enhanced the accuracy and consistency of the data collected across all the studies.

2.7. Risk of Bias and Quality Assessment

Initially, the quality of clinical trials was evaluated using the Cochrane tool [22]. This comprehensive tool evaluates several crucial factors including random sequence generation, allocation concealment, blinding of participants and personnel, blinding of outcome assessment, incomplete outcome data, selective reporting, and other biases. Each trial was independently reviewed by two researchers to mitigate any potential assessment bias, ensuring a rigorous and objective evaluation of the trial quality.

3. Results

3.1. Study Selection and Study Characteristics

A total of 767 articles were identified according to the initial search, of which 72 duplicate entries were eliminated, 607 records were excluded before screening based on their titles and abstracts, and 72 articles were excluded after a full read for not matching the inclusion criteria or having no available data. The systematic review included a total of six studies in the final analysis, delineated in Figure 1, spanning a period from 2017 to 2023.

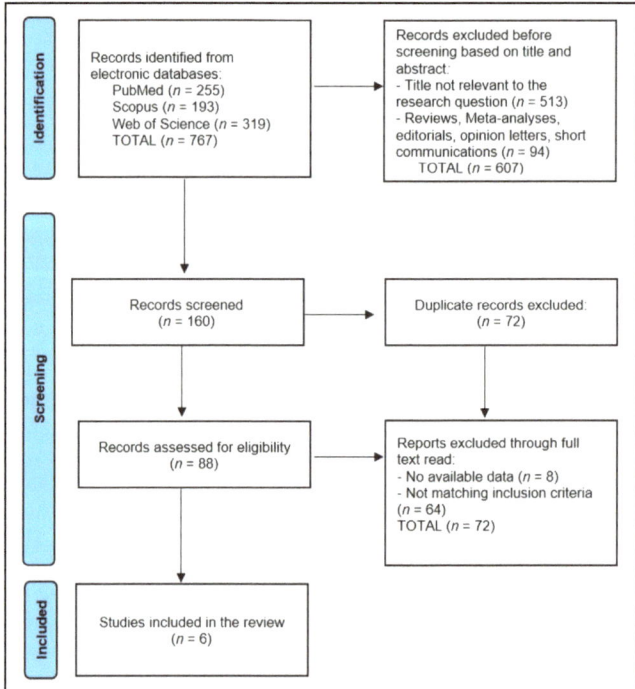

Figure 1. PRISMA flow diagram.

The analysis of the study characteristics from six research articles [23–28], as detailed in Table 1, covered a period from 2020 to 2023. These studies assessed the impact of the SGLT2 inhibitor ertugliflozin on cardiovascular outcomes in patients with type 2 diabetes. All six studies employed a randomized clinical trial design and were recognized for their high-quality methodology. The studies conducted by Cannon et al. [23] and Cosentino et al. [27] in 2020; Dagogo et al. [24] and Segar et al. [26] in 2022; and Cherney et al. [25] and Pratley et al. [28] in 2023 were all international trials, highlighting the global interest and applicability of ertugliflozin in diverse patient populations.

Table 1. Study characteristics.

Number	First Author	Reference	Country	Study Year	Study Design	Study Quality
1	Cannon et al.	[23]	International	2020	Randomized clinical trial	High
2	Dagogo et al.	[24]	International	2022	Randomized clinical trial	High
3	Cherney et al.	[25]	International	2023	Randomized clinical trial	High
4	Segar et al.	[26]	International	2022	Randomized clinical trial	High
5	Cosentino et al.	[27]	International	2020	Randomized clinical trial	High
6	Pratley et al.	[28]	International	2023	Randomized clinical trial	High

3.2. Results of Individual Studies

The existing studies performed subgroup and post hoc analyses of patients from the VERTIS-CV trial, each author focusing on the effect of ertugliflozin on different population features. Cannon et al. [21] examined a large subgroup of 5499 patients treated with ertugliflozin versus a placebo group of 2747 patients. The mean age of the participants was 64.4 years, and the average body mass index (BMI) was 32.0 ± 5.5, suggesting a predominately overweight population. A notable 87.8% of the cohort was white. This analysis focused on overall cardiovascular safety and efficacy, providing a baseline comparison for the effectiveness of ertugliflozin in reducing major adverse cardiovascular events (MACEs) compared to a placebo.

Dagogo et al. [24] analyzed the effects of combining ertugliflozin with metformin versus ertugliflozin alone in a total of 6286 participants. This comparison aimed to identify any additional benefits of metformin in the management of cardiovascular risks. The patients had a similar age profile (mean age of 64.0 years) and BMI (32.0 ± 5.4), with 87.6% of the participants being white. The study quantitatively assessed the reduction in MACEs and hospitalizations for heart failure, comparing the combined treatment versus ertugliflozin alone.

Cherney et al. [25] focused specifically on patients with heart failure, analyzing 1605 patients treated with ertugliflozin. This group was further subdivided based on heart failure presence and absence, comparing these to their respective placebo groups. This detailed examination helped elucidate the drug's impact on heart failure outcomes, a critical aspect given the high cardiovascular risk in the diabetic population. Segar et al. [24] also investigated a large cohort of 5499 patients, mirroring the subgroup in Cannon et al. [21], but potentially focusing on different aspects, such as the progression of renal impairment or specific cardiovascular outcomes over time. Detailed statistical results from this study would help in understanding the long-term benefits or risks associated with ertugliflozin treatment.

Cosentino et al. [27] divided their analysis among patients based on the severity of heart failure and ejection fraction (EF), creating subgroups such as HF with EF ≤ 45% and HF with EF >45%, compared against similar placebo groups. This stratification provided insights into how varying degrees of cardiac function might influence the efficacy of ertugliflozin, particularly in reducing hospitalizations or improving survival rates. Pratley et al. [26] explored age-related effects by dividing their cohort into segments based on age thresholds (≥65 years and <65 years) and further stratified by elderly (≥75 years) versus younger segments (<75 years). By detailing BMI across these age groups and comparing them to corresponding placebo groups, the study highlighted how age and body composition interact with treatment outcomes, particularly in terms of renal function and cardiovascular risk reduction, as seen in Table 2.

Table 2. Characteristics of patients.

Number	First Author	Reference	Sample Size	Age (Years)	Comparison Group	BMI	Race
1	Cannon et al.	[23]	Ertugliflozin (n = 5499)	64.4	Placebo (n = 2747)	32.0 ± 5.5	White 87.8%
2	Dagogo et al.	[24]	Ertugliflozin + Metformin (n = 6286)	64.0	Without Metformin (n = 1960)	32.0 ± 5.4	White 87.6%
3	Cherney et al.	[25]	Ertugliflozin HF (n = 1605)	64.6	Ertugliflozin no HF (n = 3894), Placebo HF (n = 834), Placebo no HF (n = 1913)	32.5 ± 5.4	NR
4	Segar et al.	[26]	Ertugliflozin (n = 5499)	64.4	Placebo (n = 2747)	NR	NR
5	Cosentino et al.	[27]	Ertugliflozin HF, EF ≤ 45% (n = 319), Ertugliflozin HF, EF > 45% (n = 680)	HF, EF ≤ 45%: 64.2 years, HF, EF > 45% 64.7 years	Placebo HF, EF ≤ 45% (n = 159), Placebo HF, EF > 45% (n = 327)	HF, EF ≤ 45%: 32.1, HF, EF > 45% 32.9	NR
6	Pratley et al.	[28]	Ertugliflozin aged ≥ 65 years (n = 2775), Ertugliflozin < 65 years (n = 2722), Ertugliflozin ≥ 75 years (n = 593), Ertugliflozin < 75 years (n = 4900)	NR	Placebo aged ≥ 65 years (n = 1370), Placebo < 65 years (n = 1375), Placebo ≥ 75 years (n = 310), Placebo < 75 years (n = 2435)	Ertugliflozin aged ≥ 65 years (BMI = 31.2), Ertugliflozin < 65 years (BMI = 32.7), Ertugliflozin ≥ 75 years (BMI = 30.4), Ertugliflozin < 75 years (BMI = 32.2)	85.2–92.4%

NR—not reported; BMI—body mass index; HF—heart failure; EF—ejection fraction.

3.3. Results of Synthesis

The participants had a mean duration of diabetes ranging from 9.4 to 13.8 years, reflecting a long-term management scenario typical in type 2 diabetes populations. Across the studies, the mean or median HbA1c values were consistently around 8.2%. Lipid profiles, where reported, indicated a mixed dyslipidemia common in type 2 diabetes. For instance, Cannon et al. [21] reported a detailed lipid profile with total cholesterol at 168.9 ± 46.9 mg/dL, LDL cholesterol at 89.3 ± 38.5 mg/dL, HDL cholesterol at 43.7 ± 12.0 mg/dL, and triglycerides at 181.4 ± 119.2 mg/dL. Meanwhile, Dagogo et al. [22] noted a 75.8% prevalence of dyslipidemia in their study cohort. The eGFR showed variability across the subgroups, ranging from 46.0 to 78.1 mL/min/1.73 m^2, as presented in Table 3.

The studies consistently reported a high prevalence of coronary artery disease (CAD), ranging from 71.8% to 75.9%, emphasizing the significant cardiovascular risk in the diabetic population studied. Heart failure incidence varied, with reported percentages typically around 20% to 25%, except in specific heart failure subgroups where it was notably higher. Regarding the cardiovascular and mortality outcomes, Cannon et al. [21] observed a hazard ratio (HR) of 0.97 (95.6% CI, 0.85–1.11) for MACEs, indicating that ertugliflozin was noninferior to a placebo in preventing cardiovascular events. This study also reported more substantial results for heart failure-related outcomes, with HR for hospitalization due to heart failure at 0.70 (95% CI, 0.54–0.90), showcasing a significant reduction.

Dagogo et al. [24] found variable HRs for MACEs when ertugliflozin was used in combination with other diabetes medications. Notably, the HR was 0.92 (95% CI 0.790, 1.073) when used with metformin, and without metformin, the HR rose to 1.13 (95% CI 0.867, 1.480). This suggests a potential interaction effect between ertugliflozin and metformin that may enhance cardiovascular protection.

Table 3. Metabolic disease characteristics.

Number	First Author	Reference	Duration of Diabetes	HbA1c (Mean/Median)	Lipids *	eGFR
1	Cannon et al.	[23]	13.0 years	8.2%	Total cholesterol: 168.9 ± 46.9 mg/dL, LDL: 89.3 ± 38.5 mg/dL, HDL: 43.7 ± 12.0 mg/dL, Triglycerides: 181.4 ± 119.2 mg/dL	76.1 ± 20.9 mL/min/1.73 m^2
2	Dagogo et al.	[24]	12.5 years	8.2%	Dyslipidemia: 75.8%	78.1 ± 20.1 mL/min/1.73 m^2
3	Cherney et al.	[25]	12.5–13.4 years	8.2%	NR	73.5–76.8 mL/min/1.73 m^2
4	Segar et al.	[26]	13.0 years	8.2%	LDL: 2.3(1.0) mmol/L, HDL: 1.1 (0.3) mmol/L, Triglycerides: 181.4 ± 119.2 mg/dL	76.1 ± 20.9 mL/min/1.73 m^2
5	Cosentino et al.	[27]	HF, EF ≤ 45%: 13.4 years, HF, EF > 45% 13.0 years	HF, EF ≤ 45%: 8.2%, HF, EF > 45% 8.3%	NR	HF, EF ≤ 45%: 32.1% (30–60) years, HF, EF > 45% 26.6% (30–60)
6	Pratley et al.	[28]	9.4–13.8 years	8.0–8.4	NR	46.0–57.1% between 60 and 90 mL/min/1.73 m^2

NR—not reported; *—lipids comprise cholesterol, low-density (LDL) and high-density (HDL) lipoproteins, and triglycerides; eGFR—estimated Glomerular Filtration Rate.

Cosentino et al. [27] and Pratley et al. [28] both reported significant reductions in heart failure hospitalizations, with Pratley et al. noting HRs of 0.72 (95% CI 0.52–0.99) for older adults (≥65 years) and 0.66 (95% CI 0.43–1.02) for those younger than 65, indicating robust protective effects across age groups. Additionally, Pratley et al. [28] highlighted ertugliflozin's renal benefits, showing HRs that indicated reduced risks of significant declines in renal function, such as a sustained eGFR reduction of at least 40% from baseline with HRs of 0.71 (95% CI 0.47–1.09) for older adults and 0.62 (95% CI 0.43–0.91) for the younger subgroup, as presented in Table 4.

Table 4. Analysis of outcomes.

Number	First Author	Reference	CAD (%)	Heart Failure (%)	Other Comorbidities * (%)	Outcomes (Mortality, Major Cardiovascular Events, Hospitalization)
1	Cannon et al.	[23]	75.9%	23.7%	Peripheral arterial disease: 18.7%, Cerebrovascular disease: 22.9%	MACE: HR 0.97 (95.6% CI, 0.85–1.11), $p < 0.001$ for noninferiority. Death from cardiovascular causes or hospitalization for heart failure: HR 0.88 (95.8% CI, 0.75–1.03), $p = 0.11$ for superiority. Hospitalization for heart failure: HR 0.70 (95% CI, 0.54–0.90).

Table 4. Cont.

Number	First Author	Reference	CAD (%)	Heart Failure (%)	Other Comorbidities * (%)	Outcomes (Mortality, Major Cardiovascular Events, Hospitalization)
2	Dagogo et al.	[24]	75.8%	22.5%	Diabetic microvascular disease: 36.5%; Hypertension 91.3%	MACE with metformin: HR 0.92 (95% CI 0.790, 1.073); without metformin: HR 1.13 (95% CI 0.867, 1.480); MACE with insulin: HR 0.91 (95% CI 0.765, 1.092); without insulin: HR 1.06 (95% CI 0.867, 1.293); MACE with SUs: HR 1.11 (95% CI 0.890, 1.388); without SUs: HR 0.90 (95% CI 0.761, 1.060); MACE with DPP-4 inhibitors: HR 0.77 (95% CI 0.502, 1.173); without DPP-4 inhibitors: HR 1.00 (95% CI 0.867, 1.147). Hospitalization for heart failure (HHF) with metformin: HR 0.69 (95% CI 0.503, 0.940); without metformin: HR 0.71 (95% CI 0.449, 1.117).
3	Cherney et al.	[25]	NR	29.6%	NR	HF subgroup: HR 0.53 (95% CI 0.33–0.84); No-HF subgroup: HR 0.76 (95% CI 0.53–1.08).
4	Segar et al.	[26]	75.9%	23.7%	NR	Hemoglobin mediated 63.33% (95% CI 26.08–231.35) of the effect on the risk of hospitalization for heart failure when considering weighted average changes. Hematocrit mediated 40.0% (95% CI 10.61–151.17) of the effect on the risk of hospitalization for heart failure in early time period changes.
5	Cosentino et al.	[27]	HF, EF ≤ 45%: 96.9%, HF, EF > 45% 94.4%	HF, EF ≤ 45%: 1.9%, HF, EF > 45% 3.8%	HF, EF ≤ 45%: 67.3% NYHA III, HF, EF > 45% 64.3% NYHA III	Total HHF and CV death events HR = 0.83 (0.72–0.96); ertugliflozin vs. placebo 0.70 (0.54–0.90) ertugliflozin: 0.75, placebo: 1.05; ertugliflozin 5 mg vs. placebo 0.71 (0.52–0.97); ertugliflozin 5 mg: 0.75, placebo: 1.05; ertugliflozin 15 mg vs. placebo 0.68 (0.50–0.93) ertugliflozin 15 mg: 0.72, placebo: 1.05.
6	Pratley et al.	[28]	71.8–84.4%	20.7–27.1%	36.2–40.8%	Ertugliflozin versus placebo was associated with reductions in the risk of hospitalization for heart failure (≥65 years: HR 0.72, 95% CI 0.52–0.99; <65 years: 0.66, 0.43–1.02), the prespecified kidney composite outcome of a doubling in serum creatinine (≥65 years: 0.84, 0.60–1.17; <65 years: 0.78, 0.55–1.10), and the prespecified exploratory kidney composite outcome of sustained eGFR reduction of at least 40% from baseline (≥65 years: 0.71, 0.47–1.09; <65 years: 0.62, 0.43–0.91).

NR—not reported; CI—confidence interval; HR—hazard ratio; *—other comorbidities include cerebrovascular disease, peripheral arterial disease, coronary revascularization; stroke; CAD—coronary arterial disease; MACEs—major adverse cardiovascular events; SU—sulfonylureas; DPP—dipeptidyl peptidase.

4. Discussion

4.1. Summary of Evidence

The current systematic review has elucidated several critical findings that bear significant implications for clinical practice. The most notable outcome is the substantial reduction in hospitalization for heart failure when patients are treated with ertugliflozin, as evidenced by a hazard ratio (HR) of 0.70. This finding is particularly important, considering the high morbidity and healthcare costs associated with heart failure hospitalizations in diabetic populations. The ability of ertugliflozin to reduce these hospitalizations supports its role not only as a glycemic control agent but also as a preventive strategy against one of the most severe cardiovascular complications in diabetes.

Furthermore, the additional cardiovascular benefits observed when ertugliflozin was combined with metformin, although not reaching statistical significance (HR 0.92), suggest a potential synergistic effect that could be clinically relevant. This outcome prompts consideration of the combination of these therapies as a standard approach in managing patients with T2DM who are at elevated cardiovascular risk. Given that metformin is already a first-line treatment in T2DM, the integration of ertugliflozin could be readily implemented, enhancing patient outcomes through combined metabolic and cardiovascular risk reduction.

The subgroup analysis revealing significant benefits in patients with existing heart failure (HR 0.53) further highlights ertugliflozin's role in this high-risk subgroup. These findings suggest that ertugliflozin not only prevents the worsening of heart failure but may also contribute to its management, marking a pivotal shift in how clinicians approach the intersection of diabetes and heart failure. As heart failure and diabetes frequently coexist, introducing ertugliflozin could significantly alter clinical outcomes for these patients, potentially reducing both hospitalization rates and overall cardiovascular morbidity.

Moreover, the consistent effect across diverse international populations emphasizes the global applicability of ertugliflozin, reinforcing its potential benefit in various ethnic and racial groups. This widespread efficacy supports broad clinical adoption and suggests that ertugliflozin can be an essential component of cardiovascular risk management in diabetes care protocols globally, offering a unified strategy to tackle the dual challenges of glucose control and cardiovascular risk mitigation.

The ERASe trial from Austria [29], a phase III study, explored the use of ertugliflozin in reducing the ventricular arrhythmic burden in patients with reduced or midrange ejection fraction, irrespective of their diabetic status. This multicenter, randomized, double-blind, placebo-controlled trial aimed to enroll 402 patients to investigate whether daily administration of ertugliflozin (5 mg) could decrease the total burden of ventricular arrhythmias. However, the final results are still awaiting publication. Meanwhile, the study by Croteau et al. [30] provided mechanistic insights, demonstrating how ertugliflozin might exert its cardiovascular benefits by reversing diastolic dysfunction and impaired energetics, as evidenced by improvements in phosphocreatine (PCr) and the PCr/ATP ratio, and potentially through the reduction in elevated myocardial intracellular sodium in a mouse model of diabetic cardiomyopathy, which also has possible implications in the prevention of arrythmia. However, the ERASe study does have its limitations, such as the relatively limited 52-week follow-up period, as well as its focus on patients with implantable cardiac defibrillators and/or cardiac resynchronization therapy devices. This may suggest that the results may not be applicable to all heart failure patients, particularly those without such devices. Furthermore, the results include individuals from a single geographical region; thus, the trial may have limited power to generalize the results towards broader, more diverse populations.

One study that was not included in the final analysis of the current systematic review was the post hoc analysis of the VERTIS CV trial by Pandey et al. [31], revealing that ertugliflozin significantly reduced the risk of a first HHF with a hazard ratio of 0.70 (95% CI 0.54–0.90). Notably, the reduction in HHF was more pronounced in patients with EF $\leq$ 45% (HR 0.48, 95% CI 0.30–0.75) compared to those with higher EFs, demonstrating its greater efficacy in this subgroup. Additionally, the total HHF events also showed significant reduction, particularly among those with EF $\leq$ 45% (rate ratio 0.39, 95% CI 0.26–0.57). Also, regarding mortality, the hazard ratio was 0.92 (95% CI, 0.77–1.10), which indicates no significant reduction in death from all causes with ertugliflozin compared to a placebo. Overall, ertugliflozin has been shown to provide significant benefits in reducing hospitalization for heart failure and improving glycemic control; the VERTIS CV trial data did not demonstrate a statistically significant reduction in overall cardiovascular mortality. Zhang et al.'s meta-analysis [32] of ertugliflozin in type 2 diabetes patients highlighted its effectiveness in reducing glycated hemoglobin levels (weighted mean difference −0.77%

for 5 mg and −0.82% for 15 mg), fasting plasma glucose, and body weight, with an added risk of genital mycotic infections.

The VERTIS RENAL study [33] evaluated the efficacy of ertugliflozin in T2DM patients with stage 3 CKD, showing modest A1C reductions at 26 weeks (least squares mean changes of −0.3% for both 5 mg and 15 mg doses compared to −0.4% for placebo). Despite some compliance issues with metformin use, ertugliflozin also led to significant improvements in body weight and fasting plasma glucose levels, demonstrating its utility even in patients with diminished renal function. On the other hand, the VERTIS CV trial [34] analyzed the effects on kidney composite outcomes in a broader diabetic population with established atherosclerotic cardiovascular disease, finding a notable reduction in the risk of severe renal outcomes (HR 0.66, 95% CI 0.50–0.88) for the exploratory composite renal endpoint. Furthermore, ertugliflozin was associated with a preservation of eGFR and a consistent decrease in the urinary albumin/creatinine ratio (UACR), which was particularly beneficial in patients with macroalbuminuria and high/very high-risk CKD.

The studies by Amin et al. and Cheng et al. provide a comprehensive examination of the cardiovascular and renal impacts of ertugliflozin in patients with type 2 diabetes mellitus. Amin et al. [35] focused on the blood pressure-lowering effects of ertugliflozin in T2DM patients with hypertension, demonstrating that all doses of ertugliflozin (1, 5, 25 mg) achieved significant reductions in 24 h mean systolic blood pressure by −3.0 to −4.0 mmHg over 4 weeks, which was comparable to the −3.2 mmHg reduction with hydrochlorothiazide (HCTZ; 12.5 mg). The drug also effectively decreased fasting plasma glucose and increased urinary glucose excretion without affecting plasma renin or urinary aldosterone levels. On the other hand, Cheng et al. [36] conducted a systematic review and meta-analysis assessing ertugliflozin's long-term effects on renal function and cardiovascular outcomes. Their findings indicated a statistically significant reduction in estimated glomerular filtration rate (eGFR) by 0.60 mL/min/1.73 m^2, suggesting a potential decline in renal function associated with the drug's use over no more than 52 weeks. However, ertugliflozin did not increase the risk of major cardiovascular events like acute myocardial infarction or angina pectoris, highlighting its cardiovascular safety profile. Together, these studies affirm the beneficial effects of ertugliflozin on blood pressure and cardiovascular risk in T2DM patients, while also prompting caution regarding its potential renal effects over time.

The clinical utility of this systematic review is significant, especially in enhancing cardiovascular management strategies for patients with type 2 diabetes treated with ertugliflozin. By evaluating diverse international studies, the review substantiates ertugliflozin's role in reducing major adverse cardiovascular events and heart failure-related hospitalizations. This evidence supports more informed therapeutic decisions, suggesting that ertugliflozin can be effectively integrated into treatment protocols to improve cardiovascular outcomes in this high-risk patient population, thus guiding personalized treatment approaches and improving overall care quality.

4.2. Limitations

However, this review is not without limitations. A significant concern is that all the studies conducted subgroup analyses or post hoc analyses on the same VERTIS-CV trial, which might introduce biases related to the selection and interpretation of data. Subgroup analyses, particularly when they are post hoc, can often find spurious associations that might not replicate in broader or different patient populations. Additionally, the review excluded studies that did not differentiate the effects of ertugliflozin from other SGLT2 inhibitors, potentially overlooking comparative insights that could refine the understanding of its unique benefits or risks. Critically, while our study design reflects broader clinical applications, it inherently limits our ability to definitively quantify the isolated impact of ertugliflozin. We acknowledge that this is a limitation and suggest that future research could benefit from conducting propensity analyses that adjust for the effects of additional medications. Our findings, therefore, should be interpreted with an understanding of these contextual nuances and the potential for confounding variables within the treatment regimes. A meta-analysis was not performed due to the significant heterogeneity among the included studies, which focused on various populations of diabetic patients and specific subgroups.

5. Conclusions

In conclusion, this systematic review provides robust evidence supporting the cardiovascular benefits of ertugliflozin in patients with T2DM, particularly in reducing hospitalizations for heart failure and possibly improving overall cardiovascular outcomes when combined with metformin. It is also important to take into consideration that up until this point, current findings still have yet to demonstrate a significant impact on overall cardiovascular mortality. These findings should encourage the incorporation of ertugliflozin into clinical practice for managing cardiovascular risk in T2DM patients. Clinicians are advised to consider these benefits when devising treatment plans for patients at high risk of cardiovascular events. Future research should aim to confirm these findings in larger, more diverse populations and explore the full potential of ertugliflozin within the broader spectrum of cardiovascular and metabolic diseases.

Author Contributions: Conceptualization, S.A.P. and A.E.; methodology, S.A.P. and A.E.; software, Z.L.P., S.A.P. and A.E.; validation, Z.L.P., B.N. and F.B.; formal analysis, B.N. and F.B.; investigation, B.N. and F.B.; resources, A.B. and A.N.; data curation, I.M.C., A.B. and A.N.; writing—original draft preparation, S.A.P., L.G., P.I.B. and F.B.; writing—review and editing, I.M.C., A.E, B.N., A.B. and A.N.; visualization, I.M.C., L.G. and P.I.B.; supervision, Z.L.P., L.G. and P.I.B.; project administration, Z.L.P., L.G., and P.I.B. All authors have read and agreed to the published version of the manuscript.

Funding: The costs of publication were supported by VICTOR BABES UNIVERSITY OF MEDICINE AND PHARMACY TIMISOARA. The findings, interpretations, and conclusions drawn in this study remain entirely independent of the financial support provided.

Institutional Review Board Statement: Not applicable.

Informed Consent Statement: Informed consent was obtained from all subjects involved in the study.

Data Availability Statement: The original contributions presented in the study are included in the article; further inquiries can be directed to the corresponding author.

Conflicts of Interest: The authors declare no conflicts of interest.

References

1. Mendis, S.; Graham, I.; Narula, J. Addressing the Global Burden of Cardiovascular Diseases; Need for Scalable and Sustainable Frameworks. *Glob. Heart* **2022**, *17*, 48. [CrossRef] [PubMed]
2. De Rosa, S.; Arcidiacono, B.; Chiefari, E.; Brunetti, A.; Indolfi, C.; Foti, D.P. Type 2 Diabetes Mellitus and Cardiovascular Disease: Genetic and Epigenetic Links. *Front. Endocrinol.* **2018**, *9*, 2. [CrossRef]
3. Ma, C.X.; Ma, X.N.; Guan, C.H.; Li, Y.D.; Mauricio, D.; Fu, S.B. Cardiovascular Disease in Type 2 Diabetes Mellitus: Progress Toward Personalized Management. *Cardiovasc. Diabetol.* **2022**, *21*, 74. [CrossRef] [PubMed]
4. Standl, E.; Khunti, K.; Hansen, T.B.; Schnell, O. The Global Epidemics of Diabetes in the 21st Century: Current Situation and Perspectives. *Eur. J. Prev. Cardiol.* **2019**, *26*, 7–14. [CrossRef] [PubMed]
5. Saeedi, P.; Petersohn, I.; Salpea, P.; Malanda, B.; Karuranga, S.; Unwin, N.; Colagiuri, S.; Guariguata, L.; Motala, A.A.; Ogurtsova, K.; et al. Global and Regional Diabetes Prevalence Estimates for 2019 and Projections for 2030 and 2045: Results from the International Diabetes Federation Diabetes Atlas, 9th Edition. *Diabetes Res. Clin. Pract.* **2019**, *157*, 107843. [CrossRef] [PubMed]
6. Galicia-Garcia, U.; Benito-Vicente, A.; Jebari, S.; Larrea-Sebal, A.; Siddiqi, H.; Uribe, K.B.; Ostolaza, H.; Martín, C. Pathophysiology of Type 2 Diabetes Mellitus. *Int. J. Mol. Sci.* **2020**, *21*, 6275. [CrossRef]
7. Chakraborty, S.; Verma, A.; Garg, R.; Singh, J.; Verma, H. Cardiometabolic Risk Factors Associated With Type 2 Diabetes Mellitus: A Mechanistic Insight. *Clin. Med. Insights Endocrinol. Diabetes* **2023**, *16*, 11795514231220780. [CrossRef] [PubMed]
8. Høilund-Carlsen, P.F.; Piri, R.; Madsen, P.L.; Revheim, M.E.; Werner, T.J.; Alavi, A.; Gerke, O.; Sturek, M. Atherosclerosis Burdens in Diabetes Mellitus: Assessment by PET Imaging. *Int. J. Mol. Sci.* **2022**, *23*, 10268. [CrossRef] [PubMed]
9. Davies, M.J.; D'Alessio, D.A.; Fradkin, J.; Kernan, W.N.; Mathieu, C.; Mingrone, G.; Rossing, P.; Tsapas, A.; Wexler, D.J.; Buse, J.B. Management of Hyperglycemia in Type 2 Diabetes, 2018. A Consensus Report by the American Diabetes Association (ADA) and the European Association for the Study of Diabetes (EASD). *Diabetes Care* **2018**, *41*, 2669–2701. [CrossRef] [PubMed]

10. Davies, M.J.; Aroda, V.R.; Collins, B.S.; Gabbay, R.A.; Green, J.; Maruthur, N.M.; Rosas, S.E.; Del Prato, S.; Mathieu, C.; Mingrone, G.; et al. Management of Hyperglycaemia in Type 2 Diabetes, 2022. A Consensus Report by the American Diabetes Association (ADA) and the European Association for the Study of Diabetes (EASD). *Diabetologia* **2022**, *65*, 1925–1966. [CrossRef] [PubMed]
11. Aftab, S.; Vetrivel Suresh, R.; Sherali, N.; Daniyal, M.; Tsouklidis, N. Sodium-Glucose Cotransporter-2 (SGLT-2) Inhibitors: Benefits in Diabetics With Cardiovascular Disease. *Cureus* **2020**, *12*, e10783. [CrossRef] [PubMed]
12. Fatima, A.; Rasool, S.; Devi, S.; Talha, M.; Waqar, F.; Nasir, M.; Khan, M.R.; Ibne Ali Jaffari, S.M.; Haider, A.; Shah, S.U.; et al. Exploring the Cardiovascular Benefits of Sodium-Glucose Cotransporter-2 (SGLT2) Inhibitors: Expanding Horizons Beyond Diabetes Management. *Cureus* **2023**, *15*, e46243. [CrossRef] [PubMed]
13. Brown, E.; Wilding, J.P.H.; Alam, U.; Barber, T.M.; Karalliedde, J.; Cuthbertson, D.J. The Expanding Role of SGLT2 Inhibitors Beyond Glucose-Lowering to Cardiorenal Protection. *Ann. Med.* **2021**, *53*, 2072–2089. [CrossRef] [PubMed]
14. Blanco, C.A.; Garcia, K.; Singson, A.; Smith, W.R. Use of SGLT2 Inhibitors Reduces Heart Failure and Hospitalization: A Multicenter, Real-World Evidence Study. *Perm. J.* **2023**, *27*, 77–87. [CrossRef] [PubMed]
15. Keller, D.M.; Ahmed, N.; Tariq, H.; Walgamage, M.; Walgamage, T.; Mohammed, A.; Chou, J.T.; Kałużna-Oleksy, M.; Lesiak, M.; Straburzyńska-Migaj, E. SGLT2 Inhibitors in Type 2 Diabetes Mellitus and Heart Failure-A Concise Review. *J. Clin. Med.* **2022**, *11*, 1470. [CrossRef] [PubMed]
16. Palmiero, G.; Cesaro, A.; Vetrano, E.; Pafundi, P.C.; Galiero, R.; Caturano, A.; Moscarella, E.; Gragnano, F.; Salvatore, T.; Rinaldi, L.; et al. Impact of SGLT2 Inhibitors on Heart Failure: From Pathophysiology to Clinical Effects. *Int. J. Mol. Sci.* **2021**, *22*, 5863. [CrossRef] [PubMed]
17. Nevola, R.; Alfano, M.; Pafundi, P.C.; Brin, C.; Gragnano, F.; Calabrò, P.; Adinolfi, L.E.; Rinaldi, L.; Sasso, F.C.; Caturano, A. Cardiorenal Impact of SGLT-2 Inhibitors: A Conceptual Revolution in The Management of Type 2 Diabetes, Heart Failure and Chronic Kidney Disease. *Rev. Cardiovasc. Med.* **2022**, *23*, 106. [CrossRef] [PubMed]
18. Marrs, J.C.; Anderson, S.L. Ertugliflozin in the Treatment of Type 2 Diabetes Mellitus. *Drugs Context* **2020**, *9*, 2020-7-4. [CrossRef] [PubMed]
19. Moellmann, J.; Mann, P.A.; Kappel, B.A.; Kahles, F.; Klinkhammer, B.M.; Boor, P.; Kramann, R.; Ghesquiere, B.; Lebherz, C.; Marx, N.; et al. The Sodium-Glucose Co-Transporter-2 Inhibitor Ertugliflozin Modifies the Signature of Cardiac Substrate Metabolism and Reduces Cardiac mTOR Signalling, Endoplasmic Reticulum Stress and Apoptosis. *Diabetes Obes. Metab.* **2022**, *24*, 2263–2272. [CrossRef]
20. Fediuk, D.J.; Nucci, G.; Dawra, V.K.; Cutler, D.L.; Amin, N.B.; Terra, S.G.; Boyd, R.A.; Krishna, R.; Sahasrabudhe, V. Overview of the Clinical Pharmacology of Ertugliflozin, a Novel Sodium-Glucose Cotransporter 2 (SGLT2) Inhibitor. *Clin. Pharmacokinet.* **2020**, *59*, 949–965. [CrossRef]
21. Page, M.J.; McKenzie, J.E.; Bossuyt, P.M.; Boutron, I.; Hoffmann, T.C.; Mulrow, C.D.; Shamseer, L.; Tetzlaff, J.M.; Akl, E.A.; Brennan, S.E.; et al. The PRISMA 2020 Statement: An Updated Guideline for Reporting Systematic Reviews. *Syst. Rev.* **2021**, *10*, 89. [CrossRef] [PubMed]
22. Higgins, J.P.; Altman, D.G.; Gøtzsche, P.C.; Jüni, P.; Moher, D.; Oxman, A.D.; Savovic, J.; Schulz, K.F.; Weeks, L.; Sterne, J.A.; et al. The Cochrane Collaboration's Tool for Assessing Risk of Bias in Randomised Trials. *BMJ* **2011**, *343*, d5928. [CrossRef] [PubMed]
23. Cannon, C.P.; Pratley, R.; Dagogo-Jack, S.; Mancuso, J.; Huyck, S.; Masiukiewicz, U.; Charbonnel, B.; Frederich, R.; Gallo, S.; Cosentino, F.; et al. Cardiovascular Outcomes with Ertugliflozin in Type 2 Diabetes. *N. Engl. J. Med.* **2020**, *383*, 1425–1435. [CrossRef] [PubMed]
24. Dagogo-Jack, S.; Cannon, C.P.; Cherney, D.Z.I.; Cosentino, F.; Liu, J.; Pong, A.; Gantz, I.; Frederich, R.; Mancuso, J.P.; Pratley, R.E. Cardiorenal Outcomes with Ertugliflozin Assessed According to Baseline Glucose-Lowering Agent: An Analysis from VERTIS CV. *Diabetes Obes. Metab.* **2022**, *24*, 1245–1254. [CrossRef] [PubMed]
25. Cherney, D.Z.I.; Cosentino, F.; McGuire, D.K.; Kolkailah, A.A.; Dagogo-Jack, S.; Pratley, R.E.; Frederich, R.; Maldonado, M.; Liu, C.C.; Cannon, C.P.; et al. Effects of Ertugliflozin on Kidney Outcomes in Patients With Heart Failure at Baseline in the Evaluation of Ertugliflozin Efficacy and Safety Cardiovascular Outcomes (VERTIS CV) Trial. *Kidney Int. Rep.* **2023**, *8*, 746–753. [CrossRef] [PubMed]
26. Segar, M.W.; Kolkailah, A.A.; Frederich, R.; Pong, A.; Cannon, C.P.; Cosentino, F.; Dagogo-Jack, S.; McGuire, D.K.; Pratley, R.E.; Liu, C.C.; et al. Mediators of Ertugliflozin Effects on Heart Failure and Kidney Outcomes Among Patients with Type 2 Diabetes Mellitus. *Diabetes Obes. Metab.* **2022**, *24*, 1829–1839. [CrossRef] [PubMed]
27. Cosentino, F.; Cannon, C.P.; Cherney, D.Z.I.; Masiukiewicz, U.; Pratley, R.; Dagogo-Jack, S.; Frederich, R.; Charbonnel, B.; Mancuso, J.; Shih, W.J.; et al. Efficacy of Ertugliflozin on Heart Failure-Related Events in Patients With Type 2 Diabetes Mellitus and Established Atherosclerotic Cardiovascular Disease: Results of the VERTIS CV Trial. *Circulation* **2020**, *142*, 2205–2215. [CrossRef] [PubMed]
28. Pratley, R.E.; Cannon, C.P.; Cherney, D.Z.I.; Cosentino, F.; McGuire, D.K.; Essex, M.N.; Lawrence, D.; Jones, P.L.S.; Liu, J.; Adamsons, I.; et al. Cardiorenal Outcomes, Kidney Function, and Other Safety Outcomes with Ertugliflozin in Older Adults with Type 2 Diabetes (VERTIS CV): Secondary Analyses from a Randomised, Double-Blind Trial. *Lancet Healthy Longev.* **2023**, *4*, e143–e154. [CrossRef]
29. von Lewinski, D.; Tripolt, N.J.; Sourij, H.; Pferschy, P.N.; Oulhaj, A.; Alber, H.; Gwechenberger, M.; Martinek, M.; Seidl, S.; Moertl, D.; et al. Ertugliflozin to Reduce Arrhythmic Burden in ICD/CRT Patients (ERASe-Trial)—A Phase III Study. *Am. Heart J.* **2022**, *246*, 152–160. [CrossRef]

30. Croteau, D.; Baka, T.; Young, S.; He, H.; Chambers, J.M.; Qin, F.; Panagia, M.; Pimentel, D.R.; Balschi, J.A.; Colucci, W.S.; et al. SGLT2 Inhibitor Ertugliflozin Decreases Elevated Intracellular Sodium, and Improves Energetics and Contractile Function in Diabetic Cardiomyopathy. *Biomed. Pharmacother.* **2023**, *160*, 114310. [CrossRef]
31. Pandey, A.; Kolkailah, A.A.; Cosentino, F.; Cannon, C.P.; Frederich, R.C.; Cherney, D.Z.I.; Dagogo-Jack, S.; Pratley, R.E.; Cater, N.B.; Gantz, I.; et al. Ertugliflozin and Hospitalization for Heart Failure Across the Spectrum of Pre-Trial Ejection Fraction: Post-Hoc Analyses of the VERTIS CV Trial. *Eur. Heart J.* **2023**, *44*, 5163–5166. [CrossRef] [PubMed]
32. Zhang, F.; Wang, W.; Hou, X. Effectiveness and Safety of Ertugliflozin for Type 2 Diabetes: A Meta-Analysis of Data from Randomized Controlled Trials. *J. Diabetes Investig.* **2022**, *13*, 478–488. [CrossRef] [PubMed]
33. Grunberger, G.; Camp, S.; Johnson, J.; Huyck, S.; Terra, S.G.; Mancuso, J.P.; Jiang, Z.W.; Golm, G.; Engel, S.S.; Lauring, B. Ertugliflozin in Patients with Stage 3 Chronic Kidney Disease and Type 2 Diabetes Mellitus: The VERTIS RENAL Randomized Study. *Diabetes Ther.* **2018**, *9*, 49–66. [CrossRef] [PubMed]
34. Cherney, D.Z.I.; Charbonnel, B.; Cosentino, F.; Dagogo-Jack, S.; McGuire, D.K.; Pratley, R.; Shih, W.J.; Frederich, R.; Maldonado, M.; Pong, A.; et al. Effects of Ertugliflozin on Kidney Composite Outcomes, Renal Function and Albuminuria in Patients with Type 2 Diabetes Mellitus: An Analysis from the Randomised VERTIS CV Trial. *Diabetologia* **2021**, *64*, 1256–1267. [CrossRef] [PubMed]
35. Amin, N.B.; Wang, X.; Mitchell, J.R.; Lee, D.S.; Nucci, G.; Rusnak, J.M. Blood Pressure-Lowering Effect of the Sodium Glucose Co-Transporter-2 Inhibitor Ertugliflozin, Assessed Via Ambulatory Blood Pressure Monitoring in Patients with Type 2 Diabetes and Hypertension. *Diabetes Obes. Metab.* **2015**, *17*, 805–808. [CrossRef] [PubMed]
36. Cheng, Q.; Zou, S.; Feng, C.; Xu, C.; Zhao, Y.; Shi, X.; Sun, M. Effect of Ertugliflozin on Renal Function and Cardiovascular Outcomes in Patients with Type 2 Diabetes Mellitus: A Systematic Review and Meta-Analysis. *Medicine* **2023**, *102*, e33198. [CrossRef]

Disclaimer/Publisher's Note: The statements, opinions and data contained in all publications are solely those of the individual author(s) and contributor(s) and not of MDPI and/or the editor(s). MDPI and/or the editor(s) disclaim responsibility for any injury to people or property resulting from any ideas, methods, instructions or products referred to in the content.

Systematic Review

Meta-Analysis of the Safety and Efficacy of Direct Oral Anticoagulants for the Treatment of Left Ventricular Thrombus

Mounica Vorla [1] and Dinesh K. Kalra [2,*]

[1] Department of Internal Medicine, Carle Foundation Hospital, Urbana, IL 61822, USA; mounica.vorla@gmail.com
[2] Division of Cardiology, University of Louisville, Louisville, KY 40292, USA
* Correspondence: dinesh.kalra@louisville.edu; Tel.: +1-502-852-9505; Fax: +1-502-852-6233

Abstract: Background: Literature on the preferred anticoagulant for treating left ventricular thrombus (LVT) is lacking. Thus, our objective was to compare the efficacy of DOACs versus warfarin in treating LVT. Methods: Databases were searched for RCTs and adjusted observational studies that compared DOAC versus warfarin through March 2024. The primary efficacy outcomes of interest were LVT resolution, systemic embolism, composite of stroke, and TIA. The primary safety outcomes encompassed all-cause mortality and bleeding events. Results: Our meta-analysis including 31 studies demonstrated that DOAC use was associated with higher odds of thrombus resolution (OR: 1.08, 95% CI: 0.86–1.31, *p*: 0.46). A statistically significant reduction in the risk of stroke/TIA was observed in the DOAC group versus the warfarin group (OR: 0.65, 95% CI: 0.48–0.89, *p*: 0.007). Furthermore, statistically significant reduced risks of all-cause mortality (OR: 0.68, 95% CI: 0.47–0.98, *p*: 0.04) and bleeding events (OR: 0.70, 95% CI: 0.55–0.89, *p*: 0.004) were observed with DOAC use as compared to warfarin use. Conclusion: Compared to VKAs, DOACs are noninferior as the anticoagulant of choice for LVT treatment. However, further studies are warranted to confirm these findings.

Keywords: anticoagulants; left ventricular thrombus; embolism; prescription; major adverse cardiac events

1. Introduction

Left ventricular thrombus (LVT) is a dreaded complication in patients with myocardial infarction (MI) and dilated cardiomyopathy (DCM). Despite notable progress in managing these conditions, the occurrence of LVT persists at a considerable rate, varying between 4 and 39% in patients with acute MI [1] and 11–44% in those with DCM [2,3]. Depending on thrombus size and progression, LVT carries a risk of embolization of up to 22% [3–6] and a 37% risk of major adverse cardiovascular events (MACEs) [7].

To reduce the risk of thromboembolic (TE) events, clinical guidelines recommend anticoagulation for a duration of 3–6 months in patients with LVT. However, there seems to lack consensus among different societies regarding the choice of anticoagulation regimen. The 2013 American College of Cardiology/American Heart Association (ACC/AHA) ST segment elevation MI (STEMI) guideline recommends consideration of vitamin K antagonist (VKA) therapy for 3 months in patients with or at risk of LVT (e.g., those with anteroapical akinesis or dyskinesis) (Class IIb indication, level of evidence C) [8]. The 2023 European Society of Cardiology (ESC) guideline states that "the choice of (anticoagulant) therapy should be tailored to the patient's clinical status and the results of follow-up investigations" but does not comment on the specific type of anticoagulant [9].

VKAs, predominantly warfarin, have been traditionally used for the prevention and treatment of LVT. However, difficulty in monitoring INR, drug–food and drug–drug interactions, and suboptimal times in therapeutic range (TTR) make warfarin a challenging therapeutic option for both providers and patients. Direct oral anticoagulant (DOAC) therapy, on the other hand, seems like an attractive option with fewer side effects while

providing a more predictable and steady state of anticoagulation with enhanced patient compliance and fewer drug–drug interactions. Moreover, since inception, the cost of these drugs has fallen considerably. The 2022 AHA statement on the management of LVT indicates that DOAC therapy as a reasonable alternative to VKAs but does not comment on whether either anticoagulant is preferred [10]. In this context, our meta-analysis (meta-analysis) aimed to pool results from randomized clinical trials (RCTs) and observational studies to provide a more comprehensive understanding of the safety and efficacy of DOACs in LVT patients.

2. Methods

Our meta-analysis was conducted in accordance with the Preferred Reporting Items for Systematic Reviews and Meta-Analyses (PRISMA) 2020 guideline [11]. This study was registered with PROSPERO database (registration ID 550050) [12].

2.1. Data Sources and Searches

We conducted a literature search using the following Medical Subject Headings (MeSH) terms: "Direct Oral Anticoagulants", "warfarin", "Vitamin K antagonist", and "Left ventricular thrombus". PubMed, Cochrane, Google scholar, and ClinicalTrials.gov databases were systematically queried for all RCTs and observational studies comparing DOACs versus warfarin in patients with LVT and published between 1 January 1990 and 1 March 2024. Additionally, two investigators (MV and DK) independently reviewed the reference lists of identified studies and relevant reviews to identify additional pertinent studies.

2.2. Study Selection

Our meta-analysis encompassed all RCTs and adjusted observational studies comparing DOACs with warfarin in patients diagnosed with LVT. The following criteria were employed for study inclusion: confirmation of LVT diagnosis via cardiac imaging modalities such as transthoracic echocardiography (TTE) or cardiac magnetic resonance imaging (CMRi), a median follow-up period of at least 1 month, and the reporting of at least one clinical endpoint related to treatment approach. Excluded from our analysis were case reports, case series, cross-sectional studies, and single-arm investigations. Additionally, studies involving patients with intracardiac, ventricular mural, and right ventricular thrombus were excluded from our analysis.

2.3. Outcome Measures and Quality Assessment

The primary efficacy outcomes in our study included LVT resolution, systemic embolism, composite of stroke, and transient ischemic attack (TIA). Primary safety outcomes encompassed all-cause mortality and bleeding events. Additionally, major bleeding as defined by categories 3–5 according to The Bleeding Academic Research Consortium (BARC) [13] criteria or moderate–severe bleeding according to the Global Use of Streptokinase and t-PA for Occluded Coronary Arteries (GUSTO) criteria were also included in the safety outcome [13].

To assess the quality of included observational studies and RCTs, we employed the Newcastle–Ottawa Scale (NOS) [14] and the Cochrane Collaboration Risk-of-Bias 2 (RoB 2) [15] tools. The NOS is a 9-point scoring system comprising f variables such as study selection, comparability of groups, ascertainment of exposure, and outcome measurement in observational studies, each allocated individual scores. Scores ranging from 0 to 3 indicate a very high risk of bias, 4 to 6 indicate a high risk of bias, and 7 to 9 indicate a low risk of bias (Table 1). On the other hand, the RoB 2 is a web-based tool developed in collaboration with Cochrane to assess the overall quality of RCTs based on variables such as randomization, deviation from intended intervention, outcome measurement, and selection of reported results (Figure 1).

Table 1. Assessing the risk of bias using Newcastle–Ottawa scale in observational studies.

Study	Selection				Comparability		Outcome			Overall Score
	Selection of Subjects Truly Representative/Not	Selection of Controls Drawn from the Same Cohort/Not	Ascertainment of Exposure Drawn from Secure Record/Self-Report	Demonstration of Outcome of Interest Absent/Present	Controlled for Baseline Characteristics Yes/No	Controlled for Other Factors Yes/No	Assessment of Outcome Drawn from Secure Record/Self-Report	Follow Up Length >3/<3 Months	Adequacy of Follow-Up <20%/>80% Lost to Follow-Up	
Abdi, 2022 [16]	1	1	1	1	1	0	1	1	0, 30% lost to follow-up	Good
Albahtain, 2021 [17]	1	1	1	1	1	0	1	1	1	Good
Ali, 2020 [18]	1	1	1	1	1	1	1	1	1	Good
Bass, 2022 [19]	1	1	1	1	1	0	1	NR	1	Good
Cochran, 2021 [20]	1	1	1	1	1	0	1	1	1	Good
Daher, 2020 [21]	1	1	1	0	0	0	1	NR	1	Fair
Gama, 2019 [22]	1	1	1	1	1	1	1	NR	1	Good
Gaddeti, 2020 [23]	1	1	1	0	1	1	1	1	1	Good
Herald, 2022 [24]	1	1	1	1	1	1	1	1	1	Good
Hofer, 2021 [25]	1	1	1	1	1	0	1	1	1	Good
Huang, 2023 [26]	0, included only DCM patients	1	1	1	1	1	1	1	1	Good
Iqbal, 2020 [27]	1	1	1	1	1	1	1	1	1	Good
Isa, 2020 [28]	1	1	1	1	1	1	1	1	1	Good
Jaidka, 2018 [29]	0, only AMI patients	1	1	1	1	1	1	1	0	Good
Jones, 2021 [30]	0, only AMI patients	1	1	1	1	1	1	1	1	Good
Kim, 2022 [31]	1	1	1	1	1	1	1	1	1	Good
Liang, 2022 [32]	0, only AMI patients	1	1	1	1	1	1	1	1	Good
Mihm, 2021 [33]	1	1	1	1	1	1	1	1	0	Good
Rahunathan, 2023 [34]	1	1	1	1	1	1	1	1	1	Good
Ratnayake, 2020 [35]	0, only AMI patients	1	1	1	0	0	1	1	1	Fair
Robinson, 2020 [36]	1	1	1	1	1	1	1	1	1	Good
Seiler, 2023 [37]	1	1	1	1	1	1	1	1	1	Good
Varwani, 2021 [38]	1	1	1	1	1	1	1	1	1	Good
Willeford, 2021 [39]	1	1	1	1	1	1	1	1	1	Good
Xu, 2021 [40]	1	1	1	1	1	1	1	1	1	Good
Zhang, 2021 [41]	0, only AMI patients	1	1	1	1	1	1	1	1	Good
Zhang, 2022 [42]	0, only HF patients	1	1	1	1	1	1	1	1	Good

NR—not reported.

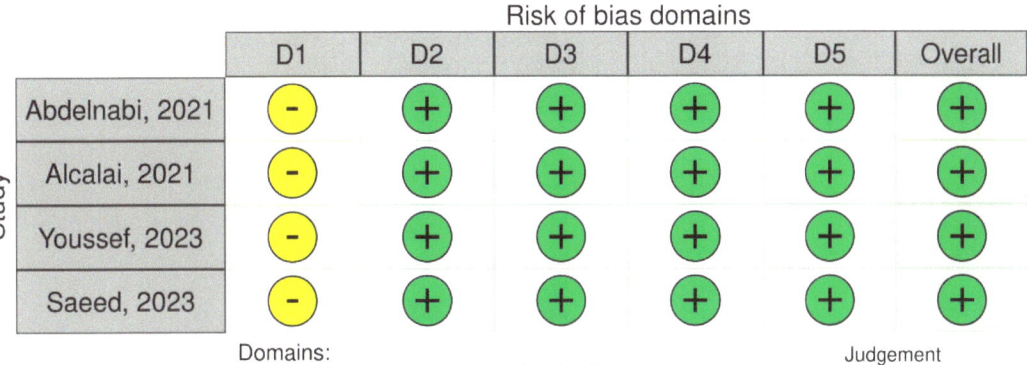

Figure 1. Assessing the risk of bias in randomized clinical trials.

2.4. Data Synthesis and Statistical Analysis

Individual study-level data extraction was independently conducted by two reviewers (MV and DK) using a predefined form, which included information on study characteristics, baseline patient characteristics, and endpoint event rates.

Our meta-analysis was conducted according to the recommendations from Cochrane Collaboration using Review Manager, version 5.3 [43]. Pooled odds ratios (ORs) and 95% confidence intervals (CIs) were calculated using random-effects models with the Mantel–Haenszel method [44]. A *p*-value of less than 0.05 was deemed statistically significant for each clinical endpoint. The extent of heterogeneity among studies was assessed using the I^2 statistic, with values exceeding 50% indicating significant heterogeneity. Forest plots were generated to visually depict the relative effect size of DOAC versus warfarin for individual clinical endpoints.

3. Results

As depicted in Figure 2 the initial search yielded 424 publications. After reviewing titles and abstracts, 141 studies were excluded for lack of relevance. The remaining 283 articles underwent a comprehensive review and assessment to determine if they met the inclusion and exclusion criteria. Following a full-text review, 31 studies were included in the final analysis.

The included studies were homogeneous regarding the inclusion and exclusion criteria. Among these, 27 were observational studies, and 4 were RCTs [45–48]. Patients were followed for an average period of 16.9 months. The baseline characteristics of the patients in the included studies are summarized in Table 2. The mean age of the patients was 59 years. Of the study participants, 33% were treated with direct oral anticoagulants (DOACs) and 67% with warfarin. All studies included in the final analysis were deemed to have a low-to-intermediate risk of bias, as assessed using the Newcastle–Ottawa Scale (NOS) and Cochrane metrics for quality assessment.

Table 2. Baseline characteristics of studies included.

Author, Year	Type of Study	Total Participants, n	DOAC/VKA Group, n	DOAC/VKA Group Mean Age, Years	Women, n (DOAC/Warfarin)	CAD, % (DOAC/Warfarin)	CHF, % (DOAC/Warfarin)
Abdelnabi, 2021 [45]	RCT	79	39/40	NR	NR	NR	NR
Abdi, 2022 [16]	Observational	40	18/19	NR	NR	NR	NR
Alhabsian, 2021 [17]	Observational	63	28/35	58/59	4/1	NR	NR
Alcalai, 2021 [46]	RCT	35	18/17	56/59	5/2	18/22	NR
Ali, 2020 [18]	Observational	92	32/60	59/58	6/11	NR	78/75
Bass, 2022 [19]	Observational	949	180/769	63/62	55/224	43/57	68/75
Cochran, 2021 [20]	Observational	73	14/59	52/62	3/14	53/61	73/81
Daher, 2020 [21]	Observational	59	17/42	57/61	3/7	88/74	NR
Gama, 2019 [22]	Observational	66	13/53	69/69	NR	NR	NR
Guddeti, 2020 [23]	Observational	99	19/80	61/61	4/25	58/66	100/96
Herald, 2022 [24]	Observational	433	134/299	66/65	18/57	35/36	88/88
Hofer, 2021 [25]	Observational	43	10/33	NR	NR	NR	NR
Huang, 2023 [26]	Observational	122	47/65	49/39	9/12	NR	NR
Iqbal, 2020 [27]	Observational	84	22/62	62/62	2/7	NR	95/94
Isa, 2020 [28]	RCT	27	14/13	55/55	1/1	NR	NR
Jaidka, 2018 [29]	Observational	49	12/37	57/61	3/9	0/8	NR
Jones, 2021 [30]	Observational	111	41/60	59/67	7/9	NR	NR
Kim, 2023 [31]	Observational	205	23/182	NR	NR	NR	NR
Liang, 2022 [32]	Observational	128	56/72	55/55	5/10	NR	NR
Mihm, 2021 [33]	Observational	108	33/75	63/60	7/9	NR	NR
Rahunathan, 2023 [34]	Observational	18	14/4	59/64	2/1	NR	NR
Ratnayake, 2020 [35]	Observational	44	2/42	NR	NR	NR	NR
Robinson, 2020 [36]	Observational	357	121/236	58/58	27/66	NR	NR
Saeed, 2023 [48]	Observational	196	98/98	56/56	17/22	NR	13/11
Seiler, 2023 [37]	Observational	101	48/54	64/62	6/12	NR	6/10
Varwani, 2021 [38]	Observational	92	58/34	NR	NR	NR	NR
Willeford, 2021 [39]	Observational	151	22/129	54/56	5/25	NR	86/85
Xu, 2021 [40]	Observational	87	25/62	59/62	6/15	NR	NR
Youssef, 2023 [47]	RCT	100	25/25	52/54	NR	NR	NR
Zhang, 2021 [41]	Observational	64	33/31	60/61	9/8	NR	NR
Zhang, 2022 [42]	Observational	187	109/78	65/63	24/12	97/61	59/39

NR—not reported.

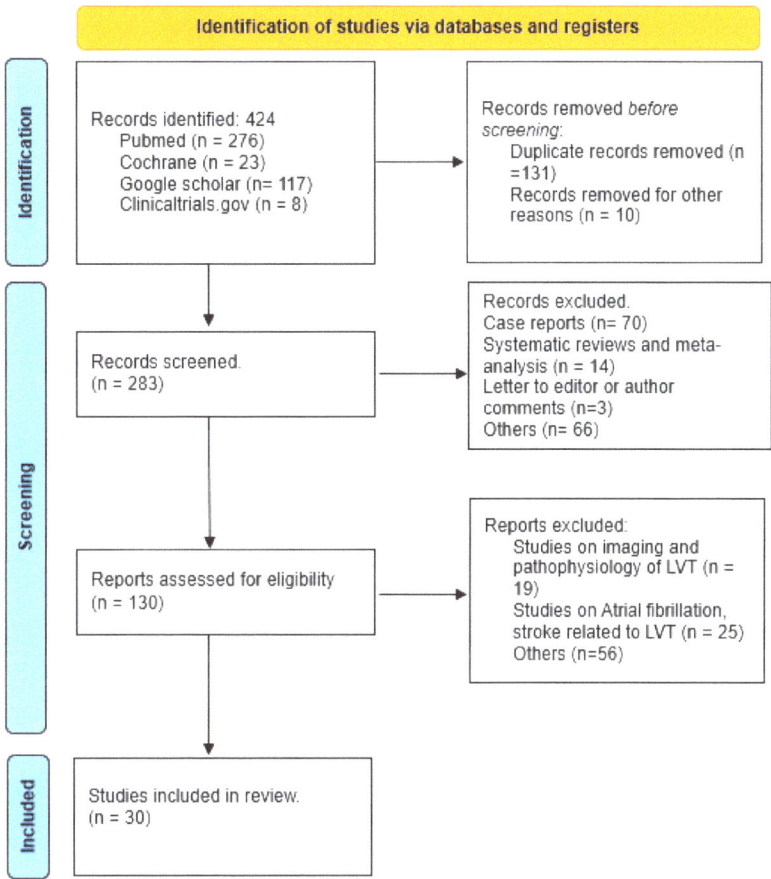

Figure 2. Preferred Reporting Items for Systematic Reviews and Meta-analyses flow sheet.

3.1. Efficacy Outcomes

LVT resolution was reported in 28 studies including 2690 patients. Compared with warfarin, DOAC use showed a trend toward higher odds of thrombus resolution (OR: 1.08, 95% CI: 0.86–1.31, *p:* 0.46) (Figure 3). The occurrence of systemic embolism was reported in 12 studies including 1508 participants. Although not statistically significant, DOAC use was associated with lowered risk of systemic embolism as compared to warfarin (OR: 0.67, 95% CI: 0.37–1.21, *p:* 0.18) (Figure 4). Additionally, 19 studies involving 2933 participants reported stroke/TIA. A statistically significant lower risk of stroke/TIA was observed in the DOAC group versus the warfarin group (OR: 0.65, 95% CI: 0.48–0.89, *p:* 0.007), with no heterogeneity (I^2-0%) among the studies included in our analysis (Figure 5).

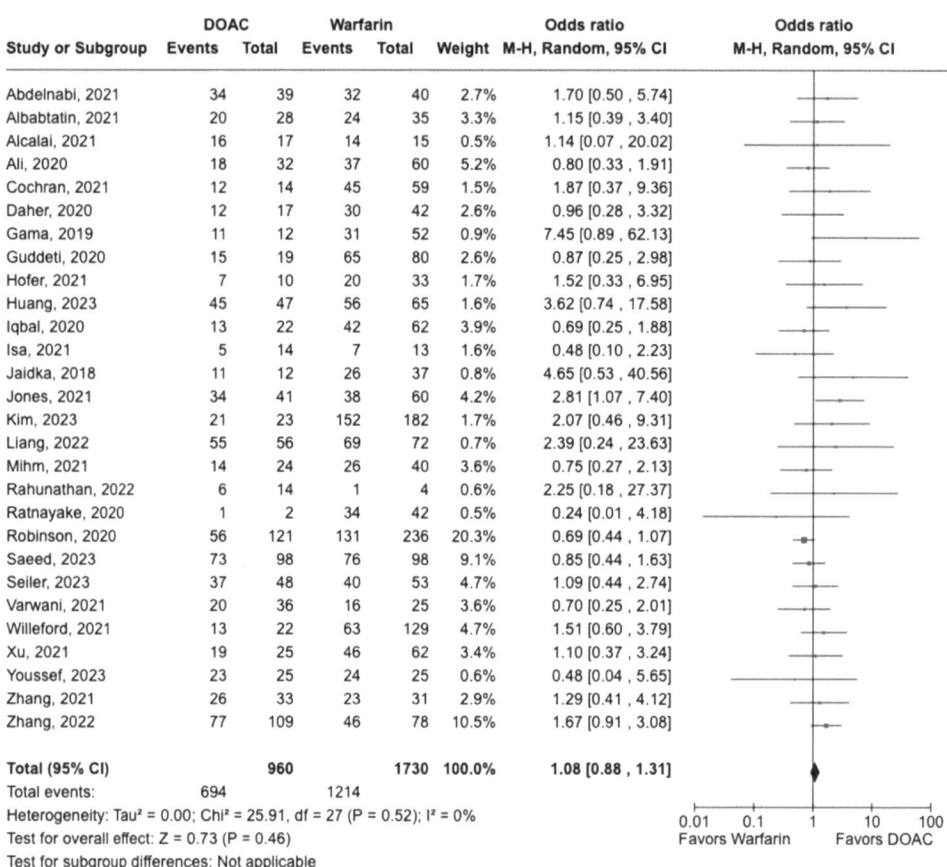

Figure 3. Forest plot of LVT resolution in trials, comparing DOAC vs. warfarin treatment groups.

Figure 4. Forest plot of systemic embolism in trials comparing DOAC vs. warfarin treatment groups.

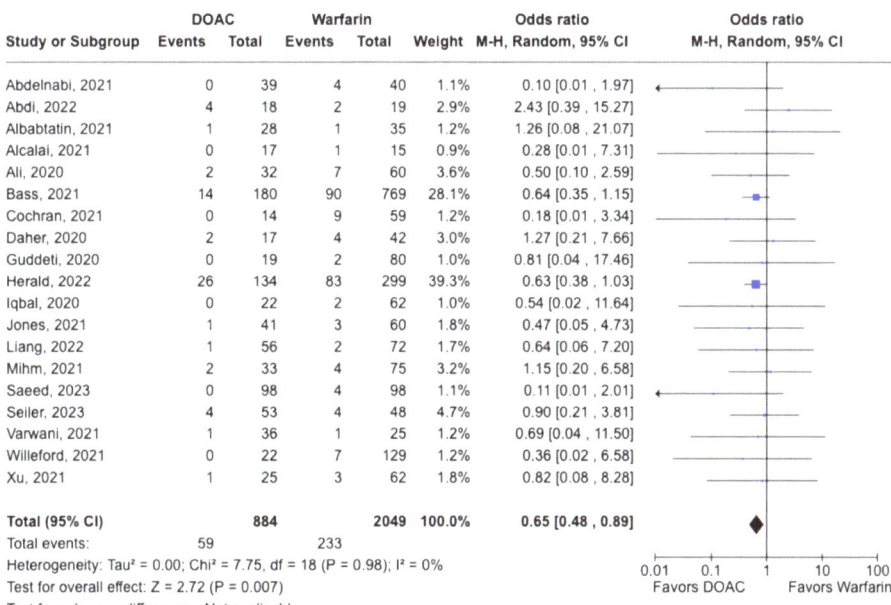

Figure 5. Forest plot comparing the occurrence of stroke/TIA in DOAC and warfarin groups.

3.2. Safety Outcomes

All-cause mortality was reported in 12 studies including 1616 patients. There was a statistically significant reduced risk of all-cause mortality with DOAC use when compared with warfarin use (OR: 0.68, 95% CI: 0.47–0.98, *p*: 0.04), with mild heterogeneity (I^2-19%) among the included studies (Figure 6). Bleeding events were reported in 21 studies including 3440 participants. DOAC use was associated with statistically significant lower odds of bleeding when compared with warfarin use (OR: 0.70, 95% CI: 0.55–0.89, *p*: 0.004), with no heterogeneity (I^2-0%) among the studies included for analysis (Figure 7). Although not statistically significant, the risk of major bleeding was also lower in the DOAC group versus warfarin group (OR: 0.75, 95% CI: 0.42–1.35, *p*: 0.34) (Figure 8).

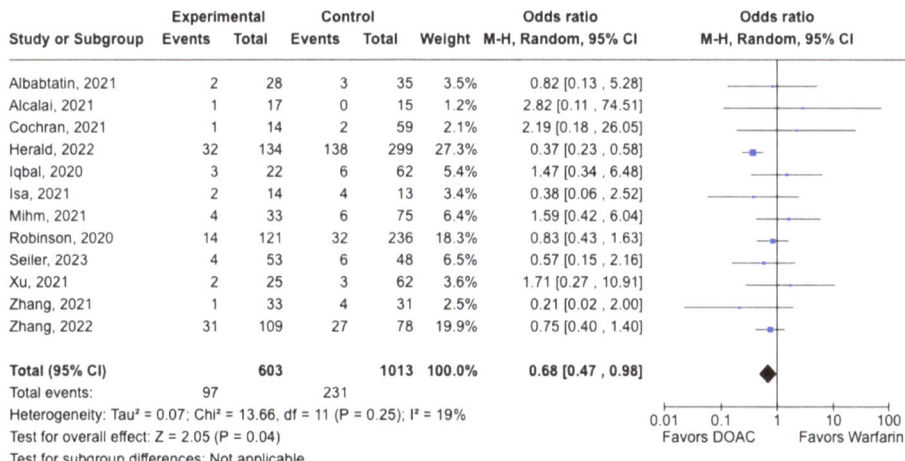

Figure 6. Forest plot comparing the occurrence of all-cause mortality in DOAC and warfarin groups.

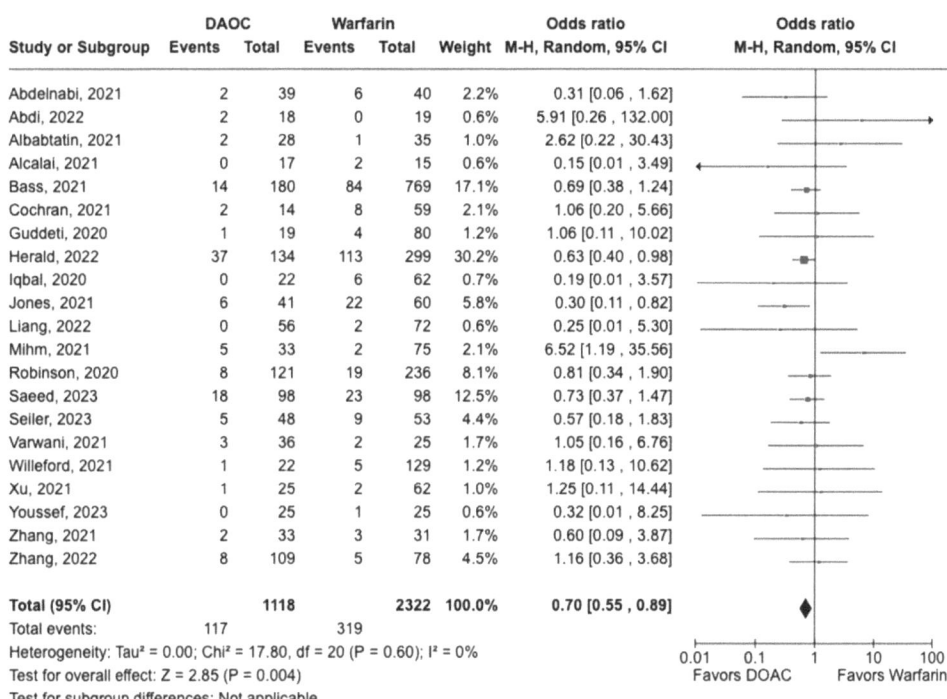

Figure 7. Forest plot comparing the occurrence of bleeding events in DOAC and warfarin groups.

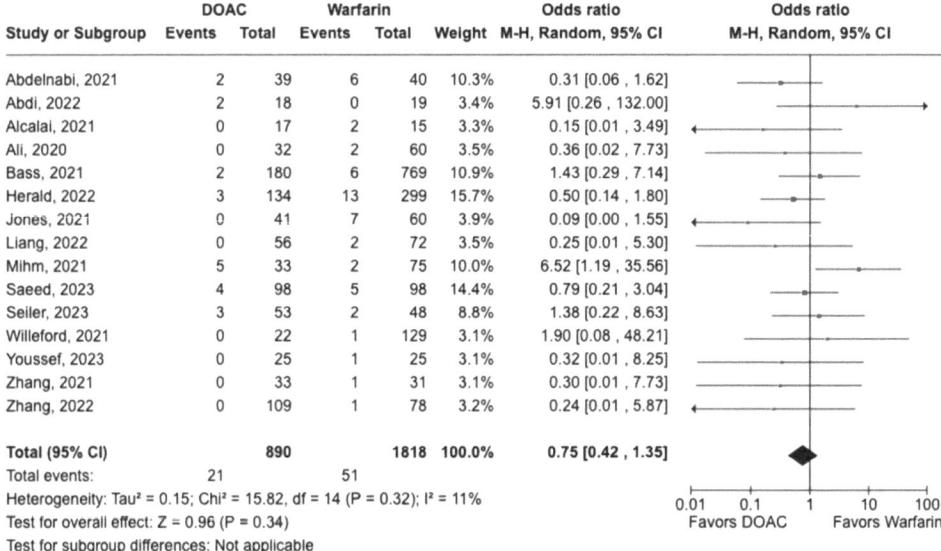

Figure 8. Forest plot comparing the occurrence of major bleeding in DOAC and warfarin groups.

4. Discussion

LVT represents a concerning complication following acute MI, with an incidence of 3.5–8% [49–51] in the postpercutaneous coronary intervention PCI era. Likewise, incidences as high as 36–44% [2,3] and 68.5% [10] have been reported in anatomic pathology studies involving patients with DCM and heart failure (HF), respectively. Because of the heightened risk of TE complications, anticoagulation is imperative for preventing stroke and systemic embolism in patients with LVT. However, due to the scarcity of robust data, DOACs are merely recommended as alternatives to warfarin in patients with LVT requiring anticoagulation. Our meta-analysis including 31 studies is the most extensive comparison to date of DOACs vs. warfarin in patients with LVT. Random-effects analysis showed that DOACs are noninferior to warfarin for pharmacological anticoagulation in patients with LVT. In fact, DOAC use was associated with a significantly lowered risk of stroke/TIA, all-cause mortality, and bleeding when compared with warfarin use.

Our meta-analysis results corroborate those of previous studies comparing DOACs vs. warfarin in patients with LVT. A subgroup analysis of seven studies [25,29,35,41,46,47] investigating the effect of DOACs with warfarin in patients after MI favored DOACs for LVT resolution (OR: 1.70, 95% CI: 0.94–3.07, *p:* 0.08). Similarly, two studies evaluated the effect of DOACs in patients with HF [42] and DCM [26]. The rates of LVT resolution were comparable between the groups but did not reach statistical significance. Given the distinct pathophysiological mechanisms in those after MI (including endocardial injury, inflammation, and blood stasis) and with HF/DCM (involving blood stasis, endothelial dysfunction, and hypercoagulability), further research exploring the impact of DOACs in different etiological contexts is warranted. Furthermore, the effect of concurrent antiplatelet therapy on LVT resolution and safety events needs investigation.

The efficacy of rivaroxaban for LVT resolution has been evaluated in five studies [17,41,42,45,52]. Similarly, apixaban was assessed in three studies [28,46,47]. LVT resolution has occurred in 75% and 79% of patients treated with rivaroxaban and apixaban, respectively. Subgroup analysis favored rivaroxaban vs. warfarin; however, this difference did not reach statistical significance (OR: 1.26, 95% CI: 0.87–1.82, *p:* 0.23). Interestingly, the subgroup analysis of studies assessing apixaban for LVT resolution favored warfarin (OR: 0.55, 95% CI: 0.17–1.82, *p:* 0.33). The difference in outcomes between rivaroxaban and apixaban could be explained by different sample sizes and study design. Additionally, differences in thrombosis based on LVT etiology may reasonably translate into differences in anticoagulant responsiveness. Therefore, further research investigating the effect of different DOACs in patients with LVT is warranted.

Finally, our study demonstrated that DOACs are no-inferior to warfarin as an anticoagulant of choice in patients with LVT. However, our study has a few limitations: (1) Most included studies are observational and nonblinded, raising concerns regarding missing data and selection bias. (2) With respect to meta-analyses, there is always the possibility of residual confounding and publication bias. (3) The imaging modality used to diagnose LVT (e.g., TTE vs. CMR) was not uniform across different studies. (4) There are no studies to date comparing the relative efficacy of different classes of DOACs (apixaban, rivaroxaban, dabigatran, etc.) with warfarin in patients with LVT. (5) We could not obtain data on adherence to DOACs or the time in the therapeutic range of warfarin treatment. (6) Finally, we were unable to standardize the dose of anticoagulants, but this also reflects the current dilemma of anticoagulation management in the LVT population in the real world.

5. Conclusions

Since their introduction for treating venous TE and atrial fibrillation, DOACs have emerged as an appealing alternative to VKAs for both patients and clinicians. They offer advantages such as reduced need for monitoring, absence of dietary restrictions, and a lower risk of bleeding. Nevertheless, adequate data are lacking regarding the efficacy and safety of DOACs in managing LVT. Our meta-analysis demonstrates that that DOACs are comparable to warfarin in terms of efficacy (LVT resolution) and are associated with a decreased incidence of adverse events (bleeding). However, dedicated randomized clinical trials will be necessary to validate our findings and inform practice guidelines.

Author Contributions: Conceptualization, M.V. and D.K.K.; methodology, M.V.; formal analysis, M.V.; data curation, M.V. and D.K.K.; writing—original draft preparation, M.V.; writing—review and editing, D.K.K.; supervision, D.K.K.; project administration, M.V. and D.K.K. All authors have read and agreed to the published version of the manuscript.

Funding: This research received no external funding.

Conflicts of Interest: The authors declare no conflict of interest.

Abbreviations

ACC	American College of Cardiology
AHA	American Heart Association
BARC	Bleeding Academic Research Consortium
CI	Confidence interval
CMRi	Cardiac magnetic resonance imaging
DCM	Dilated cardiomyopathy
DOAC	Direct oral anticoagulant
ESC	European Society of Cardiology
GUSTO	Global Use of Streptokinase and t-PA for Occluded Coronary Arteries
HF	Heart failure
LVT	Left ventricular thrombus
meta-analysis	Meta-analysis
MACE	Major adverse cardiovascular events
MeSH	Medical Subject Headings
MI	Myocardial infarction
NOS	Newcastle–Ottawa Scale
OR	Odds ratio
PRISMA	Preferred Reporting Items for Systematic Reviews and Meta-Analyses
RCT	Randomized clinical trial
RoB 2	Risk-of-Bias 2
STEMI	ST segment elevation myocardial infarction
TE	Thromboembolic events
TIA	Transient ischemic attack
TTE	Transthoracic echocardiography
TTR	Time to achieve therapeutic range
VKA	Vitamin K antagonist

References

1. McCarthy, C.P.; Vaduganathan, M.; McCarthy, K.J.; Januzzi, J.L.; Bhatt, D.L.; McEvoy, J.W. Left Ventricular Thrombus After Acute Myocardial Infarction: Screening, Prevention, and Treatment. *JAMA Cardiol.* **2018**, *3*, 642–649. [CrossRef]
2. Falk, R.H.; Foster, E.; Coats, M.H. Ventricular Thrombi and Thromboembolism in Dilated Cardiomyopathy: A Prospective Follow-up Study. *Am. Heart J.* **1992**, *123*, 136–142. [CrossRef]
3. Gottdiener, J.S.; Gay, J.A.; VanVoorhees, L.; DiBianco, R.; Fletcher, R.D. Frequency and Embolic Potential of Left Ventricular Thrombus in Dilated Cardiomyopathy: Assessment by 2-Dimensional Echocardiography. *Am. J. Cardiol.* **1983**, *52*, 1281–1285. [CrossRef]
4. Cruz Rodriguez, J.B.; Okajima, K.; Greenberg, B.H. Management of Left Ventricular Thrombus: A Narrative Review. *Ann. Transl. Med.* **2021**, *9*, 520. [CrossRef]
5. Massussi, M.; Scotti, A.; Lip, G.Y.H.; Proietti, R. Left Ventricular Thrombosis: New Perspectives on an Old Problem. *Eur. Heart J. Cardiovasc. Pharmacother.* **2021**, *7*, 158–167. [CrossRef]
6. Visser, C.A.; Kan, G.; Meltzer, R.S.; Dunning, A.J.; Roelandt, J. Embolic Potential of Left Ventricular Thrombus after Myocardial Infarction: A Two-Dimensional Echocardiographic Study of 119 Patients. *J. Am. Coll. Cardiol.* **1985**, *5*, 1276–1280. [CrossRef]
7. Lattuca, B.; Bouziri, N.; Kerneis, M.; Portal, J.-J.; Zhou, J.; Hauguel-Moreau, M.; Mameri, A.; Zeitouni, M.; Guedeney, P.; Hammoudi, N.; et al. Antithrombotic Therapy for Patients with Left Ventricular Mural Thrombus. *J. Am. Coll. Cardiol.* **2020**, *75*, 1676–1685. [CrossRef]
8. O'Gara, P.T.; Kushner, F.G.; Ascheim, D.D.; Casey, D.E.; Chung, M.K.; de Lemos, J.A.; Ettinger, S.M.; Fang, J.C.; Fesmire, F.M.; Franklin, B.A.; et al. 2013 ACCF/AHA Guideline for the Management of ST-Elevation Myocardial Infarction: A Report of the American College of Cardiology Foundation/American Heart Association Task Force on Practice Guidelines. *Circulation* **2013**, *127*, e362–e425. [CrossRef]

9. Byrne, R.A.; Rossello, X.; Coughlan, J.J.; Barbato, E.; Berry, C.; Chieffo, A.; Claeys, M.J.; Dan, G.-A.; Dweck, M.R.; Galbraith, M.; et al. 2023 ESC Guidelines for the Management of Acute Coronary Syndromes: Developed by the Task Force on the Management of Acute Coronary Syndromes of the European Society of Cardiology (ESC). *Eur. Heart J.* **2023**, *44*, 3720–3826. [CrossRef]
10. Levine, G.N.; McEvoy, J.W.; Fang, J.C.; Ibeh, C.; McCarthy, C.P.; Misra, A.; Shah, Z.I.; Shenoy, C.; Spinler, S.A.; Vallurupalli, S.; et al. Management of Patients at Risk for and With Left Ventricular Thrombus: A Scientific Statement From the American Heart Association. *Circulation* **2022**, *146*, e205–e223. [CrossRef]
11. Page, M.J.; McKenzie, J.E.; Bossuyt, P.M.; Boutron, I.; Hoffmann, T.C.; Mulrow, C.D.; Shamseer, L.; Tetzlaff, J.M.; Akl, E.A.; Brennan, S.E.; et al. The PRISMA 2020 Statement: An Updated Guideline for Reporting Systematic Reviews. *BMJ* **2021**, *372*, n71. [CrossRef]
12. Booth, A.; Clarke, M.; Dooley, G.; Ghersi, D.; Moher, D.; Petticrew, M.; Stewart, L. The Nuts and Bolts of PROSPERO: An International Prospective Register of Systematic Reviews. *Syst. Rev.* **2012**, *1*, 2. [CrossRef]
13. Mehran, R.; Rao, S.V.; Bhatt, D.L.; Gibson, C.M.; Caixeta, A.; Eikelboom, J.; Kaul, S.; Wiviott, S.D.; Menon, V.; Nikolsky, E.; et al. Standardized Bleeding Definitions for Cardiovascular Clinical Trials: A Consensus Report from the Bleeding Academic Research Consortium. *Circulation* **2011**, *123*, 2736–2747. [CrossRef]
14. Ottawa Hospital Research Institute. Available online: https://www.ohri.ca/programs/clinical_epidemiology/oxford.asp (accessed on 31 March 2024).
15. Sterne, J.A.C.; Savović, J.; Page, M.J.; Elbers, R.G.; Blencowe, N.S.; Boutron, I.; Cates, C.J.; Cheng, H.-Y.; Corbett, M.S.; Eldridge, S.M.; et al. RoB 2: A Revised Tool for Assessing Risk of Bias in Randomised Trials. *BMJ* **2019**, *366*, l4898. [CrossRef]
16. Abdi, I.A.; Karataş, M.; Öcal, L.; Elmi Abdi, A.; Farah Yusuf Mohamud, M. Retrospective Analysis of Left Ventricular Thrombus Among Heart Failure Patients with Reduced Ejection Fraction at a Single Tertiary Care Hospital in Somalia. *Open Access Emerg. Med.* **2022**, *14*, 591–597. [CrossRef]
17. Albabtain, M.A.; Alhebaishi, Y.; Al-Yafi, O.; Kheirallah, H.; Othman, A.; Alghosoon, H.; Arafat, A.A.; Alfagih, A. Rivaroxaban versus Warfarin for the Management of Left Ventricle Thrombus. *Egypt. Heart J.* **2021**, *73*, 41. [CrossRef]
18. Ali, Z.; Isom, N.; Dalia, T.; Sami, F.; Mahmood, U.; Shah, Z.; Gupta, K. Direct Oral Anticoagulant Use in Left Ventricular Thrombus. *Thrombosis J.* **2020**, *18*, 29. [CrossRef]
19. Bass, M.E.; Kiser, T.H.; Page, R.L.; McIlvennan, C.K.; Allen, L.A.; Wright, G.; Shakowski, C. Comparative Effectiveness of Direct Oral Anticoagulants and Warfarin for the Treatment of Left Ventricular Thrombus. *J. Thromb. Thrombolysis* **2021**, *52*, 517–522. [CrossRef]
20. Cochran, J.M.; Jia, X.; Kaczmarek, J.; Staggers, K.A.; Rifai, M.A.; Hamzeh, I.R.; Birnbaum, Y. Direct Oral Anticoagulants in the Treatment of Left Ventricular Thrombus: A Retrospective, Multicenter Study and Meta-Analysis of Existing Data. *J. Cardiovasc. Pharmacol. Ther.* **2021**, *26*, 173–178. [CrossRef]
21. Daher, J.; Da Costa, A.; Hilaire, C.; Ferreira, T.; Pierrard, R.; Guichard, J.B.; Romeyer, C.; Isaaz, K. Management of Left Ventricular Thrombi with Direct Oral Anticoagulants: Retrospective Comparative Study with Vitamin K Antagonists. *Arch. Cardiovasc. Dis. Suppl.* **2021**, *13*, 74. [CrossRef]
22. Gama, F.; Freitas, P.; Trabulo, M.; Ferreira, A.; Andrade, M.J.; Matos, D.; Strong, C.; Ribeiras, R.; Ferreira, J.; Mendes, M. 459 Direct Oral Anticoagulants Are an Effective Therapy for Left Ventricular Thrombus Formation. *Eur. Heart J.* **2019**, *40*, ehz747.0118. [CrossRef]
23. Guddeti, R.R.; Anwar, M.; Walters, R.W.; Apala, D.; Pajjuru, V.; Kousa, O.; Gujjula, N.R.; Alla, V.M. Treatment of Left Ventricular Thrombus With Direct Oral Anticoagulants: A Retrospective Observational Study. *Am. J. Med.* **2020**, *133*, 1488–1491. [CrossRef]
24. Herald, J.; Goitia, J.; Duan, L.; Chen, A.; Lee, M.-S. Safety and Effectiveness of Direct Oral Anticoagulants Versus Warfarin for Treating Left Ventricular Thrombus. *Am. J. Cardiovasc. Drugs* **2022**, *22*, 437–444. [CrossRef]
25. Hofer, F.; Kazem, N.; Schweitzer, R.; Horvat, P.; Winter, M.-P.; Koller, L.; Hengstenberg, C.; Sulzgruber, P.; Niessner, A. The Prognostic Impact of Left Ventricular Thrombus Resolution after Acute Coronary Syndrome and Risk Modulation via Antithrombotic Treatment Strategies. *Clin. Cardiol.* **2021**, *44*, 1692–1699. [CrossRef]
26. Huang, L.; Zhao, X.; Wang, J.; Liang, L.; Tian, P.; Chen, Y.; Zhai, M.; Huang, Y.; Zhou, Q.; Xin, A.; et al. Clinical Profile, Treatment, and Prognosis of Left Ventricular Thrombus in Dilated Cardiomyopathy. *Clin. Appl. Thromb. Hemost.* **2023**, *29*, 10760296231179683. [CrossRef]
27. Iqbal, H.; Straw, S.; Craven, T.P.; Stirling, K.; Wheatcroft, S.B.; Witte, K.K. Direct Oral Anticoagulants Compared to Vitamin K Antagonist for the Management of Left Ventricular Thrombus. *ESC Heart Fail.* **2020**, *7*, 2032–2041. [CrossRef]
28. Isa, W.Y.H.W.; Hwong, N.; Mohamed Yusof, A.K.; Yusof, Z.; Loong, N.S.; Wan-Arfah, N.; Naing, N.N. Apixaban versus Warfarin in Patients with Left Ventricular Thrombus: A Pilot Prospective Randomized Outcome Blinded Study Investigating Size Reduction or Resolution of Left Ventricular Thrombus. *J. Clin. Prev. Cardiol.* **2020**, *9*, 150. [CrossRef]
29. Jaidka, A.; Zhu, T.; Lavi, S.; Johri, A. Treatment of Left Ventricular Thrombus Using Warfarin versus Direct Oral Anticoagulants Following Anterior Myocardial Infarction. *Can. J. Cardiol.* **2018**, *34*, S143. [CrossRef]
30. Jones, D.A.; Wright, P.; Alizadeh, M.A.; Fhadil, S.; Rathod, K.S.; Guttmann, O.; Knight, C.; Timmis, A.; Baumbach, A.; Wragg, A.; et al. The Use of Novel Oral Anticoagulants Compared to Vitamin K Antagonists (Warfarin) in Patients with Left Ventricular Thrombus after Acute Myocardial Infarction. *Eur. Heart J. Cardiovasc. Pharmacother.* **2021**, *7*, 398–404. [CrossRef]

31. Kim, S.-E.; Lee, C.J.; Oh, J.; Kang, S.-M. Factors Influencing Left Ventricular Thrombus Resolution and Its Significance on Clinical Outcomes. *ESC Heart Fail.* **2023**, *10*, 1987–1995. [CrossRef]
32. Liang, J.; Wang, Z.; Zhou, Y.; Shen, H.; Chai, M.; Ma, X.; Han, H.; Shao, Q.; Li, Q. Efficacy and Safety of Direct Oral Anticoagulants in the Treatment of Left Ventricular Thrombus After Acute Anterior Myocardial Infarction in Patients Who Underwent Percutaneous Coronary Intervention. *Curr. Vasc. Pharmacol.* **2022**, *20*, 517–526. [CrossRef] [PubMed]
33. Mihm, A.E.; Hicklin, H.E.; Cunha, A.L.; Nisly, S.A.; Davis, K.A. Direct Oral Anticoagulants versus Warfarin for the Treatment of Left Ventricular Thrombosis. *Intern. Emerg. Med.* **2021**, *16*, 2313–2317. [CrossRef] [PubMed]
34. Rahunathan, N.; Hurdus, B.; Straw, S.; Iqbal, H.; Witte, K.; Wheatcroft, S. Improving the Management of Left Ventricular Thrombus in a Tertiary Cardiology Centre: A Quality Improvement Project. *BMJ Open Qual.* **2023**, *12*, e002111. [CrossRef] [PubMed]
35. Ratnayake, C.; Liu, B.; Benatar, J.; Stewart, R.A.H.; Somaratne, J.B. Left Ventricular Thrombus after ST Segment Elevation Myocardial Infarction: A Single-Centre Observational Study. *N. Z. Med. J.* **2020**, *133*, 45–54.
36. Robinson, A.A.; Trankle, C.R.; Eubanks, G.; Schumann, C.; Thompson, P.; Wallace, R.L.; Gottiparthi, S.; Ruth, B.; Kramer, C.M.; Salerno, M.; et al. Off-Label Use of Direct Oral Anticoagulants Compared With Warfarin for Left Ventricular Thrombi. *JAMA Cardiol.* **2020**, *5*, 685. [CrossRef]
37. Seiler, T.; Vasiliauskaite, E.; Grüter, D.; Young, M.; Attinger-Toller, A.; Madanchi, M.; Cioffi, G.M.; Tersalvi, G.; Müller, G.; Stämpfli, S.F.; et al. Direct Oral Anticoagulants Versus Vitamin K Antagonists for the Treatment of Left Ventricular Thrombi—Insights from a Swiss Multicenter Registry. *Am. J. Cardiol.* **2023**, *194*, 113–121. [CrossRef]
38. Varwani, M.H.; Shah, J.; Ngunga, M.; Jeilan, M. Treatment and Outcomes in Patients with Left Ventricular Thrombus—Experiences from the Aga Khan University Hospital, Nairobi-Kenya. *Pan Afr. Med. J.* **2021**, *39*, 212. [CrossRef]
39. Willeford, A.; Zhu, W.; Stevens, C.; Thomas, I.C. Direct Oral Anticoagulants Versus Warfarin in the Treatment of Left Ventricular Thrombus. *Ann. Pharmacother.* **2021**, *55*, 839–845. [CrossRef]
40. Xu, Z.; Li, X.; Li, X.; Gao, Y.; Mi, X. Direct Oral Anticoagulants versus Vitamin K Antagonists for Patients with Left Ventricular Thrombus. *Ann. Palliat. Med.* **2021**, *10*, 9427–9434. [CrossRef]
41. Zhang, Z.; Si, D.; Zhang, Q.; Qu, M.; Yu, M.; Jiang, Z.; Li, D.; Yang, P.; Zhang, W. Rivaroxaban versus Vitamin K Antagonists (Warfarin) Based on the Triple Therapy for Left Ventricular Thrombus after ST-Elevation Myocardial Infarction. *Heart Vessel.* **2022**, *37*, 374–384. [CrossRef]
42. Zhang, Q.; Zhang, Z.; Zheng, H.; Qu, M.; Li, S.; Yang, P.; Si, D.; Zhang, W. Rivaroxaban in Heart Failure Patients with Left Ventricular Thrombus: A Retrospective Study. *Front. Pharmacol.* **2022**, *13*, 1008031. [CrossRef] [PubMed]
43. Cochrane RevMan. RevMan: Systematic Review and Meta-Analysis Tool for Researchers Worldwide. Available online: https://revman.cochrane.org/info (accessed on 13 January 2024).
44. ScienceDirect Topics. Mantel Haenszel Test—An Overview. Available online: https://www.sciencedirect.com/topics/medicine-and-dentistry/mantel-haenszel-test (accessed on 31 March 2024).
45. Abdelnabi, M.; Saleh, Y.; Fareed, A.; Nossikof, A.; Wang, L.; Morsi, M.; Eshak, N.; Abdelkarim, O.; Badran, H.; Almaghraby, A. Comparative Study of Oral Anticoagulation in Left Ventricular Thrombi (No-LVT Trial). *J. Am. Coll. Cardiol.* **2021**, *77*, 1590–1592. [CrossRef] [PubMed]
46. Alcalai, R.; Butnaru, A.; Moravsky, G.; Yagel, O.; Rashad, R.; Ibrahimli, M.; Planer, D.; Amir, O.; Elbaz-Greener, G.; Leibowitz, D. Apixaban vs. Warfarin in Patients with Left Ventricular Thrombus: A Prospective Multicentre Randomized Clinical Trial. *Eur. Heart J. Cardiovasc. Pharmacother.* **2022**, *8*, 660–667. [CrossRef] [PubMed]
47. Youssef, A.A.; Alrefae, M.A.; Khalil, H.H.; Abdullah, H.I.; Khalifa, Z.S.; Al Shaban, A.A.; Wali, H.A.; AlRajab, M.R.; Saleh, O.M.; Nashy, B.N. Apixaban in Patients With Post-Myocardial Infarction Left Ventricular Thrombus: A Randomized Clinical Trial. *CJC Open* **2023**, *5*, 191–199. [CrossRef] [PubMed]
48. Use of XARELTO in Ventricular Thrombus. Available online: https://www.janssenscience.com/products/xarelto/medical-content/use-of-xarelto-in-ventricular-thrombus (accessed on 15 April 2024).
49. Rehan, A.; Kanwar, M.; Rosman, H.; Ahmed, S.; Ali, A.; Gardin, J.; Cohen, G. Incidence of Post Myocardial Infarction Left Ventricular Thrombus Formation in the Era of Primary Percutaneous Intervention and Glycoprotein IIb/IIIa Inhibitors. A Prospective Observational Study. *Cardiovasc. Ultrasound* **2006**, *4*, 20. [CrossRef]
50. Phan, J.; Nguyen, T.; French, J.; Moses, D.; Schlaphoff, G.; Lo, S.; Juergens, C.; Dimitri, H.; Richards, D.; Thomas, L. Incidence and Predictors of Left Ventricular Thrombus Formation Following Acute ST-Segment Elevation Myocardial Infarction: A Serial Cardiac MRI Study. *Int. J. Cardiol. Heart Vasc.* **2019**, *24*, 100395. [CrossRef] [PubMed]
51. Gianstefani, S.; Douiri, A.; Delithanasis, I.; Rogers, T.; Sen, A.; Kalra, S.; Charangwa, L.; Reiken, J.; Monaghan, M.; MacCarthy, P. Incidence and Predictors of Early Left Ventricular Thrombus after ST-Elevation Myocardial Infarction in the Contemporary Era of Primary Percutaneous Coronary Intervention. *Am. J. Cardiol.* **2014**, *113*, 1111–1116. [CrossRef]
52. Saeedi, R.; Johns, K.; Frohlich, J.; Bennett, M.T.; Bondy, G. Lipid Lowering Efficacy and Safety of Ezetimibe Combined with Rosuvastatin Compared with Titrating Rosuvastatin Monotherapy in HIV-Positive Patients. *Lipids Health Dis.* **2015**, *14*, 57. [CrossRef]

Disclaimer/Publisher's Note: The statements, opinions and data contained in all publications are solely those of the individual author(s) and contributor(s) and not of MDPI and/or the editor(s). MDPI and/or the editor(s) disclaim responsibility for any injury to people or property resulting from any ideas, methods, instructions or products referred to in the content.

Study Protocol

Role of Dapagliflozin in Ischemic Preconditioning in Patients with Symptomatic Coronary Artery Disease—DAPA-IP Study Protocol

Marco Alexander Valverde Akamine, Beatriz Moreira Ayub Ferreira Soares, João Paulo Mota Telles, Arthur Cicupira Rodrigues de Assis, Gabriela Nicole Valverde Rodriguez, Paulo Rogério Soares, William Azem Chalela and Thiago Luis Scudeler *

Instituto do Coração (InCor), Hospital das Clínicas da Faculdade de Medicina da Universidade de São Paulo, Av. Dr. Enéas de Carvalho Aguiar, 44, Cerqueira César, São Paulo 05403-000, Brazil; marco.akamine@hc.usp.br (M.A.V.A.); bia.soares@incor.usp.br (B.M.A.F.S.); joao.telles@fm.usp.br (J.P.M.T.); arthurcicupira@usp.br (A.C.R.d.A.); gabriela.rodriguez@fm.usp.br (G.N.V.R.); paulo.soares@hc.fm.usp.br (P.R.S.); william.chalela@incor.usp.br (W.A.C.)
* Correspondence: thiago.scudeler@fm.usp.br

Abstract: Background: Ischemic preconditioning (IP) is a powerful cellular protection mechanism. The cellular pathways underlying IP are extremely complex and involve the participation of cell triggers, intracellular signaling pathways, and end-effectors. Experimental studies have shown that sodium-glucose transport protein 2 (SGLT2) inhibitors promote activation of 5′-adenosine monophosphate (AMP)-activated protein kinase (AMPK), the main regulator of adenosine 5′-triphosphate homeostasis and energy metabolism in the body. Despite its cardioprotective profile demonstrated by numerous clinical trials, the results of studies on the action of SGLT2 inhibitors in IP are scarce. This study will investigate the effects of dapagliflozin on IP in patients with coronary artery disease (CAD). **Methods:** The study will include 50 patients with multivessel CAD, ischemia documented by stress testing, and preserved left ventricular ejection fraction (LVEF). Patients will undergo four exercise tests, the first two with a time interval of 30 min between them after washout of cardiovascular or hypoglycemic medications and the last two after 7 days of dapagliflozin 10 mg once a day, also with a time interval of 30 min between them. **Discussion:** The role of SGLT2 inhibitors on IP is not clearly established. Several clinical trials have shown that SGLT2 inhibitors reduce the occurrence cardiovascular events, notably heart failure. However, such studies have not shown beneficial metabolic effects of SGLT2 inhibitors, such as reducing myocardial infarction or stroke. On the other hand, experimental studies with animal models have shown the beneficial effects of SGLT2 inhibitors on IP, a mechanism that confers cardiac and vascular protection from subsequent ischemia–reperfusion (IR) injury. This is the first clinical study to evaluate the effects of SGLT2 inhibitors on IP, which could result in an important advance in the treatment of patients with stable CAD.

Keywords: ischemic preconditioning; SGLT2 inhibitors; myocardial ischemia; coronary artery disease

1. Background

Ischemic preconditioning (IP) is recognized as a protective cellular mechanism in which brief, recurrent episodes of myocardial ischemia followed by reperfusion can self-protect the heart from prolonged ischemic injury and thereby limit myocardial infarction size [1].

In humans, IP can be assessed in different scenarios, such as percutaneous coronary intervention (PCI), intermittent aortic clamping during coronary artery bypass graft surgery, and sequential stress tests [2–4]. The phenomenon of "warm-up" or "walk-through angina" has been related to IP and documented in several studies [5–8]. Improvement in ischemic parameters, such as the time to reach 1.0 mm ST segment deviation (ST) on electrocardiogram (ECG) and the time to develop angina in two sequential tests is considered a manifestation of IP.

The cellular pathways underlying IP are not fully understood and are undoubtedly complex, involving triggers, mediators, memory, and end-effectors. All steps appear to involve components such as adenosine, adenosine receptors, the activation of the adenosine triphosphate-sensitive potassium channels (K_{ATP} channels) and the ε-isoform of protein kinase C, and others including the paradoxical protective role of oxygen radicals [9]. Opening of the K_{ATP} channel in cardiac myocytes has been consistently observed to be important in contributing to protection against IP [10].

IP typically results in increased gene expression and cellular metabolism [11]. A central target of such changes in gene expression and metabolism is the mitochondria. The direct and indirect effects of IP on mitochondria can result in the activation of adenosine monophosphate-activated protein kinase (AMPK), a regulator of cellular metabolism [11].

It has been established that anaerobic glycolysis is important for the generation of ATP during ischemia and crucial for the attenuation of ischemic injury in the heart [12,13]. Increased glucose uptake in ischemic cardiomyocytes is achieved mainly by translocation of a glucose transporter, GLUT4, from its intracellular compartments to the sarcolemma [14,15]. It is known that the expression level of GLUT4 is increased by AMPK activation [16].

Thus, as AMPK is a protein that plays an important role in regulating myocardial energy metabolism, reducing ischemia/reperfusion (I/R) injury; it may have beneficial effects on numerous cardiovascular diseases, such as heart failure and ventricular remodeling, vascular endothelial dysfunction, chronic inflammation, apoptosis, and regulation of autophagy [17].

An increased intracellular ratio of AMP to ATP, as occurs during strenuous exercise, hypoxia, or nutritional deficiency, can phosphorylate a threonine, the 172nd amino acid of the α subunit, thereby activating AMPK [18].

Upon activation, AMPK shuts down ATP consumption pathways and activates catabolic ATP production pathways through downstream signaling and target molecules [19,20], regulating lipid and protein metabolism, fatty acid oxidation, glucose uptake, gluconeogenesis, and autophagy [21,22]. AMPK also plays an important role in reducing oxidative stress by regulating autophagy and anti-apoptosis of cardiomyocytes [23,24].

Several oral hypoglycemic agents are capable of promoting the activation of AMPK, including sodium-glucose transport protein 2 (SGLT2) inhibitors. This class of medication selectively inhibits SGLT2 in the renal proximal tubule, with a consequent decrease in renal tubular thresholds for glycosuria and an increase in urinary glucose excretion, reducing blood glucose independently of insulin.

After clinical analysis, the protective effect of SGLT2 inhibitors on the heart may be related to reduced blood pressure, weight loss, decreased serum uric acid level, osmotic diuresis, reduced volume load, and hemodynamic changes [25].

Currently, basic research has focused on the action of this class on energy metabolism, inflammation, oxidative stress, myocardial fibrosis and electrolyte homeostasis [26]. SGLT2 inhibitors change the energy metabolism of the heart from glucose to fat [27–29] and slightly increase the ketone level [30], which is beneficial for energy supply to the heart. Dapagliflozin may delay the occurrence and progress of diabetic cardiomyopathy [31].

Few studies have evaluated the effects of SGLT2 inhibitors on AMPK. Canagliflozin activates AMPK of human embryonic kidney cells (HEK-293) and hepatocytes by inhibiting complex I in the mitochondrial respiratory chain and increasing cellular AMP levels [32]. Clinically relevant concentrations of canagliflozin can directly inhibit the secretion of endothelial pro-inflammatory chemokines/cytokines by AMPK-dependent and -independent mechanisms without affecting early interleukin-1β (IL-1β) signaling [33]. Zhou et al. showed in a murine model that empagliflozin is capable of reducing injury to diabetic cardiac microvascular endothelial cells (CMECs) by inhibiting mitochondrial fission through activation of the AMPK-Drp1 (dynamin-related protein 1) signaling pathways, can preserve cardiac barrier function of CMECs by suppressing mitochondrial reactive oxygen species production, and can subsequently reduce oxidative stress by inhibiting CMEC senescence [34].

In turn, dapagliflozin has been shown in animal models to be capable of decreasing the activation of the NOD-like receptor family, pyrin domain containing protein 3/apoptosis-associated speck-like containing a CARD inflammasome (NLRP3/ASC), thereby attenuating myocardial inflammation, fibrosis, apoptosis, and diabetic remodeling likely mediated through AMPK activation [35]. Additionally, Tsai et al. showed that dapagliflozin reduced the H/R-elicited oxidative stress via modulation of AMPK [36]. Recently, Tanajak et al. showed that dapagliflozin has greater cardioprotective efficacy than vildagliptin in rats with cardiac I/R injury [37].

Despite its cardioprotective profile demonstrated by numerous clinical trials [38–40], the results of studies on the action of SGLT2 inhibitors in IP are scarce. The few studies that have addressed this question have been carried out in animal models [37,41,42]. Furthermore, in humans, there are no studies that have evaluated the effects of dapagliflozin on IP expression. Therefore, the objective of the present study will be to evaluate the effects of dapagliflozin on IP in patients with stable, symptomatic, multivessel coronary artery disease (CAD) and preserved left ventricular ejection fraction (LVEF).

2. Methods/Design

A total of 50 patients with documented multivessel CAD, ischemia documented by stress testing, and preserved left ventricular ejection fraction (LVEF) will be screened (Table 1).

Table 1. Inclusion criteria.

- Stable multivessel CAD (obstruction greater than 70% in at least two main coronary branches).
- LVEF ≥ 0.50, confirmed by transthoracic Doppler echocardiography.
- Documentation of stress-induced myocardial ischemia (horizontal or descending ST segment depression ≥ 1.0 mm)

CAD will be documented by coronary angiography by visual assessment of atherosclerotic lesions with luminal obstruction of more than 70% in at least two different coronary territories.

Left ventricular systolic function will be measured by transthoracic echocardiography or ventriculography and will be considered preserved if the LVEF is greater than 50%.

Exercise treadmill tests (ETTs) that result in depression of the ST segment during effort greater than or equal to 1.0 mm, horizontal or descending and associated or not with angina, will be considered positive.

Exclusion criteria are shown in Table 2.

Table 2. Exclusion criteria.

- Kidney failure (creatinine clearance < 60 mL/min)
- Severe liver failure
- Single-vessel CAD
- Myocardial infarction in the last 3 months
- LVEF < 50%
- Presence of any non-ischemic cardiomyopathy
- Moderate or severe valve disease
- Morphological changes in the QRS of the ECG and conduction defects that may interfere with the interpretation of changes in the ST segment
- Recent and negative exercise test for myocardial ischemia

Table 2. *Cont.*

-	Positive exercise test for myocardial ischemia, with signs of high risk
-	Limiting anginal symptoms or recent worsening
-	Arrhythmias that make it difficult to characterize myocardial ischemia during exercise stress (atrial fibrillation or flutter)
-	Patient refusal to participate in the study

2.1. Patient Preparation

After clinical and cardiological evaluation, patients will be instructed to stop medications with cardiovascular effects that potentially interfere with IP before sequential exercise tests, depending on the half-life of the drug. Diabetic patients will be instructed to suspend, in addition to medications with cardiovascular effects, oral hypoglycemic agents for a similar period before the tests. Only nitrates will be maintained, when necessary, up to 24 h before testing.

Patients will be instructed not to perform physical activities during the period and to control their salt intake, and patients with diabetes will be advised to strictly control their carbohydrate intake. They will also be instructed to contact the study team by telephone, who will be available 24 h a day, in case of questions or worsening of symptoms. On the day of the exams, the symptoms will be reassessed by the medical team before carrying out the sequential exercise tests.

2.2. Sequential Ergometric Tests

The study protocol will include 2 phases (Figure 1). The criterion used to evaluate the IP took into account the methodology used in previous studies [43,44]. In Phase 1, after the washout period of cardiovascular or hypoglycemic medications, all patients will undergo 2 consecutive ETT (ETT1 and ETT2), with a 30 min interval between them to identify the ischemia and document the magnitude of IP by the difference in ischemia parameters between the 2 tests. The protocol will be adopted according to the assessment of each patient's functionality (Bruce or modified Bruce). The ergometer used will be the GE T2100 Ergometric Treadmill coupled with a GE Case V6.73 system/software and a Tango M2 blood pressure monitor.

The recording system used will be 12 leads, including the classic leads of the Mason and Likar systems. Electrocardiographic recordings will be carried out in a standardized way, pre-exertion, every 5 to 10 s at a time close to T-1.0 mm, at the peak of exercise, at the time of the worst electrocardiographic change, at the time of arrhythmias, and at every minute of the recovery, which will last for 6 min.

Exercise tests that result in depression of the ST segment during effort greater than or equal to 1.0 mm, horizontal or descending, associated or not with chest pain will be considered positive.

Heart rate will be continuously monitored and documented every 15 s. Blood pressure measurement will be performed every 90 s, at the moment of T-1.0 mm, at peak effort and every minute of the recovery phase. The double product or rate pressure product (RPP) will be calculated by multiplying the heart rate in beats per minute (bpm) by the blood pressure in millimeters of mercury (mmHg), this variable being measured at the time of T-1.0 mm.

The criteria for interrupting exams will be those adopted by the recommendations of the Brazilian Society of Cardiology Guidelines [43].

After Phase 1, all patients will receive dapagliflozin at a dose of 10 mg once a day for 6 days. On the seventh day, patients will receive dapagliflozin 10 mg and will again undergo 2 consecutive ETTs (ETT3 and ETT4) 2 h after medication administration (time to reach peak plasma concentration) [Phase 2]. The time interval between ETT3 and ETT4 will be similar to that of Phase 1, that is, 30 min.

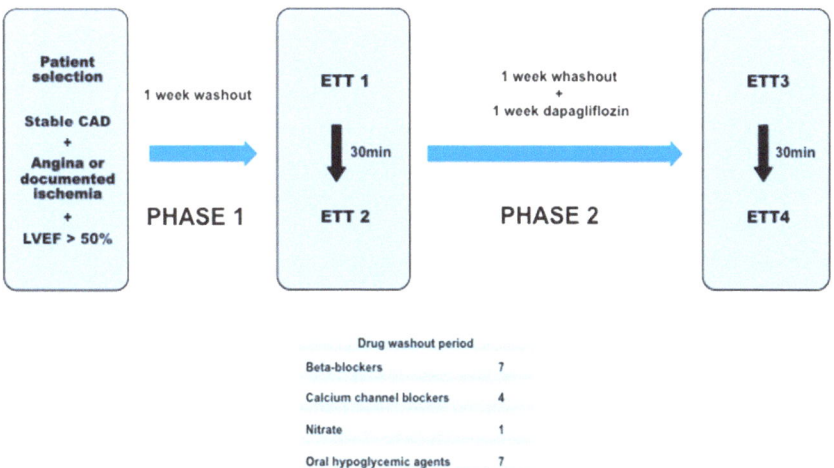

Figure 1. Study design. Legend: CAD: coronary artery disease; ETT: exercise treadmill test; LVEF: left ventricular ejection fraction.

2.3. Characterization of IP

During sequential ETTs, confirmation of T-1.0 mm will be carried out by two experienced cardiologists, independently and blind to the sequence of tests. Situations of disagreement will be resolved by consensus. An improvement equal to or greater than 30 s in the time to reach depression in 1.0 mm of the ST segment in the second sequential test compared to T-1.0 mm in the first test will be classified as IP present. The second criterion adopted to characterize the IP will be used based on the analysis of the RPP result. Therefore, when the improvement of T-1.0 mm is borderline, the RPP at the time of T-1.0 mm will be considered. If this is higher in the second test compared to the first, the IP will be considered present.

2.4. Statistical Analysis

A two-way ANOVA with repeated measures followed by the Bonferroni test will be used to compare T-1.0 mm and RPP data. Comparisons of the remaining continuous or discrete variables between the two phases will be performed using an unpaired Student's t-test or χ^2, respectively. Fisher's test will be used when appropriate. The data will be expressed as mean standard deviation and in the figures as median and interquartile ranges. A value of $p < 0.05$ will be considered significant. SPSS version 20 software will be used for all statistical analyses.

2.5. Sample Size Calculation

The primary outcome of the study will be the difference in the time required to achieve 1.0 mm of ST segment depression (T-1.0 mm) between the ETT3 and ETT1 [ETT3-ETT1] and ETT4 and ETT2 [ETT4-ETT2]. A sample of 50 patients will be necessary in order to detect a 33% increase in the time to reach T-1.0 mm with dapagliflozin compared to not using this medication, with a power of 80% and a two-sided significance level of 0.05 [45].

3. Discussion

Myocardial IP is an intracellular protective mechanism in which short periods of ischemia followed by reperfusion trigger cell signaling processes and cascades, which culminate in greater tissue resistance to subsequent ischemic insult.

The role of SGLT2 inhibitors on IP is not clearly established. Several clinical trials have shown that SGLT2 inhibitors reduce cardiovascular events, notably heart failure. However, such studies have not shown beneficial metabolic effects of SGLT2 inhibitors, such as reducing myocardial infarction or stroke. On the other hand, experimental studies with animal models have shown beneficial effects of SGLT2 inhibitors on IP, a mechanism that confers cardiac and vascular protection from subsequent ischemia–reperfusion (IR) injury. A recent meta-analysis of 16 independent animal models experiments that compared SGLT2i to a control showed that independent of diabetes, SGLT2is were significantly associated with fewer myocardial ischemia–reperfusion injuries and smaller infarct size [46]. The probable cellular mechanism that justifies the effects of dapagliflozin on cardiac ischemia/reperfusion (I/R) injury is through the modulation of AMPK.

4. Results

The trial is ongoing. Recruitment for the main trial commenced on 15 December 2022. The expected recruitment completion date is July 2025.

5. Conclusions

The DAPA-IP trial will evaluate the effects of dapagliflozin on IP and may establish a new treatment option for stable CAD patients.

Author Contributions: T.L.S. contributed to the design of the work; M.A.V.A., B.M.A.F.S. and T.L.S. wrote the main manuscript text; M.A.V.A., B.M.A.F.S., T.L.S. and J.P.M.T. contributed to the analysis and interpretation of data; A.C.R.d.A., G.N.V.R., P.R.S. and W.A.C. have substantively revised the work. All authors have read and agreed to the published version of the manuscript.

Funding: This research received no external funding.

Institutional Review Board Statement: The trial protocol was approved by the ethics committee of Faculty of Medicine, University of São Paulo (CAAE: 65199022.4.0000.0068, 16 February 2023). The study protocol is registered on the ISRCTN Registry under ISRCTN69498919 (https://doi.org/10.1186/ISRCTN12413923, accessed on 1 December 2023).

Informed Consent Statement: Written informed consent has been obtained from all participating patients.

Data Availability Statement: The data in this study are available from the corresponding author on reasonable request.

Conflicts of Interest: The authors declare no conflicts of interest.

References

1. Murry, C.E.; Jennings, R.B.; Reimer, K.A. Preconditioning with ischemia: A delay of lethal cell injury in ischemic myocardium. *Circulation* **1986**, *74*, 1124–1136. [CrossRef] [PubMed]
2. Waters, D.D.; McCans, J.L.; Crean, P.A. Serial exercise testing in patients with effort angina: Variable tolerance, fixed threshold. *J. Am. Coll. Cardiol.* **1985**, *6*, 1011–1015. [CrossRef] [PubMed]
3. MacAlpin, R.N.; Kattus, A.A. Adaptation to exercise in angina pectoris. The electrocardiogram during treadmill walking and coronary angiographic findings. *Circulation* **1966**, *33*, 183–201. [CrossRef] [PubMed]
4. Ylitalo, K.; Jama, L.; Raatikainen, P.; Peuhkurinen, K. Adaptation to myocardial ischemia during repeated dynamic exercise in relation to findings at cardiac catheterization. *Am. Heart J.* **1996**, *131*, 689–697. [CrossRef] [PubMed]
5. Jaffe, M.D.; Quinn, N.K. Warm-up phenomenon in angina pectoris. *Lancet* **1980**, *2*, 934–936. [CrossRef] [PubMed]
6. Stewart, R.A.; Simmonds, M.B.; Williams, M.J. Time course of "warm-up" in stable angina. *Am. J. Cardiol.* **1995**, *76*, 70–73. [CrossRef] [PubMed]
7. Maybaum, S.; Ilan, M.; Mogilevsky, J.; Tzivoni, D. Improvement in ischemic parameters during repeated exercise testing: A possible model for myocardial preconditioning. *Am. J. Cardiol.* **1996**, *78*, 1087–1091. [CrossRef] [PubMed]
8. Tomai, F. Warm up phenomenon and preconditioning in clinical practice. *Heart* **2002**, *87*, 99–100. [CrossRef] [PubMed]

9. Yellon, D.M.; Downey, J.M. Preconditioning the myocardium: From cellular physiology to clinical cardiology. *Physiol. Rev.* **2003**, *83*, 1113–1151. [CrossRef]
10. Gross, G.J.; Peart, J.N. KATP channels and myocardial preconditioning: An update. *Am. J. Physiol. Heart Circ. Physiol.* **2003**, *285*, H921–H930. [CrossRef]
11. Jackson, C.W.; Escobar, I.; Xu, J.; Perez-Pinzon, M.A. Effects of ischemic preconditioning on mitochondrial and metabolic neuroprotection: 5′ adenosine monophosphate-activated protein kinase and sirtuins. *Brain Circ.* **2018**, *4*, 54–61. [PubMed]
12. Opie, L.H.; Sack, M.N. Metabolic plasticity and the promotion of cardiac protection in ischemia and ischemic preconditioning. *J. Mol. Cell. Cardiol.* **2002**, *34*, 1077–1089. [CrossRef]
13. Jennings, R.A.; Kloner, R.B. Consequences of brief ischemia: Stunning, preconditioning, and their clinical implications: Part 1. *Circulation* **2001**, *104*, 2981–2989.
14. Sun, D.; Nguyen, N.; Degrado, T.R.; Schwaiger, M.; Brosius, F.C., 3rd. Ischemia induces translocation of the insulin-responsive glucose transporter GLUT4 to the plasma membrane of cardiac myocytes. *Circulation* **1994**, *89*, 793–798. [CrossRef] [PubMed]
15. Egert, S.; Nguyen, N.; DeGrado, T.R.; Schwaiger, M.; Brosius, F.C., 3rd. Myocardial glucose transporter GLUT1: Translocation induced by insulin and ischemia. *J. Mol. Cell. Cardiol.* **1999**, *31*, 1337–1344. [CrossRef] [PubMed]
16. Holmes, B.F.; Kurth-Kraczek, E.J.; Winder, W.W. Chronic activation of 5′-AMP-activated protein kinase increases GLUT4, hexokinase, and glycogen in muscle. *J. Appl. Physiol.* **1999**, *87*, 1990–1995. [CrossRef] [PubMed]
17. Lu, Q.; Li, X.; Liu, J.; Sun, X.; Rousselle, T.; Ren, D.; Tong, N.; Li, J. AMPK is associated with the beneficial effects of antidiabetic agents on cardiovascular diseases. *Biosci. Rep.* **2019**, *39*, BSR20181995. [CrossRef]
18. Hardie, D.G.; Carling, D. The AMP-activated protein kinase-fuel gauge of the mammalian cell? *Eur. J. Biochem.* **1997**, *246*, 259–273. [CrossRef]
19. Ruderman, N.B.; Carling, D.; Prentki, M.; Cacicedo, J.M. AMPK, insulin resistance, and the metabolic syndrome. *J. Clin. Investig.* **2013**, *123*, 2764–2772. [CrossRef]
20. Steinberg, G.R.; Kemp, B.E. AMPK in health and disease. *Physiol. Rev.* **2009**, *89*, 1025–1078. [CrossRef]
21. Woods, A.; Johnstone, S.R.; Dickerson, K.; Leiper, F.C.; Fryer, L.G.; Neumann, D.; Schlattner, U.; Wallimann, T.; Carlson, M.; Carling, D. LKB1 is the upstream kinase in the AMP-activated protein kinase cascade. *Curr. Biol.* **2003**, *13*, 2004–2008. [CrossRef] [PubMed]
22. Hurley, R.L.; Anderson, K.A.; Franzone, J.M.; Kemp, B.E.; Means, A.R.; Witters, L.A. The Ca^{2+}/calmodulin-dependent protein kinase kinases are AMP-activated protein kinase kinases. *J. Biol. Chem.* **2005**, *280*, 29060–29066. [CrossRef] [PubMed]
23. Bertrand, L.; Ginion, A.; Beauloye, C.; Hebert, A.D.; Guigas, B.; Hue, L.; Vanoverschelde, J.-L. AMPK activation restores the stimulation of glucose uptake in an in vitro model of insulin-resistant cardiomyocytes via the activation of protein kinase B. *Am. J. Physiol. Heart Circ. Physiol.* **2006**, *291*, H239–H250. [CrossRef]
24. He, C.; Zhu, H.; Li, H.; Zou, M.H.; Xie, Z. Dissociation of Bcl-2-Beclin1 complex by activated AMPK enhances cardiac autophagy and protects against cardiomyocyte apoptosis in diabetes. *Diabetes* **2013**, *62*, 1270–1281. [CrossRef]
25. Lopaschuk, G.D.; Verma, S. Mechanisms of Cardiovascular Benefits of Sodium Glucose Co-Transporter 2 (SGLT2) Inhibitors: A State-of-the-Art Review. *JACC Basic Transl. Sci.* **2020**, *5*, 632–644. [CrossRef]
26. Bertero, E.; Prates Roma, L.; Ameri, P.; Maack, C. Cardiac effects of SGLT2 inhibitors: The sodium hypothesis. *Cardiovasc. Res.* **2018**, *114*, 12–18. [CrossRef]
27. Merovci, A.; Solis-Herrera, C.; Daniele, G.; Eldor, R.; Fiorentino, T.V.; Tripathy, D.; Xiong, J.; Perez, Z.; Norton, L.; Abdul-Ghani, M.A.; et al. Dapagliflozin improves muscle insulin sensitivity but enhances endogenous glucose production. *J. Clin. Investig.* **2014**, *124*, 509–514. [CrossRef] [PubMed]
28. Ferrannini, E.; Mark, M.; Mayoux, E. CV Protection in the EMPA-REG OUTCOME trial: A "Thrifty Substrate" hypothesis. *Diabetes Care* **2016**, *3*, 1108–1114. [CrossRef]
29. Ferrannini, E.; Baldi, S.; Frascerra, S.; Astiarraga, B.; Heise, T.; Bizzotto, R.; Mari, A.; Pieber, T.R.; Muscelli, E. Shift to fatty substrate utilization in response to sodium-glucose cotransporter 2 inhibition in subjects without diabetes and patients with type 2 diabetes. *Diabetes* **2016**, *65*, 1190–1195. [CrossRef]
30. Taylor, S.I.; Blau, J.E.; Rother, K.I. SGLT2 Inhibitors May Predispose to Ketoacidosis. *J. Clin. Endocrinol. Metab.* **2015**, *100*, 2849–2852. [CrossRef]
31. Joubert, M.; Jagu, B.; Montaigne, D.; Marechal, X.; Tesse, A.; Ayer, A.; Dollet, L.; Le May, C.; Toumaniantz, G.; Manrique, A.; et al. The sodium-glucose cotransporter 2 inhibitor dapagliflozin prevents cardiomyopathy in a diabetic lipodystrophic mouse model. *Diabetes* **2017**, *66*, 1030–1040. [CrossRef] [PubMed]
32. Hawley, S.A.; Ford, R.J.; Smith, B.K.; Gowans, G.J.; Mancini, S.J.; Pitt, R.D.; Day, E.A.; Salt, I.P.; Steinberg, G.R.; Hardie, D.G. The Na^+/glucose cotransporter inhibitor canagliflozin activates AMPK by inhibiting mitochondrial function and increasing cellular AMP levels. *Diabetes* **2016**, *65*, 2784–2794. [CrossRef] [PubMed]
33. Mancini, S.J.; Boyd, D.; Katwan, O.J.; Strembitska, A.; Almabrouk, T.A.; Kennedy, S.; Palmer, T.M.; Salt, I.P. Canagliflozin inhibits interleukin-1 beta-stimulated cytokine and chemokine secretion in vascular endothelial cells by AMP-activated protein kinase-dependent and -independent mechanisms. *Sci. Rep.* **2018**, *8*, 5276. [CrossRef] [PubMed]
34. Zhou, H.; Wang, S.; Zhu, P.; Hu, S.; Chen, Y.; Ren, J. Empagliflozin rescues diabetic myocardial microvascular injury via AMPK-mediated inhibition of mitochondrial fission. *Redox. Biol.* **2018**, *15*, 335–346. [CrossRef] [PubMed]

35. Ye, Y.; Bajaj, M.; Yang, H.C.; Perez-Polo, J.R.; Birnbaum, Y. SGLT-2 inhibition with dapagliflozin reduces the activation of the Nlrp3/ASC inflammasome and attenuates the development of diabetic cardiomyopathy in mice with type 2 diabetes. further augmentation of the effects with saxagliptin, a DPP4 inhibitor. *Cardiovasc. Drugs Ther.* **2017**, *31*, 119–132. [CrossRef] [PubMed]
36. Tsai, K.L.; Hsieh, P.L.; Chou, W.C.; Cheng, H.C.; Huang, Y.T.; Chan, S.H. Dapagliflozin attenuates hypoxia/reoxygenation-caused cardiac dysfunction and oxidative damage through modulation of AMPK. *Cell Biosci.* **2021**, *11*, 44. [CrossRef]
37. Tanajak, P.; Sa-Nguanmoo, P.; Sivasinprasasn, S.; Thummasorn, S.; Siri-Angkul, N.; Chattipakorn, S.C.; Chattipakorn, N. Cardioprotection of dapagliflozin and vildagliptin in rats with cardiac ischemia-reperfusion injury. *J. Endocrinol.* **2018**, *236*, 69–84. [CrossRef] [PubMed]
38. Wiviott, S.D.; Raz, I.; Bonaca, M.P.; Mosenzon, O.; Kato, E.T.; Cahn, A.; Silverman, M.G.; Zelniker, T.A.; Kuder, J.F.; Murphy, S.A.; et al. Dapagliflozin and Cardiovascular Outcomes in Type 2 Diabetes. *N. Engl. J. Med.* **2019**, *380*, 347–357. [CrossRef]
39. Neal, B.; Perkovic, V.; Mahaffey, K.W.; de Zeeuw, D.; Fulcher, G.; Erondu, N.; Shaw, W.; Law, G.; Desai, M.; Matthews, D.R.; et al. Canagliflozin and Cardiovascular and Renal Events in Type 2 Diabetes. *N. Engl. J. Med.* **2017**, *377*, 644–657. [CrossRef] [PubMed]
40. Zinman, B.; Wanner, C.; Lachin, J.M.; Fitchett, D.; Bluhmki, E.; Hantel, S.; Mattheus, M.; Devins, T.; Johansen, O.E.; Woerle, H.J.; et al. Empagliflozin, Cardiovascular Outcomes, and Mortality in Type 2 Diabetes. *N. Engl. J. Med.* **2015**, *373*, 2117–2128. [CrossRef]
41. Lahnwong, S.; Palee, S.; Apaijai, N.; Sriwichaiin, S.; Kerdphoo, S.; Jaiwongkam, T.; Chattipakorn, S.C.; Chattipakorn, N.; Lahnwong, S. Acute dapagliflozin administration exerts cardioprotective effects in rats with cardiac ischemia/reperfusion injury. *Cardiovasc. Diabetol.* **2020**, *19*, 91. [CrossRef]
42. Lee, T.-M.; Chang, N.-C.; Lin, S.-Z. Dapagliflozin, a selective SGLT2 Inhibitor, attenuated cardiac fibrosis by regulating the macrophage polarization via STAT3 signaling in infarcted rat hearts. *Free Radic. Biol. Med.* **2017**, *104*, 298–310. [CrossRef]
43. Costa, L.M.A.; Rezende, P.C.; Garcia, R.M.R.; Uchida, A.H.; Seguro, L.F.B.C.; Scudeler, T.L.; Bocchi, E.A.; Krieger, J.E.; Hueb, W.; Ramires, J.A.F.; et al. Role of Trimetazidine in Ischemic Preconditioning in Patients With Symptomatic Coronary Artery Disease. *Medicine* **2015**, *94*, e1161. [CrossRef]
44. Rahmi, R.M.; Uchida, A.H.; Rezende, P.C.; Lima, E.G.; Garzillo, C.L.; Favarato, D.; Strunz, C.M.; Takiuti, M.; Girardi, P.; Hueb, W.; et al. Effect of hypoglycemic agents on ischemic preconditioning in patients with type 2 diabetes and symptomatic coronary artery disease. *Diabetes Care* **2013**, *36*, 1654–1659. [CrossRef]
45. Meneghelo, R.S.; Araújo, C.G.S.; Stein, R.; Mastrocolla, L.E.; Albuquerque, P.F.; Serra, S.M.; Sociedade Brasileira de Cardiologia. III Diretrizes da Sociedade Brasileira de Cardiologia sobre Teste Ergométrico. *Arq. Bras. Cardiol.* **2010**, *95*, 1–26.
46. Sayour, A.A.; Celeng, C.; Oláh, A.; Ruppert, M.; Merkely, B.; Radovits, T. Sodium-glucose cotransporter 2 inhibitors reduce myocardial infarct size in preclinical animal models of myocardial ischaemia-reperfusion injury: A meta-analysis. *Diabetologia* **2021**, *64*, 737–748. [CrossRef]

Disclaimer/Publisher's Note: The statements, opinions and data contained in all publications are solely those of the individual author(s) and contributor(s) and not of MDPI and/or the editor(s). MDPI and/or the editor(s) disclaim responsibility for any injury to people or property resulting from any ideas, methods, instructions or products referred to in the content.

Article

Autoimmune Thyroiditis Mitigates the Effect of Metformin on Plasma Prolactin Concentration in Men with Drug-Induced Hyperprolactinemia

Robert Krysiak *, Marcin Basiak, Witold Szkróbka and Bogusław Okopień

Department of Internal Medicine and Clinical Pharmacology, Medical University of Silesia, Medyków 18, 40-752 Katowice, Poland; mbasiak@sum.edu.pl (M.B.); wszkrobka@sum.edu.pl (W.S.); bokopien@sum.edu.pl (B.O.)
* Correspondence: rkrysiak@sum.edu.pl; Tel./Fax: +48-322523902

Abstract: Metformin inhibits the secretory function of overactive anterior pituitary cells, including lactotropes. In women of childbearing age, this effect was absent if they had coexisting autoimmune (Hashimoto) thyroiditis. The current study was aimed at investigating whether autoimmune thyroiditis modulates the impact of metformin on the plasma prolactin concentration in men. This prospective cohort study included two groups of middle-aged or elderly men with drug-induced hyperprolactinemia, namely subjects with concomitant Hashimoto thyroiditis (group A) and subjects with normal thyroid function (group B), who were matched for baseline prolactin concentration and insulin sensitivity. Titers of thyroid peroxidase and thyroglobulin antibodies, levels of C-reactive protein, markers of glucose homeostasis, concentrations of pituitary hormones (prolactin, thyrotropin, gonadotropins, and adrenocorticotropic hormone), free thyroxine, free triiodothyronine, testosterone, and insulin growth factor-1 were measured before and six months after treatment with metformin. Both study groups differed in titers of both antibodies and concentrations of C-reactive protein. The drug reduced the total and monomeric prolactin concentration only in group B, and the impact on prolactin correlated with the improvement in insulin sensitivity and systemic inflammation. There were no differences between the follow-up and baseline levels of the remaining hormones. The results allow us to conclude that autoimmune thyroiditis mitigates the impact of metformin on prolactin secretion in men.

Keywords: autoimmune thyroid disease; insulin sensitivity; lactotropes; men; prolactin excess

1. Introduction

Despite maintaining anatomical and functional connections with the brain, the pituitary is situated outside the blood–brain barrier [1]. This fact may explain why metformin, a drug of unquestionable importance in the treatment of type-2 diabetes and other insulin-resistant states [2], preferentially accumulates in this structure [3] and attenuates the secretory function of different anterior pituitary cells, including thyrotropes [4,5], gonadotropes [6,7], and lactotropes [8–12]. Interestingly, and probably importantly from the clinical point of view, the inhibitory effect on thyroid-stimulating hormone (TSH), follicle-stimulating hormone (FSH), luteinizing hormone (LH), and prolactin in these studies was observed only if their baseline levels were elevated. There is no evidence that metformin administration, even at high doses, results in a deficiency of pituitary hormones. The decrease in prolactin secretion has been reported irrespective of the reason for prolactin excess, including in subjects with prolactinomas [8], empty sella syndrome [8], traumatic brain injury [8], drug-induced hyperprolactinemia [9–12], and in hyperprolactinemia of unknown origin [8]. Unlike most cases of prolactin excess, which may be effectively treated with dopamine agonists (mainly bromocriptine and cabergoline), the use of these agents in antipsychotics-induced hyperprolactinemia is much more controversial because they may aggravate psychiatric disability [13]. Considering doubts concerning the use of dopaminergic agents and the functional basis of this disorder, metformin is regarded by some research groups as a potential, emerging treatment for drug-induced prolactin excess [14]. It should be kept in mind that even moderate long-term prolactin excess, irrespective of gender, predisposes to insulin resistance, prediabetes, atherogenic dyslipidemia, excessive and/or subnormal fat accumulation, subclinical atherosclerosis, and endothelial dysfunction [15–20], and therefore, it should be avoided. It seems that metformin is characterized by differences in the action of lactotrope function in different populations of subjects with hyperprolactinemia. The impact of this agent on circulating prolactin levels was found to be determined by sex because in young adults the drug reduced prolactin levels only in women but not in men [21]. Moreover, metformin reduces prolactin levels in men if they have low, but not if they have normal, testosterone concentrations [22]. These discrepancies suggest that the impact of metformin on prolactin secretion may be more prominent in individuals with low testosterone production. Testosterone deficiency may make patients with prolactin excess particularly prone to cardiometabolic complications because low testosterone production in men is often associated with metabolic comorbidities, such as obesity, insulin resistance, metabolic syndrome, and type-2 diabetes mellitus [23,24].

Recently, our research team reported that coexisting autoimmune (Hashimoto) thyroiditis mitigated the impact of metformin on the secretory function of the anterior pituitary cells of women, including attenuation of the prolactin-lowering effect [25,26]. This finding seems to be clinically relevant because Hashimoto thyroiditis is one of the most common human disorders and the most prevalent organ-specific autoimmune disorder worldwide [27,28]. Despite the female preponderance of autoimmune thyroiditis, more and more males are being diagnosed with this clinical entity [29]. To the best of our knowledge, no previous study investigated the association between metformin and autoimmune thyroiditis at the level of male lactotropes. Few studies conducted so far suggest that low testosterone concentrations and autoimmune thyroiditis are, in men, reciprocally related. Males with primary hypothyroidism, caused in the majority of patients by autoimmune thyroid disease, were characterized by gonadal hypofunction [30]. A higher ratio of estradiol to testosterone in males was associated with autoimmune thyroid disease [31]. Shorter AR (CAG)n repeats, increasing the activity of the androgen receptor, were found to predispose to a younger onset of autoimmune thyroid disease [32]. Moreover, Hashimoto thyroiditis is highly prevalent in individuals with Klinefelter's syndrome, the most prevalent sex-chromosome disorder resulting in male hypogonadism [33]. Lastly, exogenous testosterone reduced thyroid antibody titers in euthyroid men with testosterone deficiency and Hashimoto thyroiditis [34]. In addition to the association with testosterone deficiency, particularly in individuals with markedly elevated prolactin levels [35]; increased prolactin production may predispose to the development and progression of autoimmune disorders, including autoimmune thyroiditis [36]. Thus, the present study investigated whether coexisting euthyroid Hashimoto thyroiditis modulates the impact of chronic metformin treatment on plasma prolactin levels in middle-aged and elderly men with antipsychotic-induced hyperprolactinemia.

2. Results

The study groups were comparable with respect to age, smoking, percentages of patients with type-2 diabetes and prediabetes, BMI, and blood pressure (systolic and diastolic) (Table 1). Titers of thyroid antibodies (both TPOAb and TgAb, and the concentration of hsCRP were higher in group A than in group B. Levels of glucose, glycated hemoglobin, pituitary hormones (prolactin [total and monomeric], TSH, FSH, LH and ACTH, macroprolactin), free thyroid hormones, testosterone, IGF-1, and HOMA-IR, did not differ between groups A and B (Table 2).

Table 1. Baseline characteristics of both study groups.

Variable	Group A	Group B	p-Value
Number (n)	24	24	-
Age (years)	60 ± 8	62 ± 9	0.4200
Type-2 diabetes (%)/prediabetes (%)	50/50	54/46	0.8342
Smokers (%)/Number of cigarettes a day (n)/Duration of smoking (years)	42/11 ± 5/32 ± 10	46/10 ± 6/34 ± 11	0.7523
BMI (kg/m^2)	24.9 ± 4.3	24.5 ± 4.8	0.7624
Systolic blood pressure (mmHg)	130 ± 15	128 ± 14	0.6352
Diastolic blood pressure (mmHg)	85 ± 5	84 ± 5	0.4918

Group A: males with iatrogenic prolactin excess and euthyroid Hashimoto disease. Group B: males with iatrogenic prolactin excess but without thyroid disease. Except for the percentages of smokers, patients with type-2 diabetes, and individuals with prediabetes, the data have been shown as the mean ± standard deviation. Abbreviation: BMI—body mass index.

Table 2. The impact of metformin treatment on the assessed variables.

Variable	Group A	Group B	p-Value *
Glucose (mg/dL) [70–99]			
Baseline	121 ± 11	123 ± 12	0.5502
Follow-up	112 ± 10	105 ± 10	0.0193
p-value **	0.0048	<0.0001	-
HOMA1-IR [<2.0]			
Baseline	4.2 ± 1.3	4.0 ± 1.2	0.5824
Follow-up	3.2 ± 1.0	2.1 ± 0.8	0.0001
p-value **	0.0045	<0.0001	-
Glycated hemoglobin [4.0–5.6]			
Baseline	6.7 ± 0.5	6.8 ± 0.6	0.5346
Follow-up	6.3 ± 0.5	5.9 ± 0.4	0.0037
p-value **	0.0080	<0.0001	-
Total prolactin (ng/mL) [5–17]			
Baseline	55.2 ± 12.3	56.8 ± 13.8	0.6735
Follow-up	52.7 ± 13.4	44.8 ± 11.8	0.0354
p-value **	0.5702	0.0022	-
Monomeric prolactin (ng/mL) [3–15]			
Baseline	50.4 ± 11.2	52.5 ± 12.9	0.5500
Follow-up	48.6 ± 10.6	40.9 ± 11.6	0.0216
p-value **	0.5918	0.0020	-
Macroprolactin (ng/mL) [2–12]			
Baseline	4.8 ± 3.1	4.3 ± 2.9	0.5667
Follow-up	4.1 ± 3.5	3.9 ± 2.5	0.8208
p-value **	0.4670	0.6112	-
TPOAb (IU/mL) [<35]			
Baseline	840 ± 305	13 ± 12	<0.0001
Follow-up	668 ± 285	11 ± 14	<0.0001
p-value **	0.0978	0.5978	-
TgAb (IU/mL) [<35]			
Baseline	828 ± 320	17 ± 12	<0.0001
Follow-up	685 ± 276	16 ± 18	<0.0001
p-value **	0.1042	0.8219	-
TSH (mIU/L) [0.4–4.5]			
Baseline	3.2 ± 1.3	3.0 ± 1.4	0.6105
Follow-up	2.8 ± 1.2	2.3 ± 1.2	0.1557
p-value **	0.2738	0.0693	-

Table 2. Cont.

Variable	Group A	Group B	p-Value *
Free thyroxine (pmol/L) [10.2–21.4]			
Baseline	14.8 ± 2.4	15.2 ± 2.6	0.5824
Follow-up	15.1 ± 2.7	15.6 ± 2.9	0.5395
p-value **	0.6860	0.6174	-
Free triiodothyronine (pmol/L) [2.2–6.7]			
Baseline	3.6 ± 0.8	3.5 ± 0.9	0.6860
Follow-up	3.7 ± 0.8	3.7 ± 1.0	1.0000
p-value **	0.6670	0.4701	-
FSH (U/L) [1.5–9.5]			
Baseline	3.4 ± 1.0	3.2 ± 1.2	0.5346
Follow-up	3.7 ± 1.3	3.6 ± 1.4	0.7988
p-value **	0.3749	0.2934	-
LH (U/L) [1.5–8.5]			
Baseline	2.9 ± 0.8	3.0 ± 1.4	0.7626
Follow-up	3.2 ± 1.0	3.6 ± 1.1	0.1940
p-value **	0.2571	0.1056	-
Testosterone (ng/mL) [3.5–17.0]			
Baseline	4.4 ± 1.0	4.2 ± 1.3	0.5532
Follow-up	4.7 ± 1.2	4.9 ± 1.5	0.6125
p-value **	0.3517	0.908	-
ACTH (pg/mL) [15–70]			
Baseline	29 ± 14	34 ± 18	0.2883
Follow-up	35 ± 15	38 ± 16	0.5061
p-value **	0.1587	0.4200	-
IGF-1 (ng/mL) [50–180]			
Baseline	102 ± 48	97 ± 40	0.6968
Follow-up	114 ± 50	118 ± 60	0.8030
p-value **	0.4007	0.1604	-
hsCRP (mg/L) [<1.5]			
Baseline	2.8 ± 1.0	2.2 ± 0.8	0.0263
Follow-up	2.6 ± 0.8	1.5 ± 0.6	<0.0001
p-value **	0.4481	0.0013	-
Estimated glomerular filtration rate (mL/min/1.73 m^2) [<60]			
Baseline	89 ± 13	91 ± 15	0.6239
After 6 months	92 ± 14	92 ± 16	1.0000
p-value **	0.4457	0.8242	-

* Group A vs. Group B, ** Follow-up vs. baseline. Group A: males with iatrogenic prolactin excess and euthyroid Hashimoto disease. Group B: males with iatrogenic prolactin excess but without thyroid disease. Except for the percentage of smokers, the data have been shown as the mean ± standard deviation. Reference values are shown in square brackets. Abbreviations: ACTH—adrenocorticotropic hormone; FSH—follicle-stimulating hormone; HOMA-IR—the homeostatic model assessment of insulin-resistance ratio; hsCRP—high-sensitivity C-reactive protein; IGF-1—insulin growth factor-1; LH—luteinizing hormone; TgAb—thyroglobulin antibodies; TPOAb—thyroid peroxidase antibodies; TSH—thyroid-stimulating hormone.

Metformin treatment was well tolerated; side effects (decreased appetite, transient diarrhea, tiredness, and weakness) were mild and present only in the minority of patients (12.5% in group A and 8.3% in group B). Because no patient prematurely discontinued the treatment, the data of all included individuals were statistically analyzed. Over the entire study period, all patients adhered to the treatment recommendations and the recommendations concerning diet and physical activity.

Metformin did not affect BMI (group A: 24.9 ± 4.3 kg/m^2 vs. 24.3 ± 4.6 kg/m^2 [$p = 0.6428$], group B: 24.5 ± 4.8 kg/m^2 vs. 23.9 ± 4.1 kg/m^2 [$p = 0.6437$]), and follow-up BMI did not differ between both groups ($p = 0.7519$). In both treatment groups, metformin decreased glucose, HOMA-IR, and glycated hemoglobin. In group B, but not in group A, there were differences between the baseline and the follow-up plasma concentrations of both total and monomeric prolactin and in hsCRP. Titers of TPOAb and

TgAb, concentrations of macroprolactin, TSH, free thyroxine, free triiodothyronine, gonadotropins, testosterone, ACTH, and IGF-1 remained at similar levels over the entire study period. At the end of the study, there were between-group differences in glucose, HOMA-IR, glycated hemoglobin, total prolactin, monomeric prolactin, antibody titers, and hsCRP (Table 2).

The study groups differed in the percent changes from baseline for glucose, HOMA-IR, glycated hemoglobin, total prolactin, monomeric prolactin, testosterone, and hsCRP, which were greater in group B than in group A (Table 3).

Table 3. Percentage changes from baseline in the investigated variables during metformin treatment.

Variable	Group A	Group B	p-Value
Δ Glucose	-7 ± 3	-15 ± 6	<0.0001
Δ HOMA1-IR	-24 ± 20	-48 ± 23	0.0004
Δ Glycated hemoglobin	-6 ± 5	-13 ± 5	<0.0001
Δ Total prolactin	-5 ± 8	-21 ± 8	<0.0001
Δ Monomeric prolactin	-3 ± 7	-22 ± 10	<0.0001
Δ Macroprolactin	-15 ± 18	-9 ± 20	0.2803
Δ TPOAb	-20 ± 18	-15 ± 24	0.4184
Δ TgAb	-17 ± 23	-6 ± 28	0.1438
Δ TSH	-13 ± 14	-23 ± 22	0.0667
Δ Free thyroxine	2 ± 6	3 ± 7	0.5988
Δ Free triiodothyronine	3 ± 10	6 ± 14	0.3974
Δ FSH	9 ± 10	13 ± 12	0.2160
Δ LH	10 ± 12	20 ± 23	0.0653
Δ Testosterone	7 ± 13	17 ± 20	0.0457
Δ ACTH	21 ± 20	12 ± 18	0.1081
Δ IGF-1	12 ± 25	22 ± 20	0.1328
Δ hsCRP	-7 ± 14	-32 ± 20	<0.0001
Δ Estimated glomerular filtration rate	3 ± 7	1 ± 6	0.2934

Group A: males with iatrogenic prolactin excess and euthyroid Hashimoto disease. Group B: males with iatrogenic prolactin excess but without thyroid disease. Except for the percentage of smokers, the data have been shown as the mean ± standard deviation. Abbreviations: ACTH—adrenocorticotropic hormone; FSH—follicle-stimulating hormone; HOMA-IR—the homeostatic model assessment of insulin-resistance ratio; hsCRP—high-sensitivity C-reactive protein; IGF-1—insulin growth factor-1; LH—luteinizing hormone; TgAb—thyroglobulin antibodies; TPOAb—thyroid peroxidase antibodies; TSH—thyroid-stimulating hormone.

At the beginning of the study, in group A, there were positive correlations between the hsCRP concentration and the antibody titers (TPOAb: r = 0.462, p = 0.0002; TgAb: r = 0.398, p = 0.0011). The effect of treatment on the total and monomeric prolactin correlated with the (a) baseline concentrations (total prolactin—group A: r = 0.411, p = 0.0008, group B: r = 0.325, p = 0.0322; monomeric prolactin—group A: r = 0.442, p = 0.0004, group B: r = 0.346, p = 0.0226), (b) the impact of treatment on HOMA-IR (total prolactin—group A: r = 0.368, p = 0.0282, group B: r = 0.340, p = 0.0226; monomeric prolactin—group A: r = 0.394, p = 0.0014, group B: r = 0.411, p = 0.0008), and (c) in group B additionally, also with treatment-induced reduction in hsCRP concentration (r = 0.365, p = 0.0122 for total prolactin; r = 0.385, p = 0.0043 for monomeric prolactin). In group A, there were inverse correlations between the baseline antibody titers and metformin action on total prolactin (r = −0.402, p = 0.0005 for TPOAb; r = −0.375, p = 0.0049 for TgAb) and on monomeric prolactin (r = −0.387, p = 0.0023 for TPOAb; r = −0.341, p = 0.0274 for TgAb), as well as between treatment-induced changes in HOMA-IR and the baseline hsCRP (r = −0.352, p = 0.0126). In group B, treatment-induced changes in prolactin correlated with the impact of treatment on LH (r = 0.315, p = 0.0312 for total prolactin; r = 0.342, p = 0.0265 for monomeric prolactin) and inversely with the baseline testosterone (r = −0.265, p = 0.0489 for total prolactin; r = −0.269, p = 0.0411 for monomeric prolactin). The remaining correlations were insignificant.

3. Discussion

The study showed a decrease in prolactin concentrations in men aged between 50 and 75 years with elevated levels of this hormone who do not have concurrent thyroid disease. This finding contrasts with our previous observations concerning a neutral effect of metformin on prolactin levels in a younger population of hyperprolactinemic men [21]. This incongruency may be well explained by lower mean testosterone concentrations in the participants of the current study. In line with this interpretation, the reduction in the prolactin concentration is inversely correlated with the plasma testosterone concentration. Another interesting finding is that the decrease in the total prolactin level reflected a reduction in the monomeric form of prolactin. In turn, the concentration of macroprolactin, high molecular mass forms of this hormone, composed of antigen–antibody complexes of prolactin and aggregates of covalent or noncovalent polymers of monomeric prolactin [37], remained unchanged, which has been evidenced so far only in women [38]. The obtained results indicate that metformin treatment should be considered in middle-aged elderly men with diabetes or prediabetes coexisting with an elevated prolactin concentration. Because of the cardiometabolic and bone consequences of long-term prolactin excess, the risk of which increases with age [39], metformin treatment seems to bring benefits even in men with asymptomatic hyperprolactinemia. Another argument in favor of metformin is that antipsychotics seem to increase the risk of developing diabetes [40], the presence of which additionally justifies this treatment. It is worth noting that metformin treatment was well tolerated by men receiving antipsychotic drugs and did not worsen the effectiveness of antipsychotic therapy.

Metformin treatment did not significantly affect gonadotropin levels, though both groups differed in the strength of the prolactin-lowering effect, and in men without thyroid pathology, the decrease in plasma prolactin positively correlated with the impact of treatment on LH. This finding may suggest that metformin exerted a weak, and statistically insignificant, inhibitory effect on gonadotropin secretion in this population. The study included men with only mild or moderate prolactin excess but not individuals with severe hyperprolactinemia, which is often complicated by gonadal failure [41]. Thus, a relatively low mean baseline testosterone concentration (though in the majority of patients still within the reference range) was probably a consequence of age-related changes in testosterone production (late-onset hypogonadism), while the association with prolactin excess was less convincing. This explains why the improvement in lactotrope function did not result in statistically significant changes in gonadotropins and testosterone.

Undoubtedly, a novel observation of our present study is that the impact on prolactin concentration was absent in men with concurrent euthyroid Hashimoto thyroiditis. Although autoimmune thyroiditis is the major cause of hypothyroidism in developed countries, less pronounced lymphocytic infiltration and fibrotic reactions do not result in thyroid hypofunction [42]. Consequently, euthyroid Hashimoto thyroiditis is considered a milder form of autoimmune thyroiditis than hypothyroid Hashimoto thyroiditis and often precedes the hypothyroid phase [43]. Because untreated or inadequately treated hypothyroidism often causes hyperprolactinemia [44], making it difficult to differentiate prolactin excess secondary to thyroid hypofunction and antipsychotic treatment, the study included only patients with normal concentrations of TSH, free thyroxine, and free triiodothyronine, which were suggestive of an intact hypothalamic–pituitary–thyroid axis activity. Another reason for including only euthyroid men is the link between schizophrenia and hypothyroidism resulting in that all patients receiving antipsychotics on a chronic basis should be effectively substituted with thyroid hormones [45]. Thus, our findings cannot be explained by thyroid hypofunction. Moreover, because of matching, they cannot be attributed to baseline differences in prolactin concentration and insulin sensitivity. Lastly, they cannot be explained by differences in testosterone production, which was similar in both study groups, as well as by other differences between the groups. Thus, attenuation of the prolactin-lowering effect of metformin was likely associated with thyroid autoimmunity. In line with this explanation, titers of TPOAb and TgAb, which determine the severity of Hashimoto thyroiditis [28], inversely correlated with the impact of metformin on total and monomeric prolactin. Our finding is consistent with similar observations in women [25,26], which suggest that the pituitary effects of metformin are mitigated by Hashimoto thyroiditis, irrespective of gender.

In the investigated population of middle-aged and elderly men, metformin did not affect thyroid antibody titers. This finding is in disagreement with the results of a meta-analysis of four clinical studies [46], which showed a reduction in both TPOAb and TgAb. There are different possible explanations for these contradictory results. They may be attributed to differences in the assessed populations because the meta-analysis by Jia et al. included mainly women of reproductive age (the percentage of men was very low). Second, a remarkable proportion of the analyzed patients was diagnosed with subclinical hypothyroidism. Thus, the degree of inflammation and the destruction of thyrocytes was greater than in the current study. Finally, we recruited a smaller number of patients in comparison with the number of individuals included in the mentioned Chinese meta-analysis. A neutral effect on thyroid antibody titers indicates that the weak effects of metformin on prolactin levels in patients with autoimmune thyroiditis cannot be explained by a direct involvement of these antibodies.

Another finding attracting attention is the positive correlations between the impact of treatment on prolactin and insulin sensitivity, indicating that the metabolic and prolactin-lowering effects of metformin are related to each other, supporting the view that prolactin plays an important role in glucose homeostasis [15–20]. More importantly, concomitant autoimmune thyroiditis weakened metformin action on glucose homeostasis, which is supported by inverse correlations between the impact on HOMA-IR and hsCRP concentrations. Though our study was not aimed at investigating molecular mechanisms of metformin action, the obtained results may suggest that relatively small changes in plasma glucose, glycated hemoglobin, and HOMA-IR resulted from markedly elevated post-treatment prolactin levels in this population and/or from interaction between proinflammatory markers and metformin at the level of the central nervous system and/or GLUT-4, the main insulin-responsive glucose transporter, which is a key regulator of systemic glucose homeostasis [25,26,47].

The most likely explanation for a mitigatory impact of Hashimoto thyroiditis on the prolactin-lowering effect of metformin is interaction at the level of the pituitary adenosine 5′-monophosphate-activated protein kinase (AMPK) pathway. Some findings support this explanation. First, activation of AMPK is one of the main mechanisms of metformin action [48]. Second, anterior pituitary cells are characterized by an abundant expression of this enzyme [49]. Third, pituitary AMPK was found to mediate the gonadotropin-lowering effects of metformin [49], and it may mediate other pituitary effects of this drug. Lastly, hsCRP and proinflammatory cytokines, found to be overproduced in Hashimoto thyroiditis [28], exert an inhibitory effect on the activity of the AMPK pathway [50,51]. Thus, it is likely that the stimulatory effect observed in men without thyroid autoimmunity is counterbalanced by the opposite effect of the proinflammatory state. This interaction may be particularly relevant in middle-age or elderly men because low testosterone production, characterizing men in these age groups, is associated with the low activity of AMPK [52]. The alternative explanation for our findings is the opposite impact of metformin and mediators of the proinflammatory state on the activity of tuberoinfundibular dopaminergic neurons, playing a fundamental role in the regulation of prolactin secretion [53]. In line with this explanation, metformin was found to increase the central dopaminergic tone [54], where proinflammatory factors inhibited the tuberoinfundibular dopaminergic system [55]. Interestingly, increased activity of both AMPK and tuberoinfundibular dopaminergic neurons is associated with increased insulin sensitivity [56,57] and may explain correlations between the impact of metformin on prolactin and HOMA-IR.

Some other conclusions can be drawn on the basis of the obtained results. First, autoimmune thyroid disease attenuates the metabolic effects of metformin in subjects without hypothyroidism. Thus, men with Hashimoto thyroiditis and either diabetes or at high diabetes risk may gain fewer benefits from metformin treatment than their peers without thyroid disease. This finding is worth underlying because euthyroid Hashimoto thyroiditis may be complicated by insulin resistance and may predispose to type-2 diabetes and prediabetes [58]. Second, in the current study, metformin was administered at a high dose (3 g daily). In our previous ones, metformin decreased prolactin levels only if the daily dose ranged from 2.55 to 3 g [8,12]. Thus, only high-dose metformin should be recommended to reduce prolactin levels in subjects with prolactin excess. However, high doses of this agent are well tolerated even in the case of prediabetes. Finally, positive correlations between changes in total and monomeric prolactin and their baseline concentrations, as well as no statistically significant changes in the concentrations of the remaining hormones, the baseline levels of which were within the reference range, suggest that metformin does not lead to hypoprolactinemia and other pituitary hormone deficiencies. This conclusion is clinically relevant because dopaminergic agents may cause prolactin deficiency, which, at least in premenopausal women, is associated with the unfavorable metabolic profile [59].

Our study provides some therapeutic implications. First, it seems justified to assess thyroid antibodies in all men requiring antipsychotic therapy. Finding autoimmune thyroiditis should be an argument against treatment with first-generation antipsychotics, amisulpride, risperidone, and paliperidone, particularly predisposing to hyperprolactinemia [60]. Second, measurement of thyroid antibodies may be considered in all insulin-resistant men poorly responding to metformin therapy, if the reason for this resistance is unknown. Third, metformin may be considered as a treatment option for antipsychotic-induced hyperprolactinemia in men without thyroid disease if other options are not available, effective, or tolerated. However, its administration should not be recommended for patients with concomitant thyroid autoimmunity. Fourth, the treatment of patients with both prolactin excess and thyroiditis may be very difficult because the prolactin-lowering effects of cabergoline in this group are weak [61]. Last, from a pathophysiological point of view, men with diabetes or prediabetes may benefit from concomitant treatment with metformin and agents reducing thyroid antibody titers, such as vitamin D, selenomethionine, and myo-inositol [62]. Certainly, it cannot be excluded that other insulin-sensitizing agents, such as dipetidyl peptidase-4 inhibitors, glucagon-like peptide-1 agonists, and thiazolidinediones, administered in monotherapy or in combination with metformin, are superior to metformin alone in patients with autoimmune thyroiditis.

The strength of our study was a homogenous population of patients, resulting from strict inclusion and exclusion criteria. Unfortunately, this causes a situation where we can only speculate about metformin action in other groups of patients, particularly individuals with severe hyperprolactinemia or hypothyroidism. For ethical reasons, we did not include men with prolactin levels above 80 ng/mL. Severe hyperprolactinemia must always be specifically addressed, even when there is no sexual dysfunction (reduced libido and/or importance) because of the medium-/long-term risk of osteoporosis and cardiovascular issues [63]. Such patients require specific treatment, including the discontinuation or decreasing the dose of the antipsychotic drug, changing the antipsychotic drug, the addition of aripiprazole, the addition of dopamine agonists, and/or symptomatic treatment) [60]. The obtained correlations (positive between metformin action on plasma prolactin and baseline prolactin levels and negative between the impact on plasma prolactin and thyroid antibody titers) suggest, however, that metformin may reduce prolactin levels also in patients with severe hyperprolactinemia, but that this effect is probably also attenuated by coexisting autoimmune thyroiditis. Moreover, we excluded not only levothyroxine-naive patients with overt or subclinical hypothyroidism but also hypothyroid individuals in whom euthyroidism was restored by levothyroxine substitution. This enabled us to eliminate possible pharmacokinetic and pharmacodynamic interactions between metformin and exogenous levothyroxine. The results of our recent study [64] indicate that metformin reduces TSH in patients with autoimmune thyroiditis-induced hypothyroidism. However, considering that the impact of this agent on TSH in hypothyroid patients was moderate, even if metformin reduces plasma prolactin in hypothyroid men with autoimmune thyroiditis and antipsychotics-induced hyperprolactinemia, this effect is probably weak and secondary to the improvement in the hypothalamic–pituitary–thyroid axis activity but not associated with a direct effect on lactotrophic cells and/or the tuberoinfundibular dopaminergic pathway.

There are several other study limitations that should be noted. Despite being adequately powered for the primary outcome, the small sample size limits the generalizability of the results. Owing to the study design, the results might have been potentially affected by selection and confounding bias. Extrapolating previous data [65,66], it seems that the study population was characterized by sufficient iodine and insufficient selenium intake. It cannot completely exclude some differences in the impact of autoimmune thyroiditis in men with inadequate iodine and/or adequate selenium intake. Strict inclusion and exclusion criteria do not allow us to conclude whether other autoimmune disorders (including Graves' disease) modulate metformin action on antipsychotic-induced hyperprolactinemia, which justifies further studies. Precautions during study design and data analysis limited, but did not completely eliminate, the regression toward the mean [67]. Lastly, because gel-filtration chromatography is time-consuming and expensive, the macroprolactin content was measured using a less accurate polyethylene glycol precipitation method [37].

In conclusion, chronic treatment with metformin decreases elevated prolactin levels in middle-aged and elderly men receiving antipsychotic therapy. The inhibitory effect of this drug on lactotrope secretory function is not observed if hyperprolactinemia is accompanied by Hashimoto thyroiditis. This is paralleled by weaker metabolic effects in men with thyroiditis than in men without thyroid pathology. Thus, men with thyroid autoimmunity may benefit to a lesser degree from metformin treatment and may require specific therapy to lower titers of thyroid antibodies. These relationships seem similar to those in women, which suggests that the impact of autoimmune thyroiditis on metformin action is sex independent. Study limitations and the lack of other studies justify conducting a larger trial to verify the obtained results.

4. Materials and Methods

This single-center, prospective cohort study followed the principles of the Declaration of Helsinki, and its protocol was accepted by the Bioethical Committee of the Medical University of Silesia (KNW/0022/KB/207/17; 17 October 2107). All participants gave written informed consent after receiving oral and written information about the study from the investigators. Due to its character, the study did not require registration at a clinical trial registry.

4.1. Study Population

The participants of the study were recruited among middle-aged or elderly men (50–75 years old) with antipsychotic-induced hyperprolactinemia who, because of recently diagnosed type-2 diabetes or prediabetes, required metformin treatment. We included only patients with the total prolactin concentration in the range between 30 and 80 ng/mL on two different occasions and with a difference between both measurements not exceeding 10%. The widely accepted American Diabetes Association criteria were used to diagnose type-2 diabetes (fasting glucose greater than or equal to 126 mg/dL or 2 h post-challenge glucose equal to 200 mg/dL or higher) or prediabetes (fasting glucose between 100 and 125 mg/dL and/or 2 h post-challenge glucose between 140–199 mg/dL). The participants were chosen only among individuals complying with lifestyle recommendations for at least 12 weeks. The participants were allocated into one of two groups, each consisting of 24 men. The number of included patients was based on the sample-size analysis showing that 22 patients per group would be sufficient to provide significant data (25% difference in total prolactin) for 80% power with 95% confidence. Recruitment of two additional patients in each group aimed to compensate for possible dropouts. Individuals recruited to group A had to fulfill the following criteria of euthyroid Hashimoto thyroiditis: (a) titers of thyroid peroxidase antibodies (TPOAb) exceeding 100 U/mL; (b) a hypoechoic, diffusely heterogeneous echotexture with hypoechoic micronodules and/or surrounding echogenic septation on the ultrasound image; (c) plasma TSH concentration ranging from 0.4 to 4.5 mU/L; (d) free thyroxine concentration in the range between 10.2 and 21.4 pmol/L and (e) free triiodothyronine concentration in the range between 2.2 and 6.7 pmol/L. In turn, patients recruited to group B, serving as a control group, were characterized by (a) normal levels of TSH and free thyroid hormones, (b) non-elevated thyroid antibody titers, and (c) no changes in the sonographic picture of the thyroid gland. The control patients were selected from a group of 60 men who met the inclusion and exclusion criteria in order to obtain two populations similar in terms of age, body mass index (BMI), homeostatic model assessment of insulin-resistance ratio (HOMA-IR), and total prolactin concentrations. Similar proportions of men were enrolled in May or June (25 patients: group A: 13 men and group B: 12 men), and in November or December (23 patients: group A: 11 men and group B: 12 men).

The potential participants were excluded if they had a prolactin excess of another origin, macroprolactinemia, positive antibodies against thyrotropin receptor, other autoimmune or endocrine diseases, cardiovascular disorders (except for grade-1 hypertension), an estimated glomerular filtration rate less than 60 mL/min/1.73 m^2, liver failure, malabsorption syndromes, oncological disorders, any other serious disorders, received other medicines (except for antipsychotic drugs), or were poorly compliant.

4.2. Study Design

The flow of patients through the study is depicted in Figure 1. The initial dosing for metformin was 500 mg by mouth twice daily. Daily dosing was then increased in 500 mg increments weekly to a target dose of 3 g (1 g three times a day). Throughout the study, antipsychotic dosing remained unchanged, and other treatments were not allowed. Non-pharmacological interventions (diet and physical activity) from before the study began were continued. Adherence to pharmacotherapy was assessed every two months by asking the patient and counting tablets.

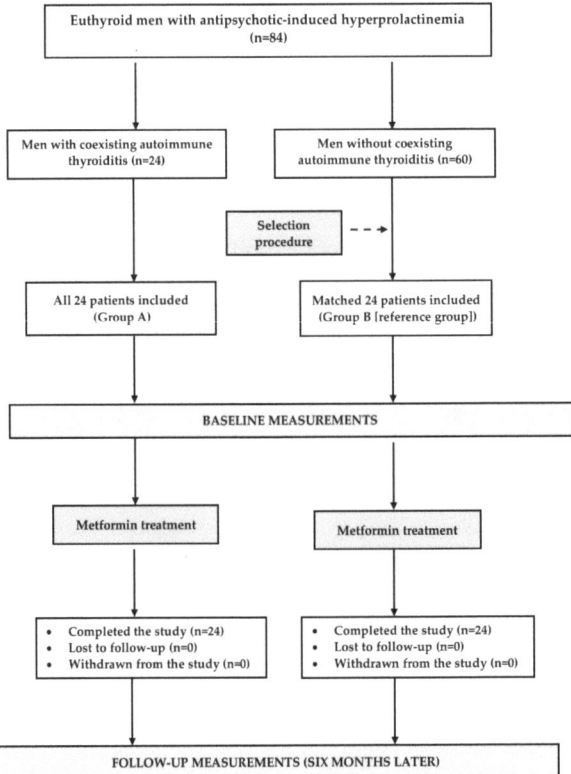

Figure 1. Flow the patients through the study.

4.3. Laboratory Assays

Blood samples for laboratory analysis were collected between 7.00 and 8.30 a.m. in a quiet and air-conditioned room (constant temperature of 23–24 °C). All assays were performed in duplicate on the first study day (before the first metformin dose) and 6 months later. Plasma levels of glucose, creatinine, and whole-blood content of glycated hemoglobin were measured using the multi-analyzer COBAS Integra 400 Plus (Roche Diagnostics, Basel, Switzerland). Titers of thyroglobulin antibodies (TgAb) and TPOAb and plasma concentrations of prolactin, insulin, thyrotropin, free thyroxine, free triiodothyronine FSH and LH were assessed using acridinium ester technology (ADVIA Centaur XP Immunoassay System, Siemens Healthcare Diagnostics, Munich, Germany). Circulating levels of adrenocorticotropic hormone (ACTH), insulin-like growth factor-1 (IGF-1), and high-sensitivity C-reactive protein (hsCRP) were measured by solid-phase enzyme-labeled chemiluminescent immunometric assays (Immulite, Siemens, Munich, Germany). Total prolactin was measured before, whereas monomeric prolactin was measured after polyethylene glycol precipitation [38]. Macroprolactin was calculated by subtracting monomeric prolactin from the total prolactin. Concentrations of adrenocorticotropic hormone (ACTH) and insulin-like growth factor-1 (IGF-1) were measured by solid-phase enzyme-labeled chemiluminescent immunometric assays (Immulite, Siemens, Munich, Germany). HOMA-IR was obtained by dividing the product of plasma glucose (in mg/dL) and insulin (in mIU/L) by 405. The estimated glomerular filtration rate was calculated as follows: $175 \times$ [plasma creatinine (μmol/L) $\times 0.0113]^{-1.154} \times$ age (years)$^{0.203}$ (the Modification Diet in the Renal Disease Study equation).

4.4. Statistical Analysis

Prior to data analysis, all variables were subjected to log transformation to meet the criteria of normality and homogeneity. The significance of differences between means was determined by Student's t-tests for independent samples (inter-group comparisons) or paired Student's t-tests (intra-group comparisons). The χ^2 test was used to compare dichotomous variables. Between-variable correlations were assessed using Pearson's correlation coefficients (r). Values of p below 0.05 indicated a significant difference.

Author Contributions: R.K.: Conceptualization, Methodology, Investigation, Data curation, Writing—original draft, Writing—review and editing; M.B.: Methodology, Investigation; W.S.: Methodology, Data curation; B.O.: Resources, Writing—review and editing, Supervision. All authors have read and agreed to the published version of the manuscript.

Funding: This work did not receive any specific grant from funding agencies in the public, commercial, or not-for-profit sectors.

Institutional Review Board Statement: The study followed the principles of the Declaration of Helsinki, and its protocol was accepted by the Bioethical Committee of the Medical University of Silesia (KNW /0022/KB/207/17; 17 October 2107).

Informed Consent Statement: All participants gave written informed consent after receiving oral and written information about the study from the investigators.

Data Availability Statement: The data that support the findings of this study are available from the corresponding author upon reasonable request.

Conflicts of Interest: The authors declare no conflicts of interest.

Abbreviations

ACTH—adrenocorticotropic hormone; AMPK—adenosine 5'-monophosphate-activated protein kinase; FSH—follicle-stimulating hormone; HOMA-IR—homeostatic model assessment of insulin-resistance ratio; hsCRP—high-sensitivity C-reactive protein; IGF-1—insulin-like growth factor-1; LH—luteinizing hormone; TgAb—thyroglobulin antibodies; TPOAb—thyroid peroxidase antibodies; TSH—thyroid-stimulating hormone.

References

1. Ueno, M. Molecular anatomy of the brain endothelial barrier: An overview of the distributional features. *Curr. Med. Chem.* **2007**, *14*, 1199–1206. [CrossRef] [PubMed]
2. Wróbel, M.P.; Marek, B.; Kajdaniuk, D.; Rokicka, D.; Szymborska-Kajanek, A.; Strojek, K. Metformin—A new old drug. *Endokrynol. Pol.* **2017**, *68*, 482–496. [CrossRef]
3. Labuzek, K.; Suchy, D.; Gabryel, B.; Bielecka, A.; Liber, S.; Okopień, B. Quantification of metformin by the HPLC method in brain regions, cerebrospinal fluid and plasma of rats treated with lipopolysaccharide. *Pharmacol. Rep.* **2010**, *62*, 956–965. [CrossRef] [PubMed]
4. Cappelli, C.; Rotondi, M.; Pirola, I.; Agosti, B.; Gandossi, E.; Valentini, U.; De Martino, E.; Cimino, A.; Chiovato, L.; Agabiti-Rosei, E.; et al. TSH-lowering effect of metformin in type 2 diabetic patients: Differences between euthyroid, untreated hypothyroid, and euthyroid on L-T4 therapy patients. *Diabetes Care* **2009**, *32*, 1589–1590. [CrossRef] [PubMed]
5. Meng, X.; Xu, S.; Chen, G.; Derwahl, M.; Liu, C. Metformin and thyroid disease. *J. Endocrinol.* **2017**, *233*, R43–R51. [CrossRef] [PubMed]
6. Krysiak, R.; Szkróbka, W.; Okopień, B. The effect of metformin on serum gonadotropin levels in postmenopausal women with diabetes and prediabetes: A pilot study. *Exp. Clin. Endocrinol. Diabetes* **2018**, *126*, 645–650. [CrossRef]
7. Velija-Ašimi, Z. Evaluation of endocrine changes in women with the polycystic ovary syndrome during metformin treatment. *Bosn. J. Basic Med. Sci.* **2013**, *13*, 180–185. [CrossRef]
8. Krysiak, R.; Okrzesik, J.; Okopień, B. The effect of short-term metformin treatment on plasma prolactin levels in bromocriptine-treated patients with hyperprolactinaemia and impaired glucose tolerance: A pilot study. *Endocrine* **2015**, *49*, 242–249. [CrossRef]
9. Wu, R.R.; Jin, H.; Gao, K.; Twamley, E.W.; Ou, J.J.; Shao, P.; Wang, J.; Guo, X.F.; Davis, J.M.; Chan, P.K.; et al. Metformin for treatment of antipsychotic-induced amenorrhea and weight gain in women with worst-episode schizophrenia: A double-blind, randomized, placebo-controlled study. *Am. J. Psychiatry* **2012**, *169*, 813–821. [CrossRef]
10. Zheng, W.; Yang, X.H.; Cai, D.B.; Ungvari, G.S.; Ng, C.H.; Wang, N.; Ning, Y.P.; Xiang, Y.T. Adjunctive metformin for antipsychotic related hyperprolactinemia: A meta-analysis of randomized controlled trials. *J. Psychopharmacol.* **2017**, *31*, 625–631. [CrossRef]
11. Bo, Q.J.; Wang, Z.M.; Li, X.B.; Ma, X.; Wang, C.Y.; de Leon, J. Adjunctive metformin for antipsychotic-induced hyperprolactinemia: A systematic review. *Psychiatry Res.* **2016**, *237*, 257–263. [CrossRef] [PubMed]
12. Krysiak, R.; Kowalcze, K.; Szkróbka, W.; Okopień, B. The effect of metformin on prolactin levels in patients with drug-induced hyperprolactinemia. *Eur. J. Intern. Med.* **2016**, *30*, 94–98. [CrossRef] [PubMed]
13. Auriemma, R.S.; Pirchio, R.; De Alcubierre, D.; Pivonello, R.; Colao, A. Dopamine agonists: From the 1970s to today. *Neuroendocrinology* **2019**, *109*, 34–41. [CrossRef] [PubMed]
14. Rusgis, M.M.; Alabbasi, A.Y.; Nelson, L.A. Guidance on the treatment of antipsychotic-induced hyperprolactinemia when switching the antipsychotic is not an option. *Am. J. Health Syst. Pharm.* **2021**, *78*, 862–871. [CrossRef] [PubMed]
15. Auriemma, R.S.; De Alcubierre, D.; Pirchio, R.; Pivonello, R.; Colao, A. The effects of hyperprolactinemia and its control on metabolic diseases. *Expert Rev. Endocrinol. Metab.* **2018**, *13*, 99–106. [CrossRef] [PubMed]
16. Auriemma, R.S.; De Alcubierre, D.; Pirchio, R.; Pivonello, R.; Colao, A. Glucose abnormalities associated to prolactin secreting pituitary adenomas. *Front. Endocrinol.* **2019**, *10*, 327. [CrossRef] [PubMed]
17. Andersen, M.; Glintborg, D. Metabolic syndrome in hyperprolactinemia. *Front. Horm. Res.* **2018**, *49*, 29–47. [PubMed]
18. Arslan, M.S.; Topaloglu, O.; Sahin, M.; Tutal, E.; Gungunes, A.; Cakir, E.; Ozturk, I.U.; Karbek, B.; Ucan, B.; Ginis, Z.; et al. Preclinical atherosclerosis in patients with prolactinoma. *Endocr. Pract.* **2014**, *20*, 447–451. [CrossRef]
19. Gierach, M.; Bruska-Sikorska, M.; Rojek, M.; Junik, R. Hyperprolactinemia and insulin resistance. *Endokrynol. Pol.* **2022**, *73*, 959–967. [CrossRef]
20. Yavuz, D.; Deyneli, O.; Akpinar, I.; Yildiz, E.; Gözü, H.; Sezgin, O.; Haklar, G.; Akalin, S. Endothelial function, insulin sensitivity and inflammatory markers in hyperprolactinemic pre-menopausal women. *Eur. J. Endocrinol.* **2003**, *149*, 187–193. [CrossRef]
21. Krysiak, R.; Szkróbka, W.; Okopień, B. Sex-dependent effect of metformin on serum prolactin levels in hyperprolactinemic patients with type 2 diabetes: A pilot study. *Exp. Clin. Endocrinol. Diabetes* **2018**, *126*, 342–348. [CrossRef]
22. Krysiak, R.; Szkróbka, W.; Okopień, B. Endogenous testosterone determines metformin action on prolactin levels in hyperprolactinaemic men: A pilot study. *Basic Clin. Pharmacol. Toxicol.* **2020**, *126*, 110–115. [CrossRef]
23. Cunningham, G.R. Testosterone and metabolic syndrome. *Asian J. Androl.* **2015**, *17*, 192–196. [CrossRef]
24. Louters, M.; Pearlman, M.; Solsrud, E.; Pearlman, A. Functional hypogonadism among patients with obesity, diabetes, and metabolic syndrome. *Int. J. Impot. Res.* **2022**, *34*, 714–720. [CrossRef]

25. Krysiak, R.; Basiak, M.; Machnik, G.; Okopień, B. Impaired gonadotropin-lowering effects of metformin in postmenopausal women with autoimmune thyroiditis: A pilot study. *Pharmaceuticals* **2023**, *16*, 922. [CrossRef]
26. Krysiak, R.; Kowalcze, K.; Madej, A.; Okopień, B. The effect of metformin on plasma prolactin levels in young women with autoimmune thyroiditis. *J. Clin. Med.* **2023**, *12*, 3769. [CrossRef]
27. Gessl, A.; Lemmens-Gruber, R.; Kautzky-Willer, A. Thyroid disorders. *Handb. Exp. Pharmacol.* **2012**, *214*, 361–386.
28. Merrill, S.J.; Minucci, S.B. Thyroid autoimmunity: An interplay of factors. *Vitam. Horm.* **2018**, *106*, 129–145.
29. Ragusa, F.; Fallahi, P.; Elia, G.; Gonnella, D.; Paparo, S.R.; Giusti, C.; Churilov, L.P.; Ferrari, S.M.; Antonelli, A. Hashimotos' thyroiditis: Epidemiology, pathogenesis, clinic and therapy. *Best Pract. Res. Clin. Endocrinol. Metab.* **2019**, *34*, 101367. [CrossRef]
30. Wortsman, J.; Rosner, W.; Dufau, M.L. Abnormal testicular function in men with primary hypothyroidism. *Am. J. Med.* **1987**, *82*, 207–212. [CrossRef] [PubMed]
31. Chen, Y.; Chen, Y.; Xia, F.; Wang, N.; Chen, C.; Nie, X.; Li, Q.; Han, B.; Zhai, H.; Jiang, B.; et al. A higher ratio of estradiol to testosterone is associated with autoimmune thyroid disease in males. *Thyroid* **2017**, *27*, 960–966. [CrossRef] [PubMed]
32. Doukas, C.; Saltiki, K.; Mantzou, A.; Cimponeriu, A.; Terzidis, K.; Sarika, L.; Mavrikakis, M.; Sfikakis, P.; Alevizaki, M. Hormonal parameters and sex hormone receptor gene polymorphisms in men with autoimmune diseases. *Rheumatol. Int.* **2013**, *34*, 575–582. [CrossRef] [PubMed]
33. Casto, C.; Pepe, G.; Li Pomi, A.; Corica, D.; Aversa, T.; Wasniewska, M. Hashimoto thyroiditis and Graves' disease in genetic syndromes in pediatric age. *Genes* **2021**, *12*, 222. [CrossRef] [PubMed]
34. Krysiak, R.; Kowalcze, K.; Okopień, B. The effect of testosterone on thyroid autoimmunity in euthyroid men with Hashimoto thyroiditis and low testosterone levels. *J. Clin. Pharm. Ther.* **2019**, *44*, 742–749. [CrossRef] [PubMed]
35. Samperi, I.; Lithgow, K.; Karavitaki, N. Hyperprolactinaemia. *J. Clin. Med.* **2019**, *8*, 2203. [CrossRef]
36. Borba, V.V.; Zandman-Goddard, G.; Shoenfeld, Y. Prolactin and autoimmunity. *Front. Immunol.* **2018**, *9*, 73. [CrossRef]
37. Fahie-Wilson, M.N.; John, R.; Ellis, A.R. Macroprolactin; high molecular mass forms of circulating prolactin. *Ann. Clin. Biochem.* **2005**, *42 Pt 3*, 175–192. [CrossRef] [PubMed]
38. Krysiak, R.; Szkróbka, W.; Okopień, B. A neutral effect of metformin treatment on macroprolactin content in women with macroprolactinemia. *Exp. Clin. Endocrinol. Diabetes* **2016**, *125*, 223–228. [CrossRef] [PubMed]
39. Cocks Eschler, D.; Javanmard, P.; Cox, K.; Geer, E.B. Prolactinoma through the female life cycle. *Endocrine* **2018**, *59*, 16–29. [CrossRef]
40. Bottai, T.; Quintin, P.; Perrin, E. Antipsychotics and the risk of diabetes: A general data review. *Eur. Psychiatry* **2005**, *20* (Suppl. S4), S349–S357. [CrossRef]
41. ThirumalaI, A.; Anawalt, B.D. Epidemiology of male hypogonadism. *Endocrinol. Metab. Clin. N. Am.* **2022**, *51*, 1–27. [CrossRef] [PubMed]
42. Caturegli, P.; De Remigis, A.; Rose, N.R. Hashimoto thyroiditis: Clinical and diagnostic criteria. *Autoimmun. Rev.* **2014**, *13*, 391–397. [CrossRef] [PubMed]
43. Orgiazzi, J. Thyroid autoimmunity. *Presse Med.* **2012**, *41*, e611–e625. [CrossRef] [PubMed]
44. Vilar, L.; Vilar, C.F.; Lyra, R.; Freitas, M.D. Pitfalls in the diagnostic evaluation of hyperprolactinemia. *Neuroendocrinology* **2019**, *109*, 7–19. [CrossRef] [PubMed]
45. Sharif, K.; Tiosano, S.; Watad, A.; Comaneshter, D.; Cohen, A.D.; Shoenfeld, Y.; Amital, H. The link between schizophrenia and hypothyroidism: A population-based study. *Immunol. Res.* **2018**, *66*, 663–667. [CrossRef]
46. Jia, X.; Zhai, T.; Zhang, J.A. Metformin reduces autoimmune antibody levels in patients with Hashimoto thyroiditis: A systematic review and meta-analysis. *Autoimmunity* **2020**, *53*, 353–361. [CrossRef] [PubMed]
47. Yaribeygi, H.; Farrokhi, F.R.; Butler, A.E.; Sahebkar, A. Insulin resistance: Review of the underlying molecular mechanisms. *J. Cell. Physiol.* **2019**, *234*, 81527–88161. [CrossRef] [PubMed]
48. Rena, G.; Hardie, D.G.; Pearson, E.R. The mechanisms of action of metformin. *Diabetologia* **2017**, *60*, 1577–1585. [CrossRef] [PubMed]
49. Tosca, L.; Froment, P.; Rame, C.; McNeilly, J.R.; McNeilly, A.S.; Maillard, V.; Dupont, J. Metformin decreases GnRH- and activin-induced gonadotropin secretion in rat pituitary cells: Potential involvement of adenosine 5′ monophosphate-activated protein kinase (PRKA). *Biol. Reprod.* **2011**, *84*, 351–362. [CrossRef] [PubMed]
50. O'Neill, L.A.; Hardie, D.G. Metabolism of inflammation limited by AMPK and pseudo-starvation. *Nature* **2013**, *493*, 346–355. [CrossRef]
51. Jeon, S.M. Regulation and function of AMPK in physiology and diseases. *Exp. Mol. Med.* **2016**, *48*, e245. [CrossRef] [PubMed]
52. Ghanim, H.; Dhindsa, S.; Batra, M.; Green, K.; Abuaysheh, S.; Kuhadiya, N.D.; Makdissi, A.; Chaudhuri, A.; Sandhu, S.; Dandona, P. Testosterone increases the expression and phosphorylation of AMP kinase α in men with hypogonadism and type 2 diabetes. *J. Clin. Endocrinol. Metab.* **2020**, *105*, 1169–1175. [CrossRef] [PubMed]
53. Grattan, D.R. 60 years of neuroendocrinology: The hypothalamo-prolactin axis. *J. Endocrinol.* **2015**, *226*, T101–T122. [CrossRef] [PubMed]
54. Ortega-González, C.; Cardoza, L.; Coutiño, B.; Hidalgo, R.; Arteaga-Troncoso, G.; Parra, A. Insulin sensitizing drugs increase the endogenous dopaminergic tone in obese insulin-resistant women with polycystic ovary syndrome. *J. Endocrinol.* **2005**, *184*, 232–239. [CrossRef] [PubMed]

55. González, M.C.; Abreu, P.; Barroso-Chinea, P.; Cruz-Muros, I.; González-Hernández, T. Effect of intracerebroventricular injection of lipopolysaccharide on the tuberoinfundibular dopaminergic system of the rat. *Neuroscience* **2004**, *127*, 251–259. [CrossRef]
56. Smith, B.K.; Steinberg, G.R. AMP-activated protein kinase, fatty acid metabolism, and insulin sensitivity. *Curr. Opin. Clin. Nutr. Metab. Care* **2017**, *20*, 248–253. [CrossRef] [PubMed]
57. Chien, H.Y.; Chen, S.M.; Li, W.C. Dopamine receptor agonists mechanism of actions on glucose lowering and their connections with prolactin actions. *Front. Clin. Diabetes Healthc.* **2023**, *4*, 935872. [CrossRef]
58. Blaslov, K.; Gajski, D.; Vucelić, V.; Gaćina, P.; Mirošević, G.; Marinković, J.; Vrkljan, M.; Rotim, K. The association of subclinical insulin resistance with thyroid autoimmunity in euthyroid individuals. *Acta Clin. Croat.* **2020**, *59*, 696–702. [CrossRef] [PubMed]
59. Krysiak, R.; Kowalcze, K.; Okopień, B. Cardiometabolic profile of young women with hypoprolactinemia. *Endocrine* **2022**, *78*, 135–141. [CrossRef] [PubMed]
60. Montejo, Á.L.; Arango, C.; Bernardo, M.; Carrasco, J.L.; Crespo-Facorro, B.; Del Pino-Montes, J.; García-Escudero, M.A.; García-Rizo, C.; González-Pinto, A.; Hernández, A.I.; et al. Multidisciplinary consensus on the therapeutic recommendations for iatrogenic hyperprolactinemia secondary to antipsychotics. *Front. Neuroendocrinol.* **2017**, *45*, 25–34. [CrossRef]
61. Krysiak, R.; Kowalcze, K.; Okopień, B. Autoimmune thyroiditis attenuates cardiometabolic effects of cabergoline in young women with hyperprolactinemia. *J. Clin. Pharmacol.* **2023**, *63*, 886–894. [CrossRef] [PubMed]
62. Krysiak, R.; Kowalcze, K.; Szkróbka, W.; Okopień, B. Sexual function and depressive symptoms in young women with euthyroid Hashimoto thyroiditis receiving vitamin D, selenomethionine and myo-inositol: A pilot study. *Nutrients* **2023**, *15*, 2815. [CrossRef] [PubMed]
63. Grigg, J.; Worsley, R.; Thew, C.; Gurvich, C.; Thomas, N.; Kulkarni, J. Antipsychotic-induced hyperprolactinemia: Synthesis of world-wide guidelines and integrated recommendations for assessment, management and future research. *Psychopharmacology* **2017**, *234*, 3279–3297. [CrossRef] [PubMed]
64. Krysiak, R.; Kowalcze, K.; Okopień, B. Impact of metformin on hypothalamic-pituitary-thyroid axis activity in women with autoimmune and non-autoimmune subclinical hypothyroidism: A pilot study. *Pharmacol. Rep.* **2024**, *76*, 195–206. [CrossRef] [PubMed]
65. Trofimiuk-Müldner, M.; Konopka, J.; Sokołowski, G.; Dubiel, A.; Kieć-Klimczak, M.; Kluczyński, Ł.; Motyka, M.; Rzepka, E.; Walczyk, J.; Sokołowska, M.; et al. Current iodine nutrition status in Poland (2017): Is the Polish model of obligatory iodine prophylaxis able to eliminate iodine deficiency in the population? *Public Health Nutr.* **2020**, *23*, 2467–2477. [CrossRef] [PubMed]
66. Kłapcińska, B.; Poprzecki, S.; Danch, A.; Sobczak, A.; Kempa, K. Selenium levels in blood of Upper Silesian population: Evidence of suboptimal selenium status in a significant percentage of the population. *Biol. Trace Elem. Res.* **2005**, *108*, 1–15. [CrossRef]
67. Streiner, D.L. Regression toward the mean: Its etiology, diagnosis, and treatment. *Can. J. Psychiatry* **2001**, *46*, 72–76. [CrossRef]

Disclaimer/Publisher's Note: The statements, opinions and data contained in all publications are solely those of the individual author(s) and contributor(s) and not of MDPI and/or the editor(s). MDPI and/or the editor(s) disclaim responsibility for any injury to people or property resulting from any ideas, methods, instructions or products referred to in the content.

MDPI AG
Grosspeteranlage 5
4052 Basel
Switzerland
Tel.: +41 61 683 77 34

Pharmaceuticals Editorial Office
E-mail: pharmaceuticals@mdpi.com
www.mdpi.com/journal/pharmaceuticals

Disclaimer/Publisher's Note: The title and front matter of this reprint are at the discretion of the Guest Editor. The publisher is not responsible for their content or any associated concerns. The statements, opinions and data contained in all individual articles are solely those of the individual Editor and contributors and not of MDPI. MDPI disclaims responsibility for any injury to people or property resulting from any ideas, methods, instructions or products referred to in the content.

www.ingramcontent.com/pod-product-compliance
Lightning Source LLC
LaVergne TN
LVHW072348090526
838202LV00019B/2505